PROVINCES

BERYL IVEY LIBRARY

N

W ← → E

S

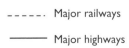

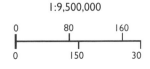

1:9,500,000

- - - - - Major railways

——— Major highways

0	80	160
0	150	30

Agricultural History

Agricultural History

HISTORY OF THE PRAIRIE WEST SERIES, VOL. 3

Edited by Gregory P. Marchildon

University of Regina

CPRC PRESS

Printed and bound in Canada at Friesens.
The text of this book is printed on 100% post-consumer recycled paper with earth-friendly vegetable-based inks.

Cover and text design: Duncan Campbell, CPRC.
Editor for the Press: David McLennan, CPRC.
Maps: Diane Perrick, CPRC.
Photographs on pages 11, 14, 56, 99, 109, 111, 112, 214 and 267 courtesy of David Mclennan, CPRC.
Index prepared by Patricia Furdek, Ottawa, Ontario.

Library and Archives Canada Cataloguing in Publication

Agricultural history / edited by Gregory P. Marchildon.

(History of the Prairie West series, ISSN 1914-864X ; 3)
Includes bibliographical references and index.
ISBN 978-0-88977-237-3

1. Agriculture—Prairie Provinces—History. 2. Agriculture—Northwest, Canadian—History. I. Marchildon, Gregory P., 1956– II. University of Regina. Canadian Plains Research Center III. Series: History of the Prairie West series ; 3

FC3209.A3A37 2011   971.2'02   C2011-902180-3

10   9   8   7   6   5   4   3   2   1

CPRC
P R E S S

Canadian Plains Research Center Press
University of Regina
Regina, Saskatchewan, Canada, s4s 0a2
TEL: (306) 585-4758 FAX: (306) 585-4699
E-MAIL: canadian.plains@uregina.ca WEB: www.cprcpress.ca

We acknowledge the financial support of the Government of Canada through the Canada Book Fund for our publishing activities.

Canadian   Patrimoine
Heritage   canadien

MIX
Paper from
responsible sources
FSC® C016245

CONTENTS

Preface

Since its inception in 1975, *Prairie Forum* has been a major repository of articles on the history of the northern Great Plains, in particular the region encompassed within the current political boundaries of the three Prairie Provinces of Manitoba, Saskatchewan and Alberta. The purpose of the History of the Prairie West Series is to make available the very best of *Prairie Forum* to as broad an audience as possible. Each volume in this series is devoted to a single, focused theme. Accompanied by dozens of new illustrations and maps as well as a searchable index, these volumes are intended to be of interest to the general reader as well as the professional historian.

The editor of the History of the Prairie West Series is Gregory P. Marchildon, Canada Research Chair in Public Policy and Economic History at the Johnson-Shoyama Graduate School of Public Policy at the University of Regina. In addition to selecting and organizing the articles based upon their quality and thematic connections to the volumes, he has chosen a cover painting for each volume which reflects the essence of each period and theme.

Acknowledgements

The editor wishes to thank CPRC Press for its support: Brian Mlazgar who has provided direction for the History of the Prairie West Series from the beginning, Duncan Campbell for his artistic input and design, Diane Perrick for the preparation of the maps and tables, and, in particular, David McLennan, who provided extensive copy-editing and image research. The editor is also grateful to Angela Scott for providing editorial commentary on an early draft of the introduction to this volume.

Introduction to *Agricultural History*

Gregory P. Marchildon

The essays selected for this volume of the *History of the Prairie West* focus on the agricultural history of the Canadian Plains. According to *The Canadian Oxford Dictionary*, agriculture is "the science or practice of cultivating the soil and rearing animals,"[1] a definition that includes farming in terms of growing crops and tending livestock within a fenced-off area of land, and ranching which involves more specialized cattle-breeding and cattle-raising operations.[2] As farming and ranching involve two very different industries and lifestyles, the essays on these subjects have been separated. In addition, since agriculture in the Prairie Provinces has been part of a global food regime since the nineteenth century, a final group of essays address the ways in which products were transported and marketed abroad.

SURVEYING THE FIELD

To begin with, however, this volume starts with four essays that provide a broad outline of agricultural history that is then explored in greater depth and detail in the sections that follow. Lewis H. Thomas (1917-83) reviews the history of agriculture from the early days of the fur trade to the First World War.[3] When Thomas completed this survey for the first issue of *Prairie Forum* in 1976, scholars were only beginning to explore the history of farming in the Canadian Plains. Though many gaps in our knowledge of this history have been filled during the past 35 years, Thomas's spirited and sure-footed narrative remains valuable.

Wayne Moodie and Barry Kaye mine the rich archival record of the Hudson's Bay Company to present a more detailed survey of indigenous agriculture during the fur trade. The authors trace the movement of maize-beans-squash agriculture from eastern areas to the Manitoba parklands and

Aboriginal settlements along the Red and the Assiniboine rivers. Predating Euro-Canadian settlement of the prairies, this form of agriculture would eventually die out with the decline of the fur trade.

The history of ranching in Canada can be divided into two main eras. The first began in the 1870s, with ranchers from the United States and investors from Great Britain beginning to fill what is now southern Alberta and southwestern Saskatchewan with cattle. This coincides with the indiscriminate slaughter of the last bison herds and the signing of the treaties that moved First Nations onto reserves. The second era began in the 1890s, as the Government of Canada began to favour fenced-in farm settlements over open range ranching. Sheilagh Jameson provides a sweeping overview of developments in ranching that are scrutinized in more depth in subsequent essays.

Since the 1970s, there has been an outpouring of research on First Nations in the context of the history of the Canadian Plains. This research has shed light on previously ignored subjects, including reserve-based farming. Summarizing this work, Bruce Dawson supplies a superb review of First Nations farming initiatives during the era of Euro-Canadian settlement. A lack of capital, substandard agricultural land on reserves, and conflicting policy objectives all served to undermine First Nations agricultural initiatives such that they had become a historical footnote by the Second World War.

FARMING

In 1812, Lord Selkirk founded the Red River Colony in what would become known as Manitoba. This was the site of the first major initiative in commercial farming in the Canadian Plains.[4] In his essay, Barry Kaye cautions against drawing too sharp a distinction between the colony's farmers and buffalo hunters. However, there was a conflict between the two groups as Red River farmers increasingly grazed livestock on the non-farm areas—the commons—that were essential to the Métis hunters for horse grazing. After 1870, except for the most poorly drained portions of the Red River lowland, the commons was surveyed and parcelled into quarter-section farms.

After southern Manitoba was settled, prospective farmers streamed further west in search of high quality agricultural land. By the early 1880s, thanks in large part to the Canadian Pacific Railway, settlers were making their way into what is now southern Saskatchewan.[5] To provide more complete understanding of what was required of these pioneers, Lyle Dick examines the cost of establishing a farm in two southeastern Saskatchewan areas—the predominantly Anglo-Canadian settlement of Abernethy, and the nearby German bloc settlement of Neudorf.[6] He explains that "farm-making was usually a gradual process" involving several years with the first phase devoted to survival and subsistence. The less cash for investment available to the settler at the beginning, the longer the process. Only after a minimum "proving-up period of three years or more" was a homesteader able to begin transforming the farm into a viable commercial operation.[7]

Although open range ranching had been part of the landscape in Alberta since the 1870s, farming would come considerably later. Warren Elofson provides a case example of how the foothill region around Pincher Creek was settled and farmed, describing how the open range was replaced by mixed livestock farms with fenced-in boundaries. In the flatter regions immediately surrounding Pincher Creek, specialized wheat farming overtook mixed farming to become the norm. By 1914, over 16,000 acres of land were seeded to wheat in an area that would eventually revert back to ranching, becoming famous as the centre of the Canadian cattle industry.

Many of the settlers moving to the Prairie West were encountering dryland farming for the first time. They travelled from regions with more precipitation, such as Ontario, Minnesota, and Europe, and they had to adjust to the semi-aridity and short growing season of the Canadian Plains. Tony Ward examines how these conditions spurred major adaptations and innovations to grain farming technology in the decades before the First World War. These early innovations include the disc plough, seed drill, self-raking reaper, and twine binder. Of course, Canadian dryland farmers benefited from the experience of dryland farmers in the United States where, for example, steam threshing had become commonplace by the 1890s. Ward estimates that the technological improvements of these years led to a substantial increase in farm size, as well as a 250 percent increase in net output, and set the stage for the wheat boom at the turn-of-the twentieth century.

Technological change only accelerated after the first phase described by Ward.[8] In his essay on the second stage of agricultural mechanization on the Canadian Plains, Bruce Shepard describes how the introduction of steam-powered (and then gasoline-powered) tractors and the development of combines for harvest, transformed grain cultivation. Mechanization also involved the extensive use of trucks—dryland farmers in the United States and Canada were central to the growth of both automobile and farm equipment manufacturing industries in the two countries. Though mechanization slowed during the Great Depression because of low wheat prices and plummeting farm income, it resumed its rapid pace after the Second World War to leave a legacy of ever-larger farms and shrinking rural populations throughout the Prairie Provinces.

RANCHING

For over a century, grain-fed Alberta beef has been a Canadian institution, a food product highly valued by consumers around the world. The heart of the cattle industry lies in the foothills of Alberta. Ranching was introduced here in the 1870s, and boomed in the decade that followed. By the census of 1891, there were approximately 128,000 head of cattle in the foothills and in the rolling country immediately east of the foothills. The majority of these cattle belonged to large herds owned by corporate ranch operations that could be established only with a large capital investment. As A. B. McCullough articulates, the bulk of this investment originated from central Canada and Great Britain, allowing the industry to grow rapidly. McCullough concludes that eastern capital so forced "the pace of development" of the industry, that it forestalled an influx of what could have been an even large number of American ranchers into the foothills.[9]

Simon Evans's first essay relates the story of the American ranchers who moved to Canada. Canadian ranchers dominated the new industry in western Alberta so these Americans had to move their herds into the short-grass prairies further east of the foothills. Evans describes the precise routes followed by American cattlemen. As farm settlement increasingly encroached on the open range in the United States, well-capitalized ranchers moved large herds over extensive distances while smaller ranchers moved their herds from Montana into eastern Alberta and southern Saskatchewan. The growth of the American cattle industry in Canada was arrested with the arrival of a "killer winter" in 1906-07, when the ranchers in the short-grass regions lost the majority of their cattle. As American cattlemen sold their surviving cattle and returned to the United States, much of the open range they had previously used for grazing was opened to farm settlement for the first time.

In his second essay, Evans thoroughly investigates the reasons for the closing of the open range in the Prairie West. According to Evans, the prosperity of open range ranching was a product of three factors: 1) the support of the Canadian government through its leasing policies; 2) the perception that the short-grass prairies of the semi-arid Palliser Triangle was unsuitable for grain farming; and 3) the almost insatiable demand by British consumers for Canadian beef. All three factors had disappeared before the winter of 1906-07 sealed the fate of the open range. At the turn of the twentieth century, an unusually wet period left the impression that large portions of the Palliser Triangle were actually suitable for farming, and the Government of Canada altered its previously generous ranch lease policy to encourage farm settlement. In addition, ranchers faced new competition when the British market

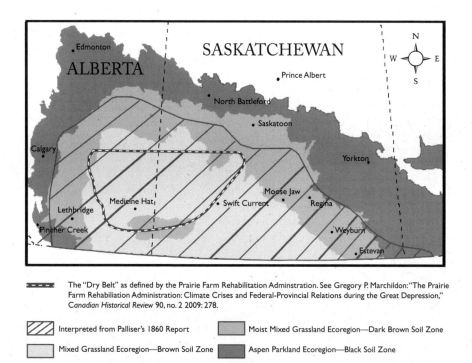

The "Dry Belt" as defined by the Prairie Farm Rehabilitation Adminstration. See Gregory P. Marchildon:"The Prairie Farm Rehabilitation Administration: Climate Crises and Federal-Provincial Relations during the Great Depression," *Canadian Historical Review* 90, no. 2 2009: 278.

Interpreted from Palliser's 1860 Report

Mixed Grassland Ecoregion—Brown Soil Zone

Moist Mixed Grassland Ecoregion—Dark Brown Soil Zone

Aspen Parkland Ecoregion—Black Soil Zone

Figure 1. The Palliser Triangle and the "Dry Belt" pictured in relation to the grasslands and parklands ecoregions.

for live Canadian cattle was altered by refrigeration, and large quantities of chilled beef were imported from Argentina.

The decision to convert the grasslands east of the foothills into croplands was not revisited until the return of drier weather after the First World War. The driest core of the Palliser Triangle, known as the Dry Belt, was the first to experience extensive drought, provoking an emergency response by the government of Alberta that encouraged a reversion to ranching (see Figure 1).[10] On the Saskatchewan side of the border, the provincial government largely continued to support grain farming. One exception was in the 1920s, when the Saskatchewan government facilitated the organization of grazing cooperatives from unused Crown lands to encourage wheat farmers in the Palliser Triangle to add livestock to their operations. Thomas Isern evaluates this policy with a case study of the Matador Community Pasture near Swift Current.

MARKETING

Whether grain or meat, the majority of what was grown or raised on the Canadian Plains was shipped to the rest of the world. Given the relatively tiny size of the home market, selling and exporting were always essential—though highly contested—aspects of Western Canadian agriculture. Until at least the 1950s, Great Britain was the largest single destination market for foodstuffs from the Canadian Plains.[11] Max Foran assesses the impact of changing British food health regulations and standards on Canadian (and American) live cattle exports to the United Kingdom. He details how Canadian exporters fought an embargo that required the slaughter of cattle at the point-of-entry into Britain in order to prevent disease from spreading to domestic cattle.

The Prairie Provinces—especially Saskatchewan—became known as "the breadbasket of the world," while vying with the United States, Argentina, and Australia to become the single largest wheat exporter in the world.[12] Private grain dealers and millers managed the export business on behalf of prairie farmers. With head offices in Winnipeg, these grain companies owned thousands of grain elevators at hundreds of railway sidings throughout the Prairie Provinces. From the beginning, many farmers concluded that these companies were colluding with the eastern-based railways and banks in order to buy wheat at artificially low prices. As a consequence, farmers formed their own organizations to fight these syndicates and receive what they considered a fair price for their product. D. J. Hall describes the agitation that led to the Manitoba Grain Act of 1900 and the accompanying struggle to make it effective in checking the worst abuses of the grain companies and the railways.

When regulation proved inadequate, farmers tried to create their own alternative to the private market by establishing their own grain companies, the largest of which was the Saskatchewan Wheat Pool established in 1924.[13] In his essay, Robert Irwin portrays how, sixteen years before this, one group of farmers launched a movement for a government-owned elevator system. Named after E. A. Partridge, a well-known farmer-radical from Sintaluta, Saskatchewan, the Partridge Plan called upon the provincial governments to own all grain storage elevators and the federal government to operate the major shipping terminal elevators in the country.[14] The Partridge Plan was launched by the Saskatchewan Grain Growers' Association, and was coolly received by provincial cabinets despite the political influence wielded by grain grower associations in Saskatchewan and Manitoba. Ultimately, a Royal Commission on Elevators appointed by the Saskatchewan government recommended that farmers dissatisfied with the private grain handlers should rely on their farmer-owned enterprises rather than government ownership to market their own grain.

Although farmers would create grain handling co-operatives in all three Prairie Provinces in the 1920s, a significant constituency of farmers held firm to the idea that only the power of government was capable of countering the

abuse of power by the private grain interests. The idea, as passed down from
E. A. Partridge and other early farm leaders, was to replace the "private chaos"
of the competitive market with the public discipline of orderly marketing. The
plummeting price of wheat during the Great Depression, and the financial
ruin faced by the prairie-based wheat pools, precipitated a crisis that led to the
first major intervention by the Government of Canada. Irwin's second essay
reviews the origins of the Canadian Wheat Board, from its first appearance
as a voluntary marketer in the depths of the Great Depression, to its development as a single-desk (compulsory) marketer of prairie grains to foreign
countries during the Second World War.

The tension between the interests that supported the free market and the
farm organizations that advocated public or co-operative alternatives to the
market continued after the Great Depression. Patrick Brennan outlines the
Winnipeg Free Press's support of free trade and its vehement opposition to
the Canada-United Kingdom Wheat Agreement of 1946. The *Free Press*'s
main target was Jimmy Gardiner, the federal minister of agriculture and chief
proponent of managed trade between Canada, the world's largest exporter of
wheat, and the United Kingdom, the world's largest importer of wheat. As
Brennan points out, the *Free Press*'s editorial opinion, though consistent with
the newspaper's long history, was not in keeping with the majority of farmers'
support for managed—as opposed to free—trade.

NOTES

1   *The Canadian Oxford Dictionary* (Toronto: Oxford University Press, 1998), 24. The word is derived from the Latin word *agricultura* in which *agri* means *field* and *cultura* means *culture*.

2   *Farm* is defined as "an area of land, and the buildings on it, used for growing crops, rearing animals, etc." (p. 502), while *ranch* is defined as a "cattle-breeding farm esp. in the western us and Canada" (p. 1194) in *The Canadian Oxford Dictionary*.

3   The Provincial Archivist of Saskatchewan from 1946 until 1957, Thomas was an associate professor of history at the Regina campus of the University of Saskatchewan from 1957 until 1964 when he joined the history department at the University of Alberta where he remained until 1982. Thomas was the author of *The Struggle for Responsible Government in the North-West Territories, 1870-97* (Toronto: University of Toronto Press, 1957) and *The North-West Territories, 1870-1905* (Ottawa: Canadian Historical Association, 1970) as well as a number of edited books on the history of the Canadian Plains.

4   See J. M. Bumsted, *Trials and Tribulations: The Red River Settlement and the Emergence of Manitoba, 1811-1870* (Winnipeg: Great Plains Publications, 2003), and *Lord Selkirk: A Life* (Winnipeg: University of Manitoba Press, 2008).

5   See volume 2 of the *History of the Prairie West* series – Gregory P. Marchildon, ed., *Immigration and Settlement, 1870-1939* (Regina: CPRC Press, 2009).

6   Dick's article generated a highly articulate response by Irene Spry in which she pointed out the importance of including labour costs—absent from Dick's analysis—in order to provide a complete picture of the costs of settlement. See Irene M. Spry, "The Costs of Making a Farm on the Prairies," *Prairie Forum* 7, no. 1 (1982): 95-99, and Lyle Dick, "A Reply to Professor Spry's Critique 'The Cost of Making a Farm on the Prairies'," *Prairie Forum* 7, no. 1 (1982): 101.

7   Lyle Dick, "Estimates of Farm-Making Costs in Saskatchewan, 1882-1914," *Prairie Forum* 6, no. 2 (1981), 194.

8   Jeremy Adelman, "The Social Bases of Technical Change: Mechanization of the Wheatlands of Argentina and Canada," *Comparative Studies in Society and History* 34, no. 2 (April 1992): 271-300.

9   A. B. McCullough, "Eastern Capital, Government Purchases and the Development of Canadian Ranching," *Prairie Forum* 22, no. 2 (1997), 234.

10   See Jack Gorman, *A Land Reclaimed: The Story of Alberta's Special Areas* (Hanna, AB: Gorman & Gorman Ltd., 1988), and Gregory P. Marchildon: "Institutional Adaptation to Drought and the Special Areas of Alberta, 1909-1939," *Prairie Forum* 32, no. 2 (2007): 251-271; and "The Prairie Farm Rehabilitation Administration: Climate Crisis and Federal-Provincial Relations during the Great Depression," *Canadian Historical Review* 90, no. 2 (2009): 275-301.

11   Gregory P. Marchildon, "Canadian-American Agricultural Trade Relations: A Brief History," *American Review of Canadian Studies* 28, no. 3 (1998): 233-252.

12   Wilfred Manenbaum, *The World Wheat Economy, 1885-1939* (Cambridge, MA: Harvard University Press, 1953).

13   Garry Fairbairn, *From Prairie Roots: The Remarkable Story of Saskatchewan Wheat Pool* (Saskatoon: Western Producer Prairie Books, 1984).

14   Murray Knuttila, *"That Man Partridge": E. A. Partridge, His Thoughts and Times* (Regina: Canadian Plains Research Center, 1994).

Surveying the Field

1. A History of Agriculture on the Prairies to 1914

Lewis H. Thomas

The period covered in this article is from fur-trade times to 1914, which is a logical division point. From 1914 to the present day there have been, firstly, a revolution in agriculture, and secondly, a change in the role of agriculture in relation to other sectors of the economy. These developments, together with the history of prairie ranching and the livestock industry, are judged to be outside the scope of this paper, which presents a general introduction to the evolution of crop growing in the region.

Agricultural activity is impossible without fertile soil and favourable climatic conditions, and unless these qualities be perceived by men, the land will remain untouched. It is therefore pertinent to begin this account with a description of the early experiments in vegetable and cereal production and with the progress of knowledge of the agricultural potential of the West (see Figure 1).

Food production during the fur-trade period, which ended in 1870, consisted primarily of the production of pemmican from buffalo meat. However, from the founding of the first posts by the Hudson's Bay Company their post managers were encouraged to grow foodstuffs to enable the posts to become self-sufficient. In 1674 the London directorate ordered that "there be provided . . . a bushel of wheat and rye, barley and oats, or a barrel of each in casks, and such sorts of garden seeds as the Governor shall advise."[1] Since the earliest company posts were on the shores of Hudson and James Bays where soil and climatic conditions were severely restrictive, no significant results were achieved in that region. But as soon as the Company moved inland the situation changed. Cumberland House, the first inland post established in 1774, had by the early nineteenth century an extensive farming operation. Travellers' accounts describe the farm, which produced field crops, vegetables and livestock.

Upstream, about twenty years earlier the French traders at Fort la Corne were producing vegetables and making this the scene of the first agricultural activity in Saskatchewan.

After the Conquest we find the Montreal traders raising garden produce at Fort Vermilion and at Peter Pond's post on the lower Athabasca. The latter, says Alexander Mackenzie, "had formed as fine a kitchen garden as I ever saw in Canada."[2] Alexander Henry, the Nor'Wester, is a major source of information, for he describes the crops which he raised during the first decade of the last century at several posts on the North Saskatchewan River. By the middle of the nineteenth century the Hudson's Bay posts at Fort Resolution on Great Slave Lake, Fort Simpson on the Mackenzie River and Fort Liard in the southwest corner of the present-day Northwest Territories had flourishing gardens. Writing in 1952, Grant MacEwan declares:

> In both the Yukon and the Northwest Territories the most promising areas for farming are along the rivers and beside the lakes. It now appears that the Yukon has at least half a million acres for which some use in an agricultural programme will be found. The Northwest Territories will have more.[3]

One hundred years before MacEwan's pronouncement, Sir John Richardson, who visited Rupert's Land as the leader of one of the Franklin search expeditions, wrote as follows on the subject of cereal production:[4]

> *Wheat is* the cereal which requires most heat of those usually cultivated in England. Its culture is said to ascend to 62° or 64° north lat. On the west side of the Scandinavian peninsula, but not to be of importance beyond the 60th. On the route of the Expedition it is raised with profit at Fort Liard in lat. 60°5' north, long. 122° 31' west, and having an altitude of between 400 and 500 feet above the sea. This locality, however, being in the vicinity of the Rocky Mountains, is subject to summer frosts; and the grain does not ripen perfectly every year, though in favourable seasons it gives a good return. At Dunvegan, on Peace River, lying in lat. 56°6' north, long. 117° 45' west, and at an altitude of 778 feet, the culture of this grain is said to be equally precarious. It grows, however, freely on the banks of the Saskatchewan, except near Hudson's Bay, where the summer temperature is too low.
>
> *Oats* are little cultivated in Rupert's Land; they require longer time than barley to ripen, and are therefore not likely to grow so far north. They have not been tried at Fort Norman, however,

which is the most advanced post in that direction where barley is cultivated... .

In good seasons *barley* ripens well at Fort Norman on the 65th parallel, ... All Mr. Bell's attempts to raise it at Fort Good Hope, two degrees further north, failed... .

Potatoes ... yield abundantly at Fort Liard, and grow, though inferior in quality, at Fort Simpson and Fort Norman. They have not succeeded at Fort Good Hope, near the 67th parallel. At the latter place *turnips* in favourable seasons attained a weight of from two to three pounds, and were generally sown in the last week of May. At Peel's River the trials made to grow culinary vegetables had no success. Nothing grew except a few cresses... .

West of Fort la Corne the chief agricultural centers were at Fort Carlton and Edmonton House. The talented writer and fur trader, Alexander Ross, writes as follows on agricultural activity at various trading posts on the Saskatchewan of which he had first-hand knowledge:[5]

Fort Edmonton is a large compact establishment, with good buildings, palisades, and bastions, pleasantly situated in a deep valley. An extensive and profitable trade is carried on with the warlike tribes of the plains-Blackfeet, Piegans, Assiniboines, and Crees. All these roving bands look up to Mr. Rowand as their common father, and he has for more than a quarter of a century taught them to love and to fear him. Attached to this place are two large parks for raising grain, and, the soil being good, it produces large crops of barley and potatoes; but the spring and fall frosts prove injurious to wheat, which, in consequence, seldom comes to maturity.

Of Carlton House, he says:

There are ... some good cultivated fields, which, with moderate industry, are said to yield abundant crops of barley and potatoes. Wheat grows here, and hops have been raised with great success; the gardens also produce good returns of onions, carrots, turnips, and cabbages. And here I noticed the best root-houses I have seen in the country. It is pleasing thus to witness the fruits of industry and progress of civilisation in the savage wilderness... .

Of Cumberland House, Ross reports:

> The trade of the place is, however, fast dwindling away to nothing; but in proportion as furs and animals of the chase are decreasing, agriculture seems to be increasing, and perhaps eventually the latter may prove to the natives more beneficial than the former.
>
> In addition to the cultivated fields, we have to notice here the cheering prospect of domestic comfort. The introduction of domestic cattle from the colony of Red River gives a new feature of civilisation to the place. Here are two fine milch cows and a bull, and more are expected. In addition to these, other proofs of industry and comfort are manifest. A neat kitchen-garden, which furnishes an ample supply of vegetables, adorns the place, in the centre of which stands a sun-dial neatly cut and figured; the latitude of the place, 53° 57' N., being marked on it.

Concluding his description of agricultural activity in the early nineteenth century, he writes:

> We have noticed on our route the commencement of agriculture and the introduction of domestic cattle at Cumberland House. From Fort Edmonton down to Carlton, and far below, a range of five hundred miles, the country and climate invite the husbandman and the plough.

The major agricultural center in the Prairie West was the Red River Settlement. There were probably gardens at the trading posts at the junction of the Red and Assiniboine before the arrival of the Selkirk settlers in 1812, but it was these hardy Scottish people who produced the first large crops of grain on the prairies. These pioneer farmers were plagued with early frosts, grasshoppers, drought and floods. But by the late 1820s agricultural produce and buffalo meat made the population self-sufficient, and the settlers had begun to produce a surplus of flour for sale to the Hudson's Bay Company. The Selkirk colonists were also able to produce a surplus of butter, pork, beef, cheese, eggs, and some vegetables, although only in small quantities at the start. By the mid-nineteenth century the Red River Settlement had reached its optimum development as an agricultural center. Further development depended on railway communication with either eastern Canada or the American midwest. Of course this also applied to the area situated west of Red River.

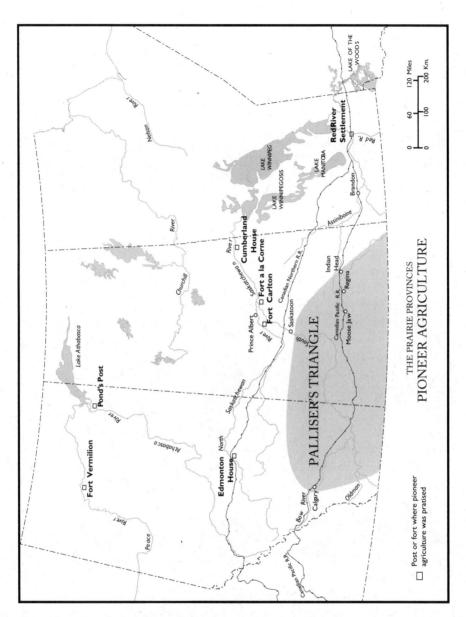

Figure 1. Locations where early agriculture was practiced.

In view of the agricultural activity at the western fur posts and Red River, it may seem strange that there were serious doubts that the plains region was suited to agriculture. This became a matter of public controversy in Britain during the middle of the nineteenth century.[6] One factor in this dispute was the reluctance of the Hudson's Bay Company to admit the possibility of success of an enterprise which would compete with and ultimately displace the fur trade. The major meat supply for the northern trade, buffalo pemmican from the plains region, would be threatened. In Canada, however, much more sanguine views prevailed. But the controversy could not be resolved without scientific investigation.

This was provided in the late 1850s. At this time the geography of the area north of the Saskatchewan was well known, but the fur traders had avoided the open plains, which was a battleground of the Crees, Assiniboines and Blackfoot. One scientific breakthrough was the publication in 1857 of Lorin Blodgett's *Climatology of the United States and of the Temperate Latitudes of the North American Continent*. The charts in that work showed that annual temperature and rainfall were similar to those prevailing in the Quebec-Montreal area. It appeared that far from being a desert, the grassland and parkland could be expected to be fertile. On-the-spot evidence was provided by two Canadians, geology professor Henry Hind and civil engineer Simon Dawson.[7] Their two-year expedition was undertaken on orders from the Government of Canada. It was designed to provide information on the transport route from Canada to the Red River Settlement, and to assess the agricultural potential of the Red River Valley and the country lying to the westward. The two men submitted separate reports. They agreed on the fertility of the Saskatchewan River valley, but differed on the fertility of the open plains. Dawson was much more optimistic than Hind, who was convinced that what was then called "the Great American Desert" extended well into the prairie region.

Meanwhile a British Parliamentary committee had been investigating the affairs of the Hudson's Bay Company and discussing the issue of uniting Canada and a portion of Rupert's Land. Its 1857 report was based on the belief that a large fertile area could be located which could be united with Canada. Probably Alexander Ross's book influenced the Committee members. Nothing came of this recommendation, but its formulation showed which way the wind was blowing.

The British-based Palliser Expedition of 1857-60 was a much more impressive scientific investigation than that of Hind and Dawson in terms of personnel and area explored.[8] It was staffed by a geologist, a botanist, a magnetic observer and a specialist in astronomy. The Hind and Dawson exploration had examined the plains country as far west as the South Saskatchewan, but Palliser's party covered not only this area but also the adjacent area west to the Rockies. It also examined possible railway routes through the mountains to the mouth of the Fraser.

Palliser's report is famous for its designation of the "fertile belt"—the crescent of tall-grass plains from Red River along the Saskatchewan River and the foothills country. It is also well known for its definition of the "Palliser Triangle," a semi-arid region which was described as an extension of the "Great American Desert," and unsuited for cultivation.

Greater in influence than Palliser or Hind was the Ontario botanist, John Macoun, who was employed by the Dominion Government in the 1870s to accompany Sandford Fleming on his railway survey, and Alfred Selwyn, the Director of the Geological Survey, on several exploratory expeditions. In 1879, 1880 and 1881 Macoun traversed the southern portion of the North-West Territories. He disputed Palliser's idea that the country between the Souris River and the Rockies was arid and unsuited to agriculture. In his reports to the Government and to the public his enthusiasm for this treeless, short-grass country was unbounded. His claim that in addition to the Saskatchewan Valley, good wheat lands extended south to the international boundary was of great influence. He undoubtedly encouraged the Canadian Pacific Railway to build due west from Winnipeg to the Rockies. And he was a tireless advocate of the agricultural potential of the entire plains region. His magnum opus, the 687-page volume entitled *Manitoba and the Great North-West*, was widely distributed and generated much enthusiasm among would-be settlers. A St. Paul newspaper later urged Minnesota readers to:

> Buy farmlands in Saskatchewan. You can leave home after Easter, sow your grain and take in the harvest and come home with your pockets full of money in time for Thanksgiving dinner.[9]

Of this boosterism Grant MacEwan quotes an unknown author: "Where everyman's a liar, that's where the West begins."[10] Certainly the record of homestead cancellations bears this out. Professor Chester Martin has written: "The cancellations in free homestead entries (more than 57 percent ... in Saskatchewan between 1911 and 1931) were the highest in any category of Dominion Lands... ."[11] The accompanying wastage of human enterprise has never been measured, but it must have been appalling.

Macoun cannot be entirely blamed for his optimistic appraisals: he visited the plains during wet years when Manitoba farmers enjoyed large crops and the country to the west was lush. The truth lies somewhere between the calculations of Macoun and Palliser.

More detailed information concerning the agricultural potential of the land in each township was provided to prospective settlers in the federal government's published reports of the surveyors in the 1880s.[12] But it is not clear how extensively these publications were distributed, and what influence they had on the decisions of the homesteaders and purchasers of lands. The files and township registers for every quarter section of homestead lands in the prairies have been preserved, and it is obvious from the number of cancellations of homestead entries for many quarter sections that the settlement of the West was a painful process for many pioneers. Untold thousands of them gave up the struggle to establish themselves on marginal land, went back east or south to the United States, or west to British Columbia. They were discouraged from moving north into the fertile belt by the absence of railway facilities, which were not constructed until after the turn of the century. But apart from the surveyors' reports and the testimony of other pioneers, the agriculturalist lacked reliable information on soil quality until the maps of the scientific Soil Surveys were published in the 1930s.

No matter how fertile the West might prove to be, it would remain an empty land, or at best a region of subsistence agriculture, without effective transportation facilities. Transportation was essential to permit the export of wheat, barley and oats, in other words, the development of an exportable cash crop. Equally important was the import of farm implements, foodstuffs, tools and household goods.

Trails traversed by Red River carts had provided overland transport during the fur-trade period, particularly over the route between Fort Garry and Edmonton. Obviously they were inadequate in serving the needs of the western farmer when large-scale settlement became a practical prospect. Nor were the paddle-wheel steamboats plying the North Saskatchewan River in the late seventies and eighties a solution. Only the railway would serve. The Canadian Pacific Railway linked the West to Central Canada and was of immense importance in serving the settlers who had arrived in the seventies, chiefly in Manitoba, and rendered Winnipeg less dependent on the line which had linked it to St. Paul in 1878. The first wheat exported from Manitoba had been shipped by flat-bottomed boat up the Red River two years earlier, in 1876.

The significance of the construction of the Railway, which was built rapidly across the plains in 1882 and 1883, was not merely that it provided a vital transportation facility, but that it became a promoter of immigration and settlement. Its 25 million-acre land grant was sold at attractive prices, for the company was primarily interested in revenues from wheat shipments and the import of commodities from Central Canada, rather than revenue from land sales.

Although the pioneer farmers recognized the value of the c.p.r., many of them were located some distance from the main line. Though new trails and ferries appeared, reaching a railway usually involved long weary hours behind a team in all sorts of weather, from dusty heat to a blinding blizzard.

Building the Canadian Northern Railway line in the Pelly district of Saskatchewan in 1909. The construction of Canadian Northern and Grand Trunk Pacific lines in the early 1900s greatly opened up regions to agricultural settlement.

An agitation for branch line railway construction soon made itself felt. Since the c.p.r. was never able to keep up with this demand, farmers favoured the construction of competing lines, such as the Manitoba and Northwestern Railway, which ran from Winnipeg to Yorkton. The reputation of the c.p.r. dipped in western opinion over issues like freight rates and the building of branches. The politics of Manitoba in the 1880s was dominated by railways, and federal-provincial relations were bedevilled by the attempts of the Manitoba Government to charter competing lines. It is perhaps not surprising that not one but two rival transcontinentals, the Canadian Northern and the Grand Trunk, were constructed after 1900.

After the North West Mounted Police had been established, the land survey system created, the Indians confined to reservations, and a government for the North West Territories organized, all in the 1870s, the West was ready to receive immigrants. The largest number came from Ontario, with smaller numbers from other parts of eastern Canada and the United States, and from the United Kingdom. As early as the 1870s non English-speaking immigrants appeared—German-speaking Mennonites and Icelanders. These were the forerunners of the great flood of continental European people who swept across the prairies seeking economic opportunity in the two decades before the outbreak of World War I. The flood ebbed and advanced, and whenever migration showed signs of decline the federal government sought to offset it by large grants of land to group settlements and colonization companies, and intensified its advertising campaign with pamphlets labelled "Homes for Millions" and "The Last, Best West." These government initiatives were backed up by the colonization departments of the railway companies.

The report of the Dominion Lands Agent at Prince Albert for the year 1898 conveys a first-hand impression of the state of affairs:

> You will observe from the accompanying tabulated statement that the number of homestead entries made shows a very marked increase over that of any recent year. This increase was mainly due to a very considerable immigration of Mennonites from Manitoba, the United States and Russia, into the Rosthern and Hague settlements, who were attracted there by the encouraging reports of the success of many of their co-religionists who had previously settled in those localities; for the same reason a still larger immigration of these people is expected there during the coming season.
>
> The year, I may truthfully say, has been a prosperous one. The harvest of 1897 was excellent; that of 1898, although not quite so luxuriant, was fairly good, notwithstanding the somewhat deleterious effects of a frost in the month of June last, a very unusual occurrence in this district, which unfortunately cut down the more tender of the growing crops; the oat crop being the heaviest sufferer from this cause.
>
> The cattle industry is steadily increasing. The export trade in livestock for the European market is assuming considerable proportions, and the prices realized are encouraging, and are having their effect upon the prosperity and comforts of the people.
>
> Upon the whole the settlers are fairly contented with their present condition and hopeful for the future. The immediate construction of another railway line giving more direct communication with the eastern markets now in prospect, will tend to allay any discontent in the remote eastern settlements. It will also open out for further settlement an immense tract of land, which for beauty, natural advantages and fertility cannot be surpassed in any country.[13]

The agent's reference to homesteaders can serve to introduce us to the land law. The so-called staple of land policy was the free homestead, but it would be misleading to suppose that all quarter sections in a township were available for this purpose. Normally a little less than one half the acreage of a township could be acquired free by homesteaders. The remainder had to be bought from the government or from the c.p.r. and other land-grant railways, or from the Hudson's Bay Company. The homesteads were usually scattered throughout the township on the even-numbered sections. Many farmers acquired homesteads along with railway lands, for it soon became apparent that a living could not be made on a 160-acre homestead.

The acquisition of a homestead involved no money transaction except for a $10.00 registration fee. The duties of a homesteader were modified in a few respects from the time the first regulations were drafted in 1871 down to the year 1930 when the federal government transferred all ungranted or unsold public lands to the three prairie provinces. The homesteader was granted title to his land after at least six months' residence each year for three years. The regulations defined the acreage to be cultivated each year, and specified that a habitable house should be constructed. These duties remained in force throughout the pioneer period, supplemented by minor modifications from time to time.[14] For example, those settlers who favoured hamlet or village settlement, such as the Mennonites, were accommodated by being excused from residing on their quarter sections.

Reference has already been made to the advertising campaigns of the federal government and the c.p.r. This may be illustrated by an extract from the 1903 report of the Superintendent of Immigration:

> ... all told, we had to deal with 114,124 requests for information by mail, in addition to many personal inquiries. In response to these we sent out 342,372 pamphlets, maps, etc. We also sent our publications to 52,653 addresses of farmers in the United States, procured through our agents, and in addition the German translation of the Atlas *of Western Canada* and the newspaper *Der Nordwesten* were sent to 52,000 special addresses of Germans in the American rural districts. We also received and sent out 72,000 papers in the Scandinavian languages, circulating them in the manner calculated to produce the best results. We shipped to our agents in the United States and Great Britain 575 cases, containing 637,578 pamphlets, etc., for distribution, our total output of literature being thus 1,313,909 separate copies or pieces.[15]

No opportunity was neglected to promote western settlement. In the same 1903 report the Superintendent noted:

> A party of Scotch curlers made a tour of Canada at the beginning of the present calendar year, and as several of them were understood to be representative agriculturists on the other side, it was felt that it would be proper for the department to show them some attention and to spend a little money in entertaining them and sending them away with a good impression of this country.[16]

Newspaper editors from the United Kingdom and the United States were transported free of charge to various districts in the hope that they would give favourable publicity to opportunities for farmers. Successful farmers were given transportation to the United Kingdom to visit their home districts with the same objective, and men of influence in Great Britain, such as Lloyd George, were given passes for visits to the Canadian West.

It is impossible to measure the effects of these public-relations stimulants. The population growth of Manitoba and the North West Territories was

On the homestead in the Eston district of Saskatchewan, circa 1915. Although settlers had begun arriving in the region after 1906, the construction of Canadian Northern's line through the area in the mid-1910s precipitated the arrival of many newcomers, such as those pictured here with their packing cases for effects still visible.

impressive. Manitoba increased from about 12,000 in 1871 to 108,000 in 1886, and 255,000 in 1901. The North West Territories farming population could be measured in hundreds in 1871, but amounted to 31,000 in 1885, and 164,000 in 1901. The decade 1901-11 exhibited the most dramatic growth in the West-from 419,000 in 1901 to 1,328,000 in 1911.

In the 1870s the frontier of settlement moved west through central Manitoba, but with the construction of the c.p.r. the Territories exhibited a marked growth paralleling the railway. The smaller number of farmers who had moved into the Saskatchewan valley in the 1870s were linked rather inadequately by the railway from Regina to Prince Albert, completed in 1890, and the Calgary-Edmonton line of 1891. The vast area between the Saskatchewan and the c.p.r. main line was largely occupied in the fifteen years preceding the War. Another extract from the Commissioner of Immigration's report for 1903 comments on the first settlements in part of this region:

> J. H. Gooderham, sub-agent of Dominion lands at Touchwood Hills, Assiniboia, reports the gradual opening up of his district for settlement, notwithstanding its distance from a railway at the present time. He says there is no more suitable tract of country than the Touchwood Hills for mixed farming, as it has all the requisites—good soil, timber, hay and water; and those settlers who have already gone into grain-raising have proved that the product is of a very superior quality. Owing to the distance from railway, this district has not been largely favoured by immigrants, but those who went there a number of years ago from Ontario and the old country, with practically nothing of commercial value, are today independent, and have secured every reasonable comfort. With the advent of the Kirkella branch of the Canadian Pacific Railway, this locality will form a very desirable location for a number of settlers from this time forward.
>
> F. J. Musgrave, sub-agent of Dominion lands at Estevan, Assiniboia, reports 523 homesteads applied for through his office, and about 10,000 acres of land under cultivation in the immediate district, the greater part of which is under flax. All settlers appear to be well pleased with the country, and the conditions of the weather have been such as to make crop prospects all that could be desired.
>
> R. M. Mitchell, sub-agent of Dominion lands at Weyburn, Assiniboia, reports the arrival of a large number of American

settlers of a thrifty class, and creditable to any nation. They
appear to be anxious to find out the laws of Canada and obey
them. It is pointed out that there is a large quantity of good
land south-west of Weyburn, yet unsurveyed, which could be
immediately settled if the new-comers could be properly located.
One thousand one hundred and ninety-five (1,195) homesteads
were taken through this sub-agency during the year.[17]

In this last great frontier area group settlements of European immigrants
were conspicuous. But there were also a few group settlements of British
origin, notably the Barr Colonists and the Patagonian Welsh.

The group settlements were served primarily by the Canadian Northern
Railway and the Grand Trunk Pacific main lines and their branches, although
C.P.R. branch lines were also in the picture. There were also many settlers in
these areas who came as individuals or families, chiefly from Eastern Canada
and the British Isles. Economic opportunity, the chance to own one's own
farm, was the dominant motive of the pioneers, although political and social
conditions in their homelands also played a part in the migration.

Many of the pioneers had previous farming experience, but many had none.
These latter were assisted by land guides and neighbours who had arrived earlier.

Having provided himself with some implements and supplies, and a few
oxen or horses, the next task for the settler was to build a house, using materials at hand—sod in the treeless plains and logs in the parkland. Only a few
could pay for lumber for a frame house and barn.

Breaking the sod was the next task, and a formidable one in the parkland,
where tree growth had to be removed. Flax was often the first crop, to break
up the sod and to clean out the wireworms, but some pioneers planted wheat
immediately. Pole fences protected the crops at the start. Barbed wire was
adopted by many from the 1880s on.

The federal government's experimental farm system was founded in 1886,
headed by William Saunders, with its headquarters in Ottawa. Branches were
soon established at Brandon and Indian Head, and the system was of immense
value to the pioneers. The tillage practice of summerfallowing was publicized
by the farm staff after its benefits were realized about 1885. Among other
ventures, experiments with early-maturing wheat were the most important.

Red Fife, an Ontario wheat, probably of Scandinavian or north Russian
origin, was the wheat of the pioneer years of western development. Its only
defect was that it involved a fairly lengthy growing season in a region subject
to early frost. Charles Saunders, William's son, was appointed Dominion

Cerealist in 1903, and in his 1909-10 report he reviewed the history of experiments to secure an earlier-maturing variety than Red Fife:

> When the Dominion Experimental Farms were first established, the settlement of the great prairie country of central and western Canada had not progressed very far, and there were various problems of vital importance connected with the growing of wheat on the plains which awaited investigation. Hence it was natural that special attention should be paid to these new sections of country. While, therefore, the needs of the older farming districts have not been overlooked, and results of value to them have been reached along various lines of research, the most interesting branches of our work have been those concerning the great wheat-growing plains. The short summer of the prairies emphasized the need for early maturing varieties of wheat, while the long distance between the farmer and the main centers of wheat consumption made it essential that only such varieties should be grown as would command an exceptionally high price in the world's markets, so that the cost of transporting the grain would be relatively low.
>
> The prairie settlers found the famous Red Fife wheat very satisfactory on the whole, except in regard to the time taken to mature the crop, which, in the less favourable seasons, was rather too long; so that the fields were sometimes touched with frost before the grain was ready to be cut, thus very seriously lessening the farmers' income. In hardness of kernel and in flour strength (the characteristics which perhaps chiefly determine the selling price of any wheat) this variety ranks at the head of its class. What was needed, therefore, for the great wheat-growing plains was an early Red Fife, a variety having all the good qualities of ordinary Red Fife with the added excellence of earliness.
>
> To meet this need, early ripening varieties of wheat were imported from various countries by the Director of the Experimental Farms and, at as early a date as possible, experiments in cross-breeding were begun for the purpose of combining in one sort all the desired qualities....[18]

From these experiments the famous Marquis wheat emerged, and Saunders goes on to say:

Sir Charles Edward Saunders (1867-1937). Saunders crossed Red Fife with Hard Red Calcutta wheat, eventually producing the Marquis strain which won universal acclaim. Marquis became the foundation crop on the prairies, enabling the large-scale commercial production of high-quality bread wheat. By 1920, approximately 90 percent of the wheat grown in western Canada was Marquis, and substantial acreages were also grown in the United States. Charles Saunders was knighted in 1934. (Courtesy of the Saskatchewan Archives Board/R-A350.)

Marquis proved remarkably successful at many points last season, the yield of over 200 bushels from a 4-acre field on the Brandon Experimental Farm being worthy of special notice. Several farmers in Northern Saskatchewan grew it with unusually good results. The best sample which reached the Cerealist's office was grown by Mr. E. B. Cay at Beatty, Sask., and showed the phenomenal weight of 66¼ pounds to the measured bushel. Other very fine samples were received from Mr. Martin Dornian, of Disley, Sask. (65 pounds per bushel), and from Mr. L. T. Symonds, of Marshall, Sask. (64¾ pounds per bushel). In addition to its earliness, Marquis wheat is very desirable in certain sections on account of its somewhat shorter straw than Red Fife. Its good appearance and excellent baking records have been discussed in previous reports.

Taking all points into consideration, Marquis wheat is recommended as the most promising sort at present available for farmers who require a hard, red wheat of high baking strength and ripening earlier than Red Fife.[19]

The experiences of the early settlers are perhaps best illustrated in their own words. The following extract from the reminiscences of Mrs. Elizabeth Ruthig[20] illustrates many of the generalizations about early prairie agriculture which have been drawn above:

> In 1907, I married Phil Ruthig, at Kitchener, Ontario. We lived a year and a half at Waterloo and Galt. Then, in 1909, came the great opportunity. My parents, Henry and Caroline Ruhl, though then in their fifties, had the fortitude to come West to

the adventure of homesteading, with my two brothers, Jacob and Henry, twenty-three and twenty-one, and a sister, Caroline, sixteen. They persuaded Phil and I to come along, and as we had hoped some day to see the Western plains, we decided this was our chance... .

Each day, from a southeast corner on Coteau Street [Moose Jaw], we watched the pageant of many families in horse and ox-drawn wagons, starting on their long journey to the south, heavily laden with supplies. Often there would be a caravan of from fifty to sixty teams. The stubborn oxen had a weakness for water holes. There was a telephone pole in the center of a large one, north of our house, east of the trail. It amused us to see oxen head for this hole and stop with one on each side of the pole. Then the driver would have to wade in and tug and pull to get them out, or unhitch them. There were anxious moments for us, wondering what lay ahead for them. We knew our turn was coming.

The Land Office was in Moose Jaw and after waiting our turn and finding where to go, father, husband Phil and brothers Jacob and Henry, left in April with two wagons well loaded with such supplies as lumber, food and bedding. They were fortunate in securing a section and a half of land... .

In the spring of 1910 we built our one room sod shack, with shingled roof and board floor. Inside, the ceiling was of V-joint and painted white. Building paper covered the mud plastered walls, then flowered wallpaper. Net curtains adorned the three small windows. We were so thrilled with our first little home in the West.

After four years of travelling back and forth to Moose Jaw by wagon—each spring going to spend six months in the city where my husband did carpenter work, each fall returning to the homestead for the winter—we decided to equip ourselves with horses and implements for farm life. Father had done the necessary breaking for homestead duties on our land. So in the fall of 1913 we left Moose Jaw with a load of furniture, making a canopy over all with the linoleum. I sat on a trunk under it, prairie schooner style, so the wind couldn't reach me. That was the last of our journeying back and forth.

In 1914 bronco horses were bought at a hundred dollars each from ranchers to the southeast. What exciting times there were breaking these horses for driving! They were of no use until properly broken. The oxen had been pretty slow going. They could not stand the heat of the day, so it meant early rising. The fields were plowed from dawn till nine in the morning, then out again from five till dusk in the evening.

The first grain was sown by hand, till seed drills were available, or could be purchased. The crop was cut with a mower, raked into long rows with a horse rake, then tied into bundles by hand. In the nice days, during the winter, the bundles, stored in granaries were placed in a wagon box and threshed with a home made flail. The chaff was removed, and to clean the grain for seed a pailful was brought into the house and poured on the table. We would all sit round and pick out the kernels for seed. It took many hours and days, but was a pastime when chill winds blew and snow and blizzards prevailed.

The first loads of grain were hauled to Vanguard, a distance of fifty miles. It usually took three days to make the trip. Soon another railway came, from Moose Jaw to Eastend, in 1913. That, too, seemed a great distance, but by leaving in the early morning and returning late at night, this trip, twenty-two miles to Meyronne, and twenty-two back, could be made in a day.

In the mechanization of agriculture the West did not lag far behind Ontario. The first mechanical thresher, using a stationary steam engine, appeared in the Red River Settlement in the 1860s. Twine binders were used to some extent on the plains in the 1880s. The first steam traction engines, used for threshing and plowing, date from the 1890s. While a harvester combine was imported from California to Saskatchewan shortly before World War I, this machine was not widely used until the 1920s.

As is well known, the advance of mechanization increased the capital cost of maintaining or acquiring a farm. The emergence of the 800-acre farm (the average size of a prairie farm today) and the drastic reduction of the number of family farms are direct consequences of mechanization. The growth of urban centers, with Regina, Saskatoon, Calgary and Edmonton rivalling the earlier dominance of Winnipeg, has produced a large class of urbanites who know as little of farm life as the average New Yorker. As the economy of the prairies has changed, the role of agriculture has undergone a relative decline. But the problems faced by farmers have been remarkably persistent.

During the pioneer period, issues of mutual concern to farmers led to formation of Grain Growers' Associations after 1900. These inevitably became pressure groups operating on provincial and federal politicians. The original organizations are the parents of the more varied and specialized organizations of today. The continuity of the problems, however, is striking: orderly marketing, freight rates (including the Crow's Nest Pass agreement), branch railway lines, elevator facilities, the maximum use of the port of Churchill and the Hudson Bay Railway. All remain subjects of debate today. Whatever the future resolution of those problems will be, Canada must recognize the special position of the West as a source of foodstuffs, and of the Western farmer as a person entitled to a stable standard of living at least equal to the national average. In a hungry world, the national responsibility for enabling the prairie farmer to make maximum use of prairie soil while conserving its fertility cannot be escaped.

NOTES

This article first appeared in *Prairie Forum* 1, no. 1 (1976): 31-46.

1   Quoted in A. S. Morton, *Under Western Skies* (Toronto, 1937), p. 135.
2   Ibid., p. 136.
3   Grant MacEwan, *Between the Red and the Rockies* (Toronto, 1952), p. 286.
4   Sir John Richardson, *Arctic Searching Expedition* ... (London, 1851), p. 269-70.
5   Alexander Ross, *The Fur Hunters of the Far West* (London, 1851), Volume II, pp. 210, 214-15, 218, 222.
6   Lewis H. Thomas, "The Mid-19th Century Debate on the Future of the North West," *Documentary Problems in Canadian History*, Vol. 1, pp. 205-27.
7   See L. H. Thomas, "The Hind and Dawson Expeditions, 1857-58", *The Beaver* (Winter, 1958), pp. 39-45.
8   See Irene M. Spry, *The Palliser Expedition* (Toronto, 1963), p. 12 ff. For an edited version of the full report see Irene M. Spry, *The Papers of the Palliser Expedition, 1857-60* (Toronto: The Champlain Society, 1968).
9   Quoted in MacEwan, *op. cit.*, p. 85.
10   Ibid.
11   Chester Martin, *"Dominion Lands" Policy* (Toronto, 1938), p. 409.
12   Descriptions of the Townships of the North-West Territories, Dominion of Canada, between the Second and Third Meridians (Ottawa, 1886).
13   *Canada Sessional Papers* (hereinafter cited as C.S.P.), 1899, Paper No. 13.
14   See J. L. Tyman, *By Section, Township and Range: Studies in Prairie Settlement* (Brandon, 1972).
15   *C.S.P.*, 1904, Paper No. 25.
16   Ibid.
17   Ibid.
18   *C.S.P.*, 1911, Paper No. 9, p. 161.
19   Ibid., p. 171.
20   Elizabeth Ruthig, "Homestead Days in the McCord District," *Saskatchewan History*, Volume VII, No. 1 (Winter, 1954), pp. 22-26.)

2. Indian Agriculture in the Fur Trade Northwest

D. Wayne Moodie and Barry Kaye

At the time of European contact, the Indian inhabitants of the lands lying to the north and west of Lake Superior did not practice agriculture, and subsisted wholly by hunting, fishing and gathering. Not until the beginning of the nineteenth century did any of the native peoples of the Northwest begin to cultivate. Within this vast region, however, only the Indians of the Manitoba parklands and of the mixed forests of adjacent northwestern Ontario and northern Minnesota took to cultivating the soil. Although attempts were also made by Indians to extend cultivation into the purely grassland environments to the west, agriculture failed to take root in this region. However, it persisted in the more humid environments to the east and was most strongly developed by the native peoples living in northwestern Ontario and adjacent Minnesota. Even in this area none of the Indians became dominantly agricultural, and their small garden plots served mainly to supplement a subsistence economy that remained based upon hunting and gathering. Despite its limited nature, the introduction of cultivation into these areas represented a significant extension of Indian agriculture beyond its traditional northern limits in native North America.[1] It also contributed to the livelihood of the Indians who adopted it during a period of rapidly depleting fur and game resources and, in so doing, became a significant component of cultural change toward the end of the fur trade era.[2]

The purpose of this article is to elucidate the nature of this expansion of Indian agriculture that took place more than a century after the first Europeans had penetrated into the Canadian Northwest. Although this development was stimulated to some extent by the European fur trade, it was essentially Indian in both initiative and character. It was also the most northerly development of Indian agriculture on the North American continent, extending its limits some three hundred miles to the north of the prehistorical agricultural frontier in central North America. This article documents the development of agriculture in the southern Manitoba lowlands and its subsequent spread into adjacent areas. It also endeavours to explain the diffusion of Indian agriculture into these areas and to describe some of its salient characteristics.

At the time of European contact, the northern limit of Indian agriculture in the western interior of North America was probably at the Knife River villages of the Hidatsa Indians, located near the confluence of the Knife and Missouri Rivers at about 47°30'north latitude.[3] The agricultural Indians of the Upper Missouri region were the Mandan, Hidatsa and Ankara, all of whom cultivated Indian corn, beans, squashes, pumpkins, sunflowers and tobacco. The first European to describe the agricultural activities of these village Indians was the French explorer La Verendrye, who, in 1738, accompanied a party of Assiniboine Indians on a trading expedition to their settlements.

There is little reason to believe that native cultivation extended beyond the Knife River at the time of La Verendrye's explorations in the 1730s. However, it is becoming increasingly apparent that the northern limit of Indian agriculture observed by the French at this time represented a southward retreat from an earlier, more poleward limit. The first to recognize this possibility was the archaeologist, Waldo R. Wedel, who wrote that "it would perhaps have been feasible to grow corn in favoured spots throughout portions of the Dakota-Manitoba mound area in prehistoric times."[4] It is apparent that the Indians were cultivating to the east of the Missouri villagers in the Sheyenne and James River valleys of North Dakota in protohistorical time,[5] while more recent archaeological research has suggested that Indian agriculture occurred prehistorically in the Red River valley as far north as present Lockport, Manitoba.[6] It might also be pointed out that tobacco, which was the most widespread of Indian cultigens in aboriginal North America, was grown in the plains well to the north of the Knife River prior to European contact. The earliest historical evidence for this occurs in the accounts of the Hudson's Bay Company trader, Matthew Cocking, who, in the course of his explorations in western Saskatchewan in 1772, described an Indian "Tobacco plantation. A small plot of ground about an hundred yards long and five wide... ."[7] Although some tobacco cultivation continued among the Indians of the Canadian Plains until the reserve period, it was largely discontinued when better quality tobacco became available through the fur trade.

The earliest historical centre of native agriculture in the Canadian Northwest was the Indian village of Netley Creek. Also known as Riviere aux Morts, Dead River, or Ne-bo-wese-be, it was established by Ottawa Indians at the turn of the nineteenth century near the junction of Netley Creek and the Red River (Figure 1). The Ottawa first began to plant at this site in 1805 and,

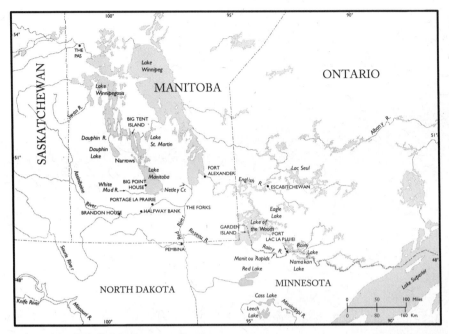

Figure 1. Indian Gardens: location map.

from there, agriculture subsequently spread among neighbouring bands of Ojibwa Indians. Neither the Ottawa nor the Ojibwa were living in the Red River valley at the time of European contact but, beginning in the 1780s, they began to replace the Cree and Assiniboine Indians to whom this territory had formerly belonged. The Ojibwa who migrated to the Red River valley were part of a general westward expansion of Ojibwa peoples into the prairie-parkland from the forests to the east. The Ottawa, in contrast, were more recent arrivals, who had come to the west from their home area in the Upper Great Lakes. According to Alexander Henry the Younger, they arrived in the Northwest about 1792, "when the prospects of great beaver hunts led them from their country."[8] Initially, the Ottawa scattered themselves widely but, by 1805, many of them hunted in the lower Red River valley and congregated each summer at the Netley Creek encampment.

It was the Red River Ottawa who first began to cultivate north of the Knife River and who played the crucial role in disseminating agriculture among the more populous, neighbouring Ojibwa. According to Lord Selkirk:

> The Indians who inhabit the country from Lake Superior to Red River are mostly of the Chippeway [Ojibwa] Nation, who have never been in the habit [of] cultivating the ground. The

Ottawas, who speak the same language & reside near Lakes Huron & Michigan have long been accustomed to plant Indian Corn, & some other vegetables tho' on a small scale. A band of these Indians, prompted by the growing scarcity of game in their own country, determined to migrate to Red River where they continued the practice of cultivating the ground.[9]

Although some agriculture had been part of the Ottawa economy in the Upper Lakes homeland, it was only when Alexander Henry gave them seed in 1805 that they first began to plant in Western Canada. As Henry noted in 1808:

The first corn and potatoes they planted here was a small quantity which I gave them in the spring of 1805, since which period they have extended their fields, and hope in a few years to make corn a regular article of traffic with us.[10]

In the same year the fur trader Peter Fidler observed that four or five Indian families had built wooden houses at Netley Creek, and that several acres of land were planted with Indian corn, potatoes and "other garden stuff."[11]

Between 1805 and the founding of the Selkirk Settlement in 1812, the Netley Creek village increased in size, and corn and potatoes raised there were sold there as provisions to passing European traders.[12] The agricultural activities of the Indians at Netley Creek soon became well known and achieved a prominence such that Lord Selkirk viewed the village as an important source of seed for the first of his Red River colonists. In his instructions of 1811 to Miles Macdonell, the first governor of the colony, Selkirk wrote that:

The Cos. [Hudson's Bay Company] establishments at Brandon House, etc., will ... supply you with seed potatoes and perhaps some seed grain ... Perhaps, however, a greater supply at least of Indian corn may be obtained from the Ottawa & Bungee [Ojibwa] Indians at Dead River near the mouth of Red River.[13]

Selkirk's letter is the first evidence to indicate that the small band of Ottawa at Netley Creek had been joined by neighbouring Ojibwa. It also suggests that some of the Red River Ojibwa had taken to cultivating alongside the Ottawa sometime prior to 1811. According to John Tanner, a whiteman who had been adopted by the Ottawa and who was living with them during this period, it was Sha-gwaw-koo-sink, an Ottawa chief at Netley Creek, who first taught the Red River Ojibwa to plant corn. In Tanner's words:

We then went down to Dead River, planted corn, and spent the summer there. Sha-gwaw-koo-sink, an Ottawwaw a friend of mine and an old man, first introduced the cultivation of corn among the Ojibbeways of the Red River country.[14]

The adoption of agriculture by the Ojibwa at Netley Creek permitted these nomadic peoples to live a more sedentary way of life than their purely hunting and gathering economy had previously allowed. The Netley Creek village also emerged as a gathering point for Ojibwa from the surrounding country. This enabled them to strengthen ties with one another and the village soon became something of a regional centre. Large numbers could assemble at this site and the Midewiwin or Grand Medicine Lodge ceremony was elaborated at seasonal meetings there. Writing retrospectively of these developments, the Reverend John West noted in 1823 that:

> There was a time when the Indians themselves had begun to collect into a kind of village towards the mouth of the Red River, had cultivated spots of ground, and had even erected something of a lodge for the purpose of performing some of their unmeaning ceremonies of ignorance and heathenism, and to which the Indians of all the surrounding country were accustomed at certain seasons to repair.[15]

On the eve of the founding of the Selkirk Settlement in 1812, events in the fur trade overtook the Indians at Netley Creek and the village fell into demise. The Ottawa abandoned the site and it would seem that several years elapsed before the Ojibwa resumed cultivating there. Tanner related that the Ottawa moved from Netley Creek to an island in Lake of the Woods, where they were observed cultivating in 1813. According to Tanner:

> After this, we started to come to an island called Me-nau-zhe-taw-naun [Garden or Plantation Island], in the Lake of the Woods, where we had concluded to plant corn, instead of our old fields at Dead River. ... we came to the Lake of the Woods, where I hunted for about a month, then went back into the country I had left, all the Indians remaining behind to clear the ground where they intended planting corn at Me-nau-zhe-tau-naung.[16]

The circumstances that led to the break-up of the Ottawa-Ojibwa village at Netley Creek are not entirely clear. Selkirk wrote that the Ottawa abandoned the site "because their corn being frequently pillaged by other Indians they thought it advisable to retire to an island in the Lake of the Woods."[17] More to the point, Selkirk indicated that the pilfering was encouraged by the North West Company which, in his opinion, was loath to see agriculture develop in the valley. The Reverend John West, however, remarked that the Hudson's Bay Company traders as well as the Nor'Westers had opposed this development as prejudicial to their interests in the fur trade, observing that "fears were entertained that the natives would be diverted from hunting furs to idle ceremonies, and an effectual stop was put to all further improvement, by the spirit of opposition that then existed in the country between the two rival Fur Companies."[18]

Following 1812, agriculture diffused widely among the Ojibwa. To the west of the Red River, a small Indian garden village, known as Grant's Village, was established in 1815 on the Assiniboine River at a place called the Half Way Bank, located midway between Brandon House and Portage la Prairie (Figure 1).[19] Lower down the Assiniboine six tents of Ojibwa were observed fishing and making gardens in 1819.[20] In the following year, Indian gardens were reported along the Whitemud River near Big Point House at the southern end of Lake Manitoba.[21] The Reverend John West, in traversing the area between Lakes Manitoba and Winnipeg in 1822, noted that a band of Indians was raising potatoes and pumpkins on the shores of Lake Manitoba.[22] By the late 1820s there is evidence that Indian cultivation had penetrated as far north as the Swan River valley, where an Indian called the Otter had gardens said to be productive of potatoes and turnips.[23]

For the most part, the agriculture that diffused into the Lake Manitoba region and northward was based upon the potato. The nature of this potato culture, and the role it played in the subsistence cycle of the Lake Manitoba Ojibwa, was graphically summarized by the Hudson's Bay Company trader, William Brown, in 1819. His account is also revealing of the quasi-sedentary living and the ceremonial gatherings that took place at the more important of these agricultural sites.

> A considerable number of the Indians particularly those of Fort Dauphin, and the Manitoba, have ground under cultivation, and raise a great many Potatoes, but that is their only crop … Those of the Manitoba … [cultivate] on an Island towards the North end of the Lake, they have erected there what they call a Big Tent, where they all assemble in spring, hold Councils and go thro' their Religious Ceremonies—The soil here is excellent and each family has a portion of it under cultivation, which

the women and old men remain, and take care of during the summer—while the young men go a hunting—In the fall of the year when they are going to abandon the place, they secure that part of the produce, under ground till spring, which they cannot carry along with them—During favourable years, they generally make a considerable quantity of maple sugar, part of which they also put in Cache—The Big Tent is constructed in the form of an arch, and consists of a slight frame of wood covered on the outside with the bark of the pine tree, and lined in the inside with bulrush mats. It is 60 ft. long—15 ft. wide—and 10 ft. high.[24]

Although some of the agricultural sites that emerged to the north and west of the Red River were relatively short-lived, agriculture persisted among the Indians of this region until the end of the fur trade period. Thus, in 1843 the Reverend Abraham Cowley recorded an instance of Indian cultivation in the vicinity of the Narrows of Lake Manitoba.[25] The Reverend James Settee observed potato fields in 1855 and 1856 on what he called the Potatoe Island, that is, the present Garden Island at the north end of Lake Manitoba.[26] Members of the Assiniboine and Saskatchewan Exploring Expedition of 1858 noted Indian potato culture on Sugar Island in Lake St. Martin in the same general area. They also observed "several places on the Dauphin River [the present Mossy River] where the Indians grow potatoes, Indian corn and melons."[27] This latter observation is noteworthy, for it appears to have been the northernmost instance of Indian corn cultivation on the continent.

Of greater significance was the expansion of agriculture to the east of Red River, a movement which began with the removal of the Ottawa to Garden Island in the southwestern corner of Lake of the Woods in 1812. Introduced by the Ottawa, it soon spread to the more numerous woodland Ojibwa around the Lake. Daniel Harmon, for example, who passed through Lake of the Woods in 1816, remarked that "the Sauteux [Ojibwa], who remain about the Lake of the Woods, now begin to plant Indian corn and potatoes, which grew well."[28] By this time, agriculture had also penetrated some distance to the north of Lake of the Woods. Thus, several Indian families about Escabitchewan House on Bali Lake were cultivating Indian corn, potatoes and beans at least as early as 1815[29] and by 1823 the Indians of Eagle Lake, south of Lac Seul, had "good gardens" described as very productive of Indian corn.[30] The expansion northward of agriculture into this region carried corn cultivation to the outer limits of the mixed forest belt, where further diffusion of this Indian cultigen was precluded by the harsh environmental conditions of the boreal forest to the north. Even at these northern margins, however, corn cultivation could play an important role in Indian subsistence, a circumstance that was appreciated by the fur traders of this country. The lack of big game, especially of moose and caribou, had reduced subsistence levels to the point where the traders came to view agriculture among the Indians as a beneficial development. Thus, lamenting the fact that his Lac Seul Indians were not raising corn, John Davis, the trader in this region, wrote:

> could the Indians be brought to dwell more at one place and employ less of their time seeking the deer [caribou] and Moose they might be induced to cultivate the soil and otherwise improve their condition ... particularly as they have an example shown them by the neighbouring Indians at Eagle Lake.[31]

From Lake of the Woods, Indian agriculture also spread along the line of the Rainy River, but this eastward thrust petered out as the climate deteriorated toward the height of land separating Lake Superior drainage from that flowing into Lake Winnipeg. Indian corn appears to have been grown only as far east as the Manitou Rapids on the Rainy River, where Major Delafield described "a small field of thriving Indian corn" in 1823.[32] Farther east, cultivation was confined to potatoes, a development that appears to have spread as far as Namakan Lake, on whose islands the Indians, according to the Reverend Peter Jacobs, "raised a good quantity of potatoes, which they barter to the traders for goods."[33]

The beginnings of agriculture to the south of the Lake of the Woods cannot be dated precisely, but, in 1821, Father S. J. N. Dumoulin, the Roman Catholic missionary at Pembina, reported that the Indians in his charge were planting at four different localities, of which the main one was on the Roseau River.[34] Although the Pembina mission was south of the 49th parallel, the Roseau site was probably in British territory, but it is likely that one or more of the three unspecified localities lay to the south of the international boundary. If so, this represents the earliest occurrence of Indian agriculture in northern Minnesota. Not until 1828, however, is there conclusive evidence of Indian agriculture in that area. In February 1829, the American Fur Company traders at Rainy Lake purchased corn grown by the Indians at present Red Lake, Minnesota.[35] In 1832 Henry Schoolcraft learned from traders at Cass Lake that the Red Lake Indians were raising considerable quantities of corn. However, travel accounts of the Upper Mississippi region to the southeast of Red Lake make no mention of Indian agriculture before 1832.[36] In that year Schoolcraft visited garden sites on Star Island in Cass Lake and along the shores of Leech Lake.[37] Although he had not observed Indian agriculture in these areas during his explorations of the Upper Mississippi country in 1820, it was well established by 1832, and had spread to the other lakeside sites, including Big Turtle Lake and Lake Winnibigoshish.

Of the different agricultural sites or complexes that emerged among the Ojibwa of the Northwest, the most important and enduring were at Lake of the Woods and Red Lake located some fifty miles to the south. It was in the Lake of the Woods-Red Lake area that Indian agriculture was most prominently developed and played its most important role both within Ojibway society and in their relations with European fur traders. Of the two complexes, that in Lake of the Woods was the earliest and it was following this development that agriculture subsequently spread to most of the groups that cultivated east of the Red River.

The earliest and most important site in Lake of the Woods was Menau-zhe-taw-naun, or Garden Island, occupied by the Ottawa in 1812. Within a few years of its founding a considerable agricultural complex had evolved on this island, which was described as follows by the Hudson's Bay Company factor at Rainy Lake in 1819:

> I visited their tents which were pitched alongside of the piece of ground which they [had] under cultivation which from the regular manner in which it was laid out would have done credit to many ... farmers, excellent Potatoes, Indian Corn, Pumpkins, Onions and Carrots. The women on whom it is a duty to do all the laborious work, were busily employed gathering....[38]

The corn culture at Garden Island was strongly commercial in character and, from the outset, part of the corn crop was sold to the fur traders of the Rainy Lake-Lake of the Woods area. As early as 1808, the Ottawa had expressed a desire to trade agricultural produce and commercial opportunities at the Lake of the Woods may well have influenced their decision to locate there. The Ottawa had a long tradition as traders and middlemen in the Michilimackinac area and elsewhere, and had raised corn commercially near parts of Lake Michigan and the southern shore of Lake Superior to provision the fur trade.[39] The "chief part" of Shaw-gwawkoo-sink's first crop at Garden Island was sold to the North West Company traders,[40] and by 1817 it was widely known that corn could be purchased from the Ottawa in the Lake of the Woods.[41]

Commercial corn production by both the Ottawa and Ojibwa in this area was initiated by the demand for provisions by the fur traders. By this time the country between Lake Superior and the Lake of the Woods had been virtually depleted of the larger food animals, and an often meager and precarious subsistence was derived by the fur traders from fish, rabbits, and wild rice (also traded from the Indians). Corn was purchased, not only to assist in feeding the men at the trading posts, but also to supply the transport brigades with essential voyaging provisions. Garden Island was strategically located in this respect, for it lay astride the main canoe route connecting the Lakehead with the Western Plains. The Indian gardens afforded the traders a small, but fairly dependable, supply of corn. The supply of corn was especially valuable when the more capricious wild rice harvests failed.[42] The traders of the North West and Hudson's Bay Companies vied with one another for the corn supply to ensure greater mobility in the competition for furs and a more assured food supply. Although the amounts traded annually by the Indians to the two companies during the period of competition are not known, the chief factor at Rainy Lake reported in the autumn of 1819 that the Hudson's Bay Company's trade at Garden Island had "been more successful than could have [been] imagined."[43] The post journal indicates that not all of the corn purchased had been transported to the company's headquarters at Rainy Lake; seventy-nine bags of corn and one bag of rice had been cached somewhere between the island and Rainy Lake.[44]

After the union of the two companies in 1821, corn production at Lake of the Woods declined, largely as a result of a drop in price brought on by the cessation of competition. The chief factor at Rainy Lake reported in 1822-23 that during the period of competition the Indians "would never give more than 2 bushels for a three point blanket, which traders, some from competition and some from necessity were obliged to give; as soon as the junction was affected the exorbitant price was reduced to a pint of powder for a bushel. On this the Indians in great parts discontinued their cultivation.[45] However, in 1824 the Hudson's Bay Company "succeeded in trading a tolerable stock of Indian corn" at Garden Island,[46] a total of seventy-six bushels.[47] In 1825 the total traded was 140 bushels[48] and in 1828 the total was one hundred fans (about seventy bushels) of very bad corn.[49]

Although the union of the companies precipitated a falling off in corn production, the decline was both short-lived and limited in effect. The Nor'Westers were quickly supplanted by the Americans, and the old rivalry was replaced by an equally intense Anglo-American rivalry. Corn regained its previous importance in the trade, and the strategic position of Garden Island along the international boundary was fully appreciated by the traders on both sides.

One of the best descriptions of the agriculture on Garden Island is contained in the reports of the Red River Exploring Expedition, which visited the island in 1857.

Garden Island is about a mile and a half long at its widest part. Its western half is thickly wooded, the greater portion of the eastern half cleared and cultivated. A field containing about 5 acres was planted with Indian corn, then nearly ripe. The corn was cultivated in hills, and kept very free from weeds ... Near the space devoted to Indian corn, were several small patches of potatoes, pumpkins, and squashes. An air of great neatness prevailed over the whole of the cultivated portion of the Island.[50]

The expedition provided the first reliable indication of Indian cultivation on islands other than Plantation Island, and a map produced by its members shows that cultivation had spread to a number of sites along the lakeshore.[51] The widespread nature of this agriculture, and evidence that it persisted into the treaty period, is corroborated by the reserve surveys which, although only partially conducted, show several agricultural islands at the north end of the Lake near Rat Portage and in Shoal Lake.[52] The number of Indians who eventually took to cultivating in the Lake of the Woods is not known, but in 1854 it was estimated that about two hundred Indians resided on Garden Island in summer and raised on it "large quantities of potatoes, Indian corn and pumpkins."[53]

Like the Lake of the Woods Ojibwa, their neighbours at nearby Red Lake in Minnesota were exceptionally predisposed towards agriculture and sold their corn to both British and American fur traders. As early as 1829, the Red Lake Indians were trading corn to the American Fur Company traders at Rainy Lake[54] and by 1832 Henry Schoolcraft related that the Red Lake band was supplying corn to "the posts on the Upper Mississippi, and even as far east as Fond du Lac."[55] Of these people it was further noted that "They are enabled to sell 3 or 400 bushels in exchange for goods and reserve to themselves a comfortable supply for the winter."[56]

The Reverend Frederic Ayer, who conducted a reconnaissance of the Upper Mississippi region in 1842-43, wrote a detailed description of the Red Lake people at this time. According to Ayer, they comprised about one hundred hunting men and from five to six hundred women and children. He described them as one of the largest bands of the Ojibwa nation and, "as a body, probably more stationary than any other band of Ojibwas." He further observed that "This band raise more corn and potatoes probably than any band of Ojibwas in this part of the country. In ordinary seasons they put up from 15 to 60 to 80 sacks of corn to a family. Their sacks contain a bushel or more each."[57]

The seasonal activities, of the Red Lake Indians, including their far-flung trading trips, were described in detail by Ayer as follows:

When the rivers open in the Spring, the men generally leave, and descend the Red River to the Colony ... They are absent about 20 or 25 days. The principal object of this visit to the Colony, is to traffick sugar with the half-breeds and others, for which they receive clothing and goods. Again in the first part of June, a considerable number start out to hunt buffalo in the plains to the West. A few also visit La Pointe and the Sault Ste Marie during the summer. With these exceptions, they spend the spring and summer at Red Lake. The men only go to the Colony and on their hunts in summer. And in their winter hunts, the men do not usually take their families with them. Fewer families than usual remain at Red Lake this winter [i.e., 1842-43], on account of the small crop of corn there last season. The hunters have taken their families with them this winter on account of the scarcity at R. Lake. There are this winter 35 lodges at the Lake. As a general thing the women and children remain here both summer and winter.[58]

The importance of corn cultivation at Red Lake, especially in relation to that of surrounding Ojibwa bands, was further underscored by Ayer's observation that "Indians from other bands in considerable numbers starving at home come here to winter and live upon the hard earned fruits of this peoples industry."[59]

The Indian agriculture that emerged in the fur trade Northwest varied greatly in extent and significance, ranging from isolated individual family plots to large village complexes. On the whole, the larger and more enduring agricultural sites were confined to the southern and eastern sections of the region, while the smaller and often ephemeral ones predominated in the extreme north and west. In some instances, agricultural produce was regularly sold to the fur traders, so that cultivation was undertaken with commercial as well as subsistence intentions by different groups or individuals within them.

Geographically, the most striking feature of this agriculture was its northern nature. Throughout most of its distribution, it was essentially confined to lakeside or insular locations, where micro-climatic conditions allowed the cultivation of corn well to the north of its prehistoric limits. Although it developed following European contact, and incorporated European crops, it was nonetheless a native agriculture derived from the maize-beans-squash complex of North American Indians. This complex remained intact at least as far north as Garden Island, located at 49°10' north. However, corn, the most adaptable of the cultigens in this complex, was cultivated significantly farther north, reaching its outer limit in northwestern Ontario at Eagle Lake and Escabitchewan and achieving its extreme northern limit on the continent in

Manitoba along the Mossy River at latitude 53°31' north. Indian agriculture thus made its farthest poleward thrust at the centre of the continent along the large north-south trending lakes of the Manitoba lowlands. Beyond this it appeared only under the auspices of church missions, and generally at a later date. It was also confined to crops introduced by Europeans.

Despite the increasing inroads of white traders, missionaries and government officials into the Northwest, the agriculture that spread among the Ojibwa remained essentially native in character. If not entirely based on Indian cultigens, it was conducted on Indian terms, a feature that for some bands persisted into the reserve period. Thus, despite dwindling game resources and persistent attempts by missionaries and representatives of government to convert them to farming, many continued to hunt and fish, and to plant the small gardens of corn and potatoes that had been their custom prior to becoming wards of the new Canadian state.

NOTES

This article first appeared in *Prairie Forum* 11, no. 2 (1986): 171-83. The authors wished to acknowledge the Hudson's Bay Company Archives, Archives of Manitoba, for permission to consult and quote from their records.

1   D.W. Moodie and Barry Kaye, "The Northern Limit of Indian Agriculture in North America," *The Geographical Review* 59 (4) (1969): 513-29.
2   D.W. Moodie, "Agriculture and the Fur Trade," in Carol M. Judd and Arthur J. Ray, eds., *Old Trails and New Directions: Papers of the Third North American Fur Trade Conference* (Toronto: University of Toronto Press, 1980), 283.
3   George F. Will, "Indian Agriculture at Its Northern Limit in the Great Plains Region of the United States," *Proceedings of the 20th International Congress of Americanists*, Vol. 1 (1924), 203-05.
4   Waldo R. Wedel, *Prehistoric Man on the Great Plains* (Norman, Oklahoma: University of Oklahoma Press, 1961), 239.
5   W. Raymond Wood, *Biesterfeldt: A Post-Contact Coalescent Site on the Northeastern Plains*, Smithsonian Contributions to Anthropology no. 15 (Washington: Smithsonian Institution, 1971). See also William D. Strong, "From History to Prehistory in the Northern Great Plains," *Smithsonian Miscellaneous Collections*, vol. 100 (Washington: Smithsonian Institution, 1940) 353-94.
6   Personal Communication, Dr. A.P. Buchner, Project Archaeologist, Manitoba Culture Heritage, Historic Resources Branch, Winnipeg, Manitoba.
7   York post journal, Journal of a Journey Inland with the Natives by Matthew Cocking Second at York Fort; commencing Saturday 27th June 1772 and ending Friday, 18 June 1773, Hudson's Bay Company Archives, Archives of Manitoba, Winnipeg (hereafter HBCA, AM), B239/a/69, fo. 9.

8   Alexander Henry, in Elliott Coues, ed., *New Light on the Early History of the Greater Northwest. The Manuscript Journals of Alexander Henry, Fur Trader of the Northwest Company, and of David Thomson, Official Geographer and Explorer of the Same Company, 1799-1814*, 2 Vols. (Minneapolis: Ross and Haines, 1965), 2: 448.

9   Manuscript by Lord Selkirk relating to Red River, 1819, Selkirk Papers, microfilm copies in the Archives of Manitoba, Winnipeg (hereafter SP), Vol. 47, p. 12, 836.

10   Alexander Henry, *New Light on the Early History*, 2: 448.

11   Peter Fidler, Journal of Exploration and Survey, 1794-1808, HBCA, AM, E3/3, fo. 58, 24 May 1808.

12   Brandon House post journal, 3 September 1807, HBCA, PAM, B22/a/15, fo. 3; Pembina post journal, 30 August 1809, HBCA, AM, B160/a/2, fo. 2.

13   Selkirk's Instructions to Miles Macdonell, 1811, SP, Vol. 1, pp. 176-77.

14   John Tanner, in Edwin James, ed., *A Narrative of the Captivity and Adventures of John Tanner during Thirty Years Residence among the Indians in the Interior of North America* (Minneapolis: Ross and Haines, 1956), 171.

15   John West, *The Substance of a Journal during a Residence at the Red River Colony, British North America: And Frequent Excursions Among the Noah West Indians in the Years 1820, 1821, 1822, 1823* (London: L.B. Seeley and Son, 1824), 129.

16   John Tanner, *A Narrative of the Captivity and Adventures of John Tanner*, 190 and 191.

17   Manuscript by Lord Selkirk relating to Red River, 1819, SP, Vol. 47, p. 12, 665.

18   John West, *The Substance of a Journal*, 130.

19   Manitoba Report on District, 1820, HBCA, AM, B51/e/1, fo. 18.

20   Journal at the Forks, Red River, 20 May 1819, HBCA, AM, B51/a/2, fo. 3.

21   Report of the Brandon House District by Peter Fidler, 1821, HBCA, AM, B51/a/2, fo. 3.

22   John West, *The Substance of a Journal*, 97.

23   Several references in the following Fort Pelly post journals, HBCA, AM, B159/a/10, 1828-29; B159/a/11, 1829-30; B159/a/13, 1831-32.

24   Report of the Manitoba District by William Brown, 1818-19, HBCA, AM, B122/e/1, fo. 8.

25   Journal of the Reverend Abraham Cowley, 13 February 1843, Church Missionary Society Records, Film A86, microfilm copies in the Archives of Manitoba, Winnipeg.

26   Journal of the Reverend James Settee, 13 October 1856, ibid., Film A95.

27   Henry Youle Hind, *North West Territory: Reports of Progress Together with a Preliminary and General Report on the Assiniboine and Saskatchewan Exploring Expedition, Made Under Instruction from the Provincial Secretary, Canada* (Toronto: J. Lovell, 1859), 7.

28   Daniel Williams Hannon, in W. Kaye Lamb, ed., *Sixteen Years in the Indian Country: His Journal, 1800-1816* (Toronto: Macmillan, 1957), 211.

29   Osnaburgh House District Report, 1815-16, I-IBCA, AM, B155/e/3, fos. 3, 7; Red Lake District Report, 1816-17, HBCA, PAM, B177/e/1, fo. 3.

30   Lac Seul District Report, 1823-24, HBCA, AM, B107/e/1, fos. 2, 5; Escabitchewan District Report, 1823-24, HBCA, PAM, B64/a/10, fos. 32-33.

31   Lac Seul District Report, 1823-24, HBCA, AM, B107/e/1, fo. 5.

32   Major Joseph Delafield, in Robert McElroy and Thomas Riggs, eds., *The Unfortified Boundary: A Diary of the First Survey of the Canadian Boundary line from St. Regis to the Lake of the Woods by Major Joseph Delafield, American Agent under Articles VI and VII of the Treaty of Ghent* (New York: Private Printing, 1943), 433.

33 Peter Jacobs, *Journal of the Reverend Peter Jacobs, Indian Wesleyan Missionary, from Rice Lake to the Hudson's Bay Territory, and Returning; Commencing May, 1852; With a Brief Account of His Life, and a Short History of the Wesleyan Mission in that Country* (New York: The Author, 1857), 30.
34 Father Dumoulin to Archbishop J.O. Plessis, 16 August 1821, in Grace Lee Nute, ed., *Documents Relating to Northwest Missions, 1815-1827* (St. Paul: Minnesota Historical Society, 1942), 324.
35 Fort Lac la Pluie post journal, 28 February 1829, HBCA, AM, B105/a/13, fo. 6.
36 None of the following accounts mentions Indian agriculture: Z.M. Pike, *An Account of Expeditions to the Sources of the Mississippi* (Philadelphia: C. and A. Conrad, 1810); Henry Rowe Schoolcraft, *Narrative Journal of Travels to the Sources of the Mississippi River, in the Year 1820*, edited by Mentor L. Williams (East Lansing, Michigan: Michigan State College Press, 1953); Ralph H. Brown, ed., "With Cass in the Northwest in 1820," *Minnesota History* Vol. 23 (1942): 126-48, 233-52, and 328-48; and J.C. Beltrami, *A Pilgrimage in Europe and America, Leading to the Discovery of the Sources of the Mississippi and Bloody River*, 2 Vols. (London: Hunt and Clarke, 1828).
37 Henry Rowe Schoolcraft, in Philip P. Mason, ed., *Expedition to Lake Itasca; the Discovery of the Source of the Mississippi* (East Lansing, Michigan: Michigan State University Press, 1958), 20-21, 209, 260, 328, 333 and 335.
38 Fort Lac la Pluie post journal, 10 September 1819, HBCA, AM, B105/a/7, fo. 30.
39 Charles M. Gates, ed., *Five Fur Traders of the Northwest* (St. Paul: Minnesota Historical Society, 1965), 32; Alexander Henry, *Travels and Adventures in Canada and the Indian Territories, Between the Years 1760 and 1776* (Toronto: G. N. Morang and Company, 1901), 48-49; W. Vernon Kinietz, *The Indians of the Western Great Lakes, 1615-1760* (Ann Arbor, Michigan: University of Michigan Press, 1940), 235-36.
40 Miles Macdonell's journal, 23 January 1814, SP, Vol. 63, p. 16, 876.
41 Miles Macdonell to Lord Selkirk, 3 August 1817, SP, Vol. 12, p. 3, 900.
42 J.D. Cameron to George Simpson, Lac la Pluie, 20 May 1829, HBCA, AM, D5/3, fo. 353.
43 Fort Lac la Pluie post journal, 5 October 1819, HBCA, AM, B105/a/7, fo. 39.
44 Ibid. The Indian bags or fans, were seven-tenths of a bushel. The bags were made of fawn skins, taken off nearly whole, and consequently they varied with the size of the animal.
45 Lac la Pluie Report on District, 1822-23, HBCA, AM, B105/e/2, fo. 3.
46 Lac la Pluie Report on District, 1824-25, HBCA, AM, B105/e/4, fo. 1.
47 Lac la Pluie Report on District, 16 October 1824, HBCA, AM, B105/a/10.
48 Lac la Pluie Report on District, 1825-26, HBCA, AM, B105/e/6, fo. 11.
49 J.D. Cameron to George Simpson, Lac la Pluie, 20 May 1829, HBCA, AM, D5/3, fo. 353.
50 *Report on the Exploration of the Country between Lake Superior and the Red River Settlement* (Legislative Assembly of Canada; Toronto, 1858), 137 (para. 134).
51 Ibid., map following Introduction.
52 Leo G. Waisberg, "The Economy of the Boundary Waters Ojibwa" (unpublished mss. in possession of Grand Council Treaty No. 3, Kenora, Ontario, 1979), 52.
53 Journal of the Reverend Robert M. McDonald, 17 February 1854, Church Missionary Society Records, Film A93.

54 Fort Lac la Pluie post journal, 28 February 1829, HBCA, AM, B105/a/13, fo. 6.

55 Schoolcraft, *Expedition to Lake Itasca*, 21.

56 Frederic Ayer to David Greene, Red Lake, 12 January 1845, in Grace Lee Nute, comp., Manuscripts Relating to Northwest Missions, Division of Archives and Manuscripts, Minnesota Historical Society, St. Paul.

57 W.T. Boutwell et. al. [Messrs. Hall, Ayer and Boutwell] to David Greene, Pokegoma, 6th March 1843, in ibid.

58 Loc. cit.

59 Frederic Ayer to David Greene, op. cit.

3. The Ranching Industry of Western Canada:
    Its Initial Epoch, 1873-1910

Sheilagh S. Jameson

The story of ranching in Western Canada is a truly important segment of our nation's history. It has been dismissed, quite wrongly, by some as a replica of America's wild west; for others it embodies something of the romanticism of bygone years—a nostalgic feeling for the freedom and appeal of the ranching era as expressed in the Calgary Stampede and other rodeos. Certainly the days when the range was open and the cattle barons ruled may truly be regarded as the most romantic era of the West's history, but it is also one of special significance. It had an important economic, political and social impact on Canada as a whole. Only in recent years has the extent of this impact been explored by historians and some recognition given to the importance of the early ranching industry.

The first breeding herd of cattle to be introduced into the West's major ranching region, southern Alberta, paradoxically came from the north. In October 1873, Methodist missionary Rev. John McDougall and his trader brother, David, drove eleven cows and a bull from Fort Edmonton to the Stoney Indian Mission at Morleyville in the Foothills region on the Bow River. During the following year the North-West Mounted Police (NWMP) brought some 235 head of cattle, beef animals and draught oxen, on their trek West. The police created the beginning of a Western market for beef and milk and foremost among those who took advantage of this opportunity was Joseph McFarland who, in 1875, drove a herd from Montana to the vicinity of the Fort Macleod post. That same year a trader named John Shaw came from the Kootenay area with a herd of 456 cattle, bringing them up the Foothills to Morleyville. On finding a police post at Fort Calgary and activity at the Hudson's Bay Company posts on the Bow and Elbow Rivers, he abandoned his original plan of trailing his herd to the Company's establishment at Fort Edmonton and wintered them at Morley.

Following this beginning, various traders, most notably George Emerson and Tom Lynch, brought herds north from the western states, the flow increasing as the 1870s progressed. During these years various obstacles to ranching were removed and prospects for development of a full-scale industry deepened. In 1877 with the signing of Treaty Seven the Indians of the Blackfoot Confederacy, no longer free roaming owners of the land, were assigned reserves. Then by 1879 the buffalo were gone and the grasslands of the West were ready for great herds of cattle. The Mounted Police presence with the establishment of law in advance of settlement was an important positive factor in the country's development at this time. In addition the police, with their increasing demand for beef and for horses for patrol work, provided a good share of the growing local market. Another source was the government's need to purchase beef for the Indians to alleviate the starvation wrought by the extermination of the buffalo. In an attempt to encourage the Indians to adopt a new way of life and at the same time to reduce the cost of purchasing beef, the government in 1879 established two farms or ranches, one near Calgary and the other in the vicinity of Fort Macleod. The enterprise met with limited success and it was still necessary to purchase large quantities of beef for delivery on the reserves.

From the beginning the relationship of the Mounted Police with the ranching industry was of particular significance. At the time of its inception and during its early formative years the police provided the infant stock business with protection and encouragement. A further stamp of approval came as a substantial proportion of members of the force, men in a unique position to assess the potential of the situation, joined the ranching fraternity. From 1877 onward those planning to make this move on the expiration of their three-year term of service purchased cattle from the herds that were driven north. Indicative of their involvement in the stock business during its primary years is the fact that the first three brands registered in the North-West Territories were issued to policemen, namely Inspectors P. R. Neale and S. B. Steele, jointly, on 29 January 1880, Superintendent W. Winder, 19 March 1880 and Inspector C. E. Denny, 22 April 1880.

Superintendent William Winder of the Mounted Police, generally known as Captain Winder, is credited with filling a special promotional role in the development of the Western cattle industry. He was a native of Lennoxville, Quebec, and when accorded a period of leave in 1879 he spent it in his home area in the Eastern Townships. Here he discussed with the wealthy professionals, merchants, businessmen and farmers who were his friends and associates, the financial possibilities and the appeal of the rolling acres of free grasslands in the West, pointing out that an incipient ranching business was already operating in the Alberta Foothills with promising success. He advocated the formation of large cattle companies, a suggestion that was in accordance with reports of the North American stock situation which were emanating from Great Britain and the United States. During the late 1870s a roaring cattle business in the American West, based to a large extent on rocketing export sales to Britain, had sprung into existence. Stories of fabulous fortunes to be made on the Western rangelands had resulted by 1879 in a range boom, and huge ranch companies, many of them financed by British investors, were vying for positions on the American plains. The Canadian government and Canadian investors, aware of this burgeoning industry, were anxiously considering such lucrative ventures in Canada's west; however, there was still some uncertainty regarding the threat of marauding Indian bands, adverse weather conditions and the stability of grass resources—certainly, Captain John Palliser's report of his western expedition, 1857-60, with its emphasis on the aridity and barren-

William Winder, photographed circa late-1860s, when he was a captain in the Cookshire Calvary in the Eastern Townships of Quebec. Winder came west with the North-West Mounted Police in 1874 and became the Superintendant at Fort Macleod. Returning to the Eastern Townships while on leave in 1879, Winder aroused the interest of wealthy businessmen and investors with his enthusiastic statements regarding the viability of ranching in the west. Upon his retirement from the force two years later, Winder formalized his existing ties in the bourgeoning Canadian ranching industry, founding the Winder Ranch. (Courtesy of the Glenbow Archives/NA-98-28.)

ness of the Canadian plains, was not encouraging. Therefore the enthusiastic firsthand accounts of Captain Winder and some of his fellow officers, reinforced as they were by the North-West Mounted Police official reports which detailed ranching potential and progress, sounded a welcome note of encouragement. Surely here was not only a golden opportunity for Canadian investors but the challenge of diverting northward the British capital which was pouring into Wyoming, Texas and Montana.

An initial requirement for a large-scale Western ranching industry was a change in land regulations. The Dominion Lands Act of 1872 had provided for development of a general settlement and although later amended did not permit the establishment of a large cattle industry. Prospective investors and ranchers pressured the government to make appropriate changes. An improved cattle market with the added incentive of a privileged position for Canadian cattle on British markets, created by the imposition in 1879 of an embargo on live cattle from the United States, gave weight to their requests. Many of those anxious to become participants in a Western ranching enterprise had influential status in business and/or political connections, a number being involved directly or indirectly with Sir John A. Macdonald's Conservative government. This fact undoubtedly was an aid in precipitating action. There were, in addition, internal national concerns that helped provide a favourable climate for Western ranch expansion. The importation of cattle from the United States to provide beef for Western Indians created a large ongoing expenditure which could be reduced sharply if a local market were established. Also, a thriving cattle business in the West would give greater credence to the building of the controversial railway across the prairies.

The man who spearheaded negotiations with the government for the creation of conditions appropriate for range development was Senator M. H. Cochrane of Compton, Quebec. Noted internationally for the purebred Shorthorns he raised on his Hillhurst stock farm near Compton, Senator Cochrane had now reached the conclusion that the greatest profits lay in mass production of good beef and that this could be achieved on the extensive ranges of the West. Meetings with government ministers and an exchange of correspondence over approximately a two-year period resulted on 23 December 1881 in the passing by Macdonald's government of an Order-in-Council permitting land leases of up to 100,000 acres on a twenty-one-year, or less, basis at the rate of one cent per acre per year; cattle for stocking such leases were to be allowed duty-free entry during 1881 and 1882. Cochrane, who had been allowed his choice of Western land, had proceeded with the establishment of his ranch. He selected his lease stretching along the Bow River with headquarters at the Big Hill (later the site of the town of Cochrane) approximately twenty-five miles west of Calgary, and in October 1881, the first great herd of Cochrane Ranche cattle, purchased in Montana, arrived on the Bow River range. Also that fall over sixty head of purebred Angus, Hereford and Shorthorn bulls, most of them British imports, were shipped by rail and river boat to Fort Benton, Montana and then trailed north. On 30 November 1881 the Cochrane Ranche cattle numbered some 6,800 head.

Thus the year 1881 marked the commencement of the era of the big ranches. Competition among hopeful prospective ranch owners and speculators for both favourably located leases and available investment became eager. The enthusiasm was further fueled by the wholehearted endorsement of Alberta's ranch country by the Governor-General, the Marquis of Lorne, and his party on their 1881 tour of the West—this gave the concept of Western ranching a certain prestige on both sides of the Atlantic.

From the scramble for lands, four particularly large leaseholders emerged. Of these Senator Cochrane, besides being first, became controller of the greatest acreage. Nevertheless his operation underwent some major setbacks. Adverse weather conditions during the first two winters, coupled with mis-management, the result of stock-handling decisions being made in the East, caused serious cattle losses. Undaunted, the Senator decided to run sheep on the Bow River land and to move his cattle further south; in 1883 he took over several leases in the area between the Waterton and Oldman Rivers and that summer the remnants of the two great Cochrane Ranche cattle drives of 1881 and 1882 travelled south. To satisfy land regulations another company, the British American Ranche Company, functioning under essentially the same controlling interests, was formed to operate the sheep enterprise. These two ranches gave Senator Cochrane rights over a total of 334,500 acres of grazing land. Some 8,200 head of sheep were brought to the Bow River range in 1883, but during the late 1880s this ranch gradually went out of business. Although sheep losses and poor wool prices were factors in its failure, the decisive cause was the increasing pressure exerted by both the government and encroaching settlers for the release of the lands for settlement. Meanwhile the cattle business in the south also faced some initial reverses. Probably the greatest potential danger that threatened it came in 1885 from the Indians on the nearby Blood Reserve. News of Métis and Indian successes in the Saskatchewan Rebellion fueled a growing unrest and a belligerent attitude towards whites among the Bloods. Some Cochrane Ranche cattle were killed but the police established an outpost at nearby Stand Off and, with increased patrol work, managed to maintain peaceful relations. In succeeding years the Cochrane Ranche pros-pered and became a profitable enterprise but after Senator Cochrane's death in 1903 the pivotal force of the operation appeared to be gone. W. F. Cochrane, one of the Senator's sons who had managed the business since 1884, shortly returned to Quebec. In 1906 the Company ceased to function and the land was sold to the Mormon Church.

Following the establishment of the Cochrane Ranche the other three moguls of the cattle industry quickly came into being. The one of these which became most widely known and proved best able to survive the onslaught of time was the Bar U. Its instigator was Fred Stimson, a successful farmer and stock raiser in the Compton area who, fired by the glowing accounts of Western ranching related by his brother-in-law, Captain William Winder, persuaded Sir Hugh

Among Alberta's black pioneers were John Ware (circa 1845-1905) and his family (left to right: Mildred Ware, Robert, Nettie and John). John Ware was born into slavery on a cotton plantation in South Carolina, and after gaining his freedom at the end of the American Civil War he headed to Texas where he became a cowboy and got his start in the ranching industry. In 1882, Ware was hired on to a cattle drive to bring 3,000 head of cattle from the U.S. to what became the Bar U Ranch. He worked for the Bar U and the Quorn Ranch, gaining the respect of his peers and a considerable reputation for his skills in horsemanship, roping, and steer-wrestling. Ware eventually established ranches of his own and became one of the best-known cattlemen of Alberta's early ranching era. John Ware, who reputedly had never been thrown from a horse, died in 1905 when his mount tripped in a badger hole and fell on top of him, killing him instantly. His funeral in Calgary on September 14 that year was one of the largest in the young city's history. Alberta's Mount Ware, John Ware Ridge, and Ware Creek honour his legacy. (Courtesy of the Glenbow Archives/NA-263-1.)

Allan, head of the Allan Steamship Line, and his brother Andrew, to provide the necessary financial backing and in March 1882 the North-West Cattle Company was formed with Fred Stimson as manager. In enthusiastic anticipation he had already registered a brand, choosing Bar U (Ū), which was issued in his own name on 20 October 1881—a brand which shortly came to identify the ranch and which eventually received international recognition. An effective manager, involved in both stockmen's concerns and the social life of the Foothills ranch community, Stimson was also a colourful character and an excellent raconteur who placed the stamp of his own unique personality on Alberta's ranching scene, and through his fund of amusing and sometimes preposterous stories, added a certain flavour to Western lore. The Bar U itself earned a special place in the annals of Western ranch history. Many of Alberta's ranchers and stockmen rode for the Bar U and so became indoctrinated into ranch life. A number of the range land's famed and distinctive characters were associated with this ranch, among whom, besides Stimson, were Herb Millar, who was hired in Chicago in the fall of 1881; John Ware, noted black cowboy and rancher; and, perhaps most importantly, George Lane. After serving as foreman from 1884 to 1889 Lane branched out on his own. When in 1902 the possibility of obtaining the Bar U arose he managed to interest the noted Winnipeg-based meat-packing firm, Gordon, Ironsides and Fares, in the deal and a new company, Lane, Gordon, Ironsides and Fares, was formed for the express purpose of purchasing the Bar U. After successfully piloting the ranch through the numerous ups and downs that beset the industry he bought out the Gordon, Ironsides and Fares firm in 1920, becoming sole owner of the Bar U.

Another ranch started in 1882 was the Oxley. During the flurry of ranch fever in 1881 John Craig, an Ontario farmer and Shorthorn breeder, with the support of several Canadian shareholders, organized the Dominion Livestock Company and obtained a 100,000-acre lease in the Porcupine Hills and up the western reaches of Willow Creek. Requiring more capital he went to England where he attracted the interest of Alexander Staveley Hill, a Conservative member of the British Parliament, and the Earl of Lathom, a prominent cattle breeder. Hill's participation was dependent on certain conditions—English capital only was subscribed and the company, formed on 29 March 1882, was named the Oxley Ranche Company after Hill's country home, Oxley Manor. Hill himself assumed the position of managing director and Craig was named ranch manager. As the original lease included some rough land unsuitable for grazing, another lease of 87,000 acres, excellent range further east and south on Willow Creek, was acquired in Hill's name. In August 1882 a herd of approximately 3,400 head of cattle purchased in Montana arrived and the operation was under way. From the outset the Oxley was plagued by internal problems, becoming a classic example of conflict between absentee owners and ranch managers. After the inevitable severance of Craig's relations with the Oxley, he produced a book, *Ranching with Lords and Commons*, in which he records the peculiar problems involved in managing a ranch for the British aristocracy. In 1886 the ranch was reorganized under the name, the New Oxley (Canada) Ranche Company. Despite further problems it achieved some success and a place of respect among the great ranches of the Alberta Foothills.

The fourth of the quartet of big ranches was the Walrond Ranche Company. Organized in 1883 by Dr. Duncan McEachran, Dominion Veterinary Surgeon, it was financed chiefly by British capital with Sir John Walrond—Walrond being the main shareholder. McEachran, vice-president of the company, acted as manager. The Walrond acquired several leases on the north fork of the Oldman River and the southern stretches of the Porcupine Hills. McEachran, a capable manager, imported good stock and developed a successful cattle export business. The Walrond for more than a decade pursued its profitable way despite the ever-increasing advance of settlement. Then came the disastrous winter of 1906-07 with terrible cattle losses and this, added to the discouraging situation which faced big ranchers, spelled the end. In 1908 the Walrond stock was sold and the lease went to W.R. Hull with Pat Burns later taking control.

During their heyday these four giants of the cattle industry, the Cochrane Ranche Company, the North-West Cattle Company (Bar U), the Oxley Ranche Company and the Walrond Ranche Company, held almost one-third of all land in the southwestern part of Alberta, of which any use was being made by ranchers or settlers. By this time practically all the Foothills region from Cochrane south had been assigned in some way. Other large, and comparatively small, companies held leases and stock ranged west to the mountain threshold and eastward onto the plains. This area was originally chosen partly for the shelter and security of the hills and for the more picturesque ranch headquarter sites they afforded, and to avoid complete dependence on the grass resources of the more arid prairie. Many of these ranches also were significant in the saga of range history. A few might be mentioned: the Quorn, 1886, financed by a Leicestershire, England, syndicate and formed primarily for raising horses for the British hunt club market; the Stewart Ranche, 1881, established by Captain John Stewart of Ottawa and with strong NWMP connections; the Winder Ranche started by Mounted Police Superintendent William Winder; the Alberta Ranche, founded by Sir Francis DeWinton who had been aide-de-camp of Lord Lorne during his 1881 tour; the Bow River Horse Ranche, which took over some of the original Cochrane Ranche range; the Glengarry Ranch, 1885, established by A.B. Macdonald, Ontario merchant and eventually owned by railway contractors William Mackenzie and Donald Mann; the Maunsell Ranch, owned by brothers Edward and George Maunsell, formerly of the NWMP—the list could continue.

The Foothills ranch community, unlike most frontier societies, was law-abiding to a marked degree, a situation reflecting not only the background of most of its components, the majority of whom were middle or upper-class British, Eastern Canadians or French, but also the pervading influence of the Mounted Police. A homogeneous relationship existed between the ranchers and the police, produced in part by similar social and cultural backgrounds and further enhanced by the substantial number of ex-members of the force who became stockmen.

Although the major part of the early development of the Western Canadian ranching industry occurred along the Alberta Foothills and their eastern fringe, some ranches of varying types were established on the prairies to the east and south as early as the 1880s. In 1886 the Powder River Cattle Company of Wyoming, under Moreton Frewen, its British manager and part owner, brought approximately eight thousand head of cattle north and established a ranch east of the Foothills leases on Mosquito Creek. The brand of this spread, the 76, later became a part of the most famed ranch/farm scheme to grace the prairies, the Canadian Agricultural, Coal and Colonization Company (CACC).

The CACC had its roots in a Canadian Pacific Railway (CPR) colonization scheme. The CPR, despite its mutually beneficial association with the ranch industry during the early 1880s, was from its inception strongly committed to settlement, for the filling up of the western lands with settlers would create a much larger volume of freight flowing both east and west—besides there was CPR land available for sale, hopefully to large numbers of settlers. So as early as 1884 the rail company established ten experimental farms along the line running across the arid prairie west of Swift Current. The plan was to grow field crops and vegetables as a test of the suitability of the country for farming. After good initial results because of an unusual amount of rainfall in 1884 and into 1885, the area reverted to its usual dry state and by 1886 the experiment was recognized as a failure. At this

Sir John Lister-Kaye (1853-1924), circa 1880s. (Courtesy of the Glenbow Archives/NA-1710-1.)

point an English promoter, Sir John Lister-Kaye, became involved. Sir John was already operating a farm of nearly seven thousand acres at Balgonie, near Major Bell's famed farming empire in the Qu'Appelle Valley but he envisaged a much more grandiose venture, large enough to overshadow Major Bell's project. He approached the CPR and the Dominion government with details of his scheme. He wished to establish ten ranch/farms patterned upon English estates located at intervals along the CPR line west of Swift Current, for which he would require ten blocks of land each ten thousand acres in size. The CPR and the government were pleased to encourage such an enterprise and provided land at reasonable rates, so after some financing problems were solved the CACC was formed in 1888 with Lister-Kaye as manager. Sir John's Balgonie farm was purchased by the company and, as planned, ten more ranch/farms were established between Swift Current and Langdon, some twenty miles east of Calgary.

Sir John was a man of action and construction of fine houses, stables and other buildings at each farm commenced immediately; young Englishmen, many of them inexperienced, arrived to man the operation and soon walking ploughs were turning the prairie sod. Stocking the establishments was a priority and true to his propensity for action on a grand scale, Sir John purchased the Powder River Cattle Company's entire Canadian herd, some 5,800 head, along with their 76 brand. This became the main brand of the new outfit and as years passed the CACC became widely known as the 76 Ranch. In addition to fine horses and purebred Polled Angus and Galloway bulls, a large number of Yorkshire swine were brought in, but Sir John's sheep enterprise was his most ambitious; he imported some three hundred pedigreed rams from England and had a flock of over ten thousand head trailed north from Montana. Some of his management schemes were unique in Western Canada. For example, during a killing drought in 1889 he attempted to save the wheat on his one thousand-acre fields by having water hauled in carts. Again he envisioned Swift Current as the dairy capital of the West and built a creamery and cheese factory there, but failed to recognize the fact that the wild 76 range cows were not milk cows; naturally, they and the cowboys who handled them were all vigorously opposed to the idea, so this plan too failed. However, his project of establishing a meat packing business in Calgary met with unqualified success.

By 1890 the company was in a serious financial condition and Sir John Lister-Kaye was forced to resign. That year proved to be a particularly disastrous one. In April, a fire started by a spark from a CPR locomotive caused the death of over one thousand sheep, the loss being compounded by the fact that the majority were ewes and the lambing season was approaching. That summer a burning July drought was followed by a fierce hailstorm with the result that no grain was harvested in the Swift Current area that fall. In 1895 the CACC sold its holdings to a new London-based organization, the Canadian Land and Ranche Company. D. H. Andrews, an Englishman respected in the stock business who had come to the CACC at the time of the purchase of the 76 herd, became manager. Consolidation policies were adopted and in 1897 the company arranged with the CPR and the government to exchange some twenty-four thousand acres of arid prairie in the Maple Creek area and west of Medicine Hat for more arable land around Swift Current, Gull Lake and Crane Lake. During much of the next decade the 76, despite some serious reverses, conducted a largely successful cattle and sheep business. However, in 1903 the combination of spring snowstorm losses and poor wool prices added to the increasing demand for farm land by settlers who were crowding into the Swift Current district led the company to dispose of its sheep operation and sell its land at Swift Current and Rush Lake. During the next few years pressure from the westward flow of settlement strengthened and the Company sold off more of its holdings, retaining sizeable cattle and horse herds which were ranged mainly at the Crane Lake ranch. The killing winter of 1906-07 dealt a final blow to the 76 and in 1909 the Winnipeg-based firm of Gordon, Ironsides and Fares acquired the remainder of the holdings of the company. This pioneering ranch/farm experiment, despite its failures, achieved some valuable successes. Although it did not fulfill its original objectives as a colonization scheme, many of the company's Scottish and English employees remained in the country, particularly in the Swift Current area, becoming a part of the district's agricultural base. In addition, the company's purebred animals, like those of a number of other ranches conceived on an ambitious scale, were major factors in the production of good stock across the West.

Another ranch/colonization enterprise considerably smaller but in operation earlier than Lister-Kaye's gigantic project, was the Military Colonization Company Ranche, founded in 1883. The instigator was a retired military man, General Thomas Bland Strange, who obtained a lease on the Bow River where he planned to raise horses for the British Army and to provide a type of haven for his British officer shareholders, a scheme that was largely unsuccessful.

On the plains in the southeastern part of the district of Alberta and across Assiniboia, a considerable number of American ranches appeared. The Circle Ranch owned by the Conrad brothers and the I. G. Baker Company of Fort Benton, Montana, were the most noted of the early American spreads coming into the country before 1890. Their cattle were wide-ranging but the main concentration was in the Medicine Hat area. As the 1890s and early 1900s passed, increasing numbers of American cattlemen, seeking to escape the overcrowded southern range and the pressure of advancing settlement, brought their herds northward. Sufficient rain fell during 1898 and the first few years of the new century to produce good grass in the normally arid areas south of Medicine Hat, Maple Creek and Swift Current so ranch prospects appeared promising. The largest American cattle business to come into this area was that of A. J. "Uncle Tony" Day, a Texas rancher whose operation had been gradually pushed northward by the rising swell of homesteaders. Leaving his last stop in South Dakota in 1902 he brought his herds of approximately twenty-five thousand head of cattle and six hundred horses north of the border and here on several leases extending from Swift Current westward to Maple Creek, around the Cypress Hills and south of Medicine Hat, he established the Turkey Track Ranch, named from the distinctive brand his animals carried. Tony Day was an expert cattleman but the two greatest enemies of the early Western ranchers, encroaching settlement and killing winter weather, proved too much for him. Almost two-thirds of the Turkey Track cattle perished during the winter of 1906-07 and in 1908 the remainder of his herd was sold to Gordon, Ironsides and Fares Company.

Another famed cattle outfit that came north during this period was the Scottish-owned Matador Ranch with headquarters in Colorado. In 1904 a large lease was obtained north of Rush Lake along the South Saskatchewan River and the following spring shipments of stocker cattle commenced arriving. The cow/calf part of the Matador operation remained in the United States and the Canadian ranch was used entirely for fattening young stock for beef shipments to Chicago. For this reason the 1906-07 winter, although causing a loss of almost 50 percent of the Matador cattle in the Saskatchewan region, did not deal the ranch the crippling blow suffered by many other spreads. After prospering for more than a decade the Matador decided to close down its Canadian operation in 1921.

In the wake of the American ranchers, droves of farmers following the same trails from the western states pushed into the southern section of the plains of Alberta and Assiniboia. Not all were Americans, some were Canadians returning northward, others were Europeans, Germans mainly, who had tried the Western American farm lands and now were looking for greener pastures. South of the border, homesteads were gone and land prices high, so the farmers' trek northward was an attempt to grasp a fast-disappearing opportunity to obtain a homestead or to buy cheaply land available from the government or the CPR. True, it was arid country, but some had had experience in farming dry land—real dryland farming techniques came later but initial attempts were under way. Added incentives were good grain prices, and Canada's welcome to the tillers of the soil. So ranchers were squeezed out and settlers won another battle in their war with the cattlemen.

Conflict with settlement had become a part of life for the cattle industry soon after its inception. As early as 1885 a group spearheaded by Calgary's first farmers, Sam Livingston and John Glenn, and called the Alberta Settlers' Rights Association, sent a petition to Sir John A. Macdonald requesting that all townships around Calgary be opened for homestead entry and that settlers already established receive patent for their land. As numbers of settlers increased the feud became bitter but at no time did it erupt into violence and bloodshed. One particularly contentious case of a settler on the Walrond range gained national publicity, becoming an issue in Ottawa where the Conservative government, in the face of strong criticism in opposition ranks and among much of the public, ruled in favour of the ranch. Nevertheless the rumblings produced by this conflict presaged to some ranchmen a coming change in the comfortable, compatible relationship between the Conservative western ranchers and their friends and supporters in government.

A further disruption for the big ranchers occurred in 1892 when the section of the Calgary and Edmonton Railway running south to Fort Macleod was constructed. Not only did this line cut through the ranges of many of the large ranches, thus interfering with grazing patterns, but land required for railway grants was included in their leases. This situation precipitated the passing in 1892 of an order-in-council to terminate closed leases in 1896. The ranchers were given options to buy comparatively large portions of their leased land at reasonable prices. An added concession was an extension of stock-watering reserves, that is the exemption from settlement and maintenance of public access rights to areas on creeks and rivers, which thus prevented homesteaders from cutting off water resources from ranch cattle.

The year 1896 heralded significant changes in the Western cattle industry. First the termination of the original no-settlement leases took effect and although blocks of land could be purchased and new leases with homestead release could be obtained, the days of the open range when the cattlemen controlled the country were over. Another blow to ranchers in 1896 was the defeat of the Conservative government as the incoming Liberals posed a further threat to the stockmen's cause. Their ranks included both Clifford Sifton, Minister of the Interior, soon famed for his immigration policy designed to fill the West with settlers, and Edmonton-based Frank Oliver, Member of Parliament for the District of Alberta, who was an ardent champion of settlers' rights and a vigorous opponent of the ranchers and their interests. Deprived of their influential connections with Ottawa, the cattlemen, seeking strength in unity, formed the Western Stock Growers' Association in 1896, an organization designed to act as the voice of the industry.

The early 1900s brought difficult years for the cattle industry. The immigration policies of the government and the CPR bore fruit; the stream of incoming settlers, small until 1896, then rose rapidly becoming a mighty flood through the first years of the new century and continuing at high levels until 1912. A spirit of farming optimism pervaded the West. Grain prices were good and farmers were encouraged by higher than average rainfall, so when the government, despite ranchers' protests that much of the country was too dry for farming, opened the dryland grazing areas of the southeast for settlement, homesteaders poured in and their plows turned under thousands of acres of native prairie wool, that short nutritious blend of grasses which could better withstand dry conditions and which made excellent feed for stock. The severe drought of 1910 forced some of the settlers to abandon their dryland farms but their grain-growing attempts had caused irreparable damage to the prairie ecology as prairie wool once destroyed does not restore itself. For the ranchers meanwhile, settlement advance was taking its toll. Then came the bad winter of 1906-07. In 1907-08 quite a number of ranchers went out of business; the losses they had sustained were a factor but in most cases the decisive reason was the agrarian advance.

So ended the days when the cattlemen ruled the West, the times when their influence was a truly dominant force in government, and when, as during the decade following 1881, the cattle export trade boomed and their industry became the most promising factor in the Canadian economy. Nevertheless the influence of this period persisted with perhaps the social elements of its character being

Saddle bronc riding at the first Calgary Stampede in September 1912. A massive crowd of over 100,000 people attended to watch hundreds of cowboys from across Western Canada, the United States and Mexico compete for prize money. The second Calgary Stampede was not held until after the First World War, in 1919, however; it was dubbed the "Victory Stampede" in celebration of the end of the War. (Courtesy of the Provincial Archives of Alberta/A-12116.)

the most enduring. This was particularly true of the Foothills area. The ranchers here with initial strong social connections in Britain and Eastern Canada soon developed a way of life which essentially was maintained until the outbreak of World War I. It featured sports, notably polo, horse racing, hunting, cricket, tennis, and included such cultural events as balls, musical evenings, and visits to the theatre in Calgary. Indeed the cultural relationship between the city and the ranch community was congenial and, generally speaking, homogeneous; one manifestation of this was the Ranchmen's Club in Calgary.

Thus the life of the early ranch community became a distinctive part of Western heritage. A segment of the cattlemen's story was over but, contrary to some contemporary assessments, the saga itself was not ending. Some felt that the rancher was being pushed out, that the grain farmer and the mixed farmer whose crops were supplemented by the raising of some stock were taking over the West. This, of course, was not so. The days of the open range were certainly gone, but the stock industry survived. The rancher changed with changing times. Obviously ownership of land was necessary for security, so many stockmen purchased as much as possible of their holdings and reduced the size of their herds to more manageable units. Although numerous ranchers operated on a smaller scale, some of the cattlemen still had large holdings and great herds. Foremost among them were Pat Burns, A. E. Cross, George Lane and A. J.

Maclean. These, the Big Four as they came to be called, underwrote the first Calgary Stampede in 1912, an attempt to keep alive memories of early ranch life in the West. They were giants in the cattle business but other ranchers, too, operated comparatively large spreads and held positions of importance in the ranch community during the first years of the new century.

There was a sense of rightness, of destiny, about the Western cattle business during its early days. The prairies and hills had been the home of buffalo and after their demise it seemed ordained that cattle should take over the range. The early ranchers did little to disrupt nature—their homes were few and scattered, the country lay largely unfenced, generally speaking the Canadian range was not over-grazed, the trails that wound across hills and plains were not very different from Indian trails and buffalo paths. There may have been a conscious or unconscious awareness on the ranchers' part of their oneness with nature, with the country—undoubtedly they felt they were the right people at the right time and place using the land as it was meant to be used. Perhaps their strength and influence rose in part from this sense of a God-given right. It was inevitable that the reign of the cattle barons should pass but their impact on the entire Canadian scene was undeniable, they provided the ranching community with a strong foundation and left a legacy of richness and colour. Their story adds to our history's uniqueness, a certain eminence, and a touch of glamour.

NOTES

This article first appeared in *Prairie Forum* 11, no. 2 (1976): 229-42. The apparent discrepancy in the spelling of the word "ranch" should be explained. During the early period it was customary for British and Canadian ranchers to spell "ranch" with a final "e" and in the official names of their cattle companies and in relative correspondence it is so spelled. In the names of American-based spreads and a few others the "e" is omitted. In the interests of accuracy and to give a greater sense of the time period I have followed the practice of spelling the various companies' names as they appeared in documents and writings during the initial ranching period. In general usage throughout the article "ranch" is spelled in the presently accepted manner.

REFERENCES

Primary

Breen, David H. "The Canadian West and the Ranching Frontier, 1875-1922" (Ph.D. dissertation, University of Alberta, 1972).

Brand Book, 1880-1890. Cattle and Horse Brands Registered by North-West Territorial Government (known as Brand Transfer Book), Glenbow-Alberta Institute.

Canada, *Sessional Papers*, 1882, vol. xv, no. 8, 45 Vic. (No. 18), "Report of Commissioner A.G. Irvine for 1881," 1 February 1882, p. 11. Also Irvine's report for 1880.

Canada, *Sessional Papers*, 1882, vol. xv, no. 8, 45 Vic. (No. 18), "Dominion Land Regulations," p. 6.

Department of the Interior, Timber and Grazing Branch, Public Archives of Canada (records re early ranches, 1880-1926, photocopies in Glenbow-Alberta Institute).

Evans, Simon, "The Passing of a Frontier: Ranching in the Canadian West, 1882-1912" (unpublished Ph.D. dissertation, University of Calgary, 1976).

Hardisty, Richard, Papers, J. Bunn to Richard Hardisty, 14 August 1875, Glenbow-Alberta Institute.

Topographical Maps, 1925-26, with leases marked by J. E. A. Macleod, Calgary. Glenbow Library.

Topographical Map showing ranch leases and related data obtained from Sessional Papers No. 53, A 1885 and No.20, A1886 by Tom Gooden, Glenbow Library.

Western Stock Growers' Association Papers, 1896-1921, Minute Book, 1896-1921.

White, Frank, Notebook—Stock Records, Contracts, Cochrane Ranche, 1881-82, Glenbow-Alberta Institute.

Secondary

Breen, David H., *The Canadian Prairie West and the Ranching Frontier, 1874-1924* (Toronto: University of Toronto Press, 1983).

Brado, Edward, *Cattle Kingdom: Early Ranching in Alberta* (Vancouver/Toronto: Douglas & McIntyre, 1984).

Craig, John R., *Ranching with Lords and Commons* (Toronto: William Briggs, 1912).

Godsal, F. W., "Old Times," *Alberta Historical Review* 12, no. 4 (Autumn 1964): 19.

Jameson, Sheilagh S., "Era of the Big Ranches," *Alberta Historical Review* 18, no. 1 (Winter 1970): 1-9.

Jameson, Sheilagh S., "The Quorn Ranch," *Canadian Cattlemen* 8, nos. 2 and 3 (September and December 1945). Jameson, Sheilagh S., "The Social Elite of the Ranch Community and Calgary," in A. W. Rasporich and H. C.

Klassen, eds., *Frontier Calgary: Town, City and Region 1875-1914* (Calgary: McClelland and Stewart West, 1975).

Jameson, Sheilagh S., "Partners and Opponents: The CPR and the Ranching Industry of the West," in Hugh A. Dempsey, ed., *The CPR West: The Iron Road and the Making of a Nation* (Vancouver: Douglas & McIntyre, 1984), 71-86.

Kelly, L. V., *The Range Men* (Toronto: William Briggs, 1913).

MacEwan, Grant, *Blazing the Old Cattle Trail* (Saskatoon: Modem Press, 1962).

McGowan, Don C., *Grassland Settlers: The Swift Current Region During the Era of the Ranching Frontier* (Regina: Canadian Plains Research Center, 1975), 57-106.

4. The Roots of Agriculture: A Historiographical Review of First Nations Agriculture and Government Indian Policy

Bruce Dawson

Historians have largely ignored the agricultural pursuits of the First Nations peoples of the Canadian prairies.[1] Despite the many explorer accounts, ethnographic recordings and archaeological suggestions of an agricultural tradition amongst a number of Aboriginal groups on the northern plains, historians have rarely mentioned First Nations farmers in their annals. More often, the historical record has dismissed the concept of Aboriginal agriculture. Occasionally these dismissals have been subtle, such as the opening lines to eminent Prairie historian W. L. Morton's article "A Century of Plain and Parkland":

> One hundred years ago the prairies, the lands rolling upward from the Red River to the foothills of the Rockies, were primitive, with little trace of human habitation. No rut scored the sod, no furrow scarred the long roll of the prairie.... The plains were as thousands of years of geological and climatic change had made them. Men had hardly touched them, for man himself was primitive, in that he had adapted himself to nature, and nature to himself.[2]

The flowing, poetic description of an empty land inhabited by peoples who could not conceive of utilizing it in an agricultural sense that Morton put forth in 1969 is matched by the blunter dismissal issued by early Saskatchewan historian John Hawkes, who stated: "The Indian was not a natural farmer. He was a born hunter and warrior. Century upon century had ingrained in him the nomadic instinct; steady labour, so many hours a day, week in and week out, was as foreign to his nature as a dog kennel to a fox."[3] The concept these

"Indian Corn," the crop most often associated with the Aboriginal peoples of North America.

authors suggested, that the First Nations peoples had no tradition, interest or aptitude for agriculture, is incorrect.

Agriculture is defined as the cultivation of soil and the rearing of animals. Archaeologists suggest that people first began to domesticate plants and animals approximately 12,000 years ago. During the succeeding 8,000 years, most peoples around the world adopted agriculture as their primary livelihood. By 7,000 years ago, the first cultivated plants, gourds, were being grown in midwestern parts of North America.[4] The crop most often associated with the Aboriginal peoples of North America, maize or corn, was first grown approximately 2,500 years ago in what is now the southwestern United States. During the succeeding 2,000 years, the growing of maize spread throughout the eastern and midwestern parts of the continent. Early European explorers noted the extensive agricultural pursuits of eastern Aboriginal peoples, such as the Huron living in the Great Lakes region and the Mandan who resided along the Missouri River in what is now North Dakota.

Agriculture was not prevalent on the Canadian prairies, but it did exist. Archaeological research indicated that corn was being grown in southern Manitoba by 1,400 AD.[5] Near the town of Lockport the remains of corn, large storage pits and hoes made from bison shoulder blades were found during the 1980s. While the evidence does not indicate who these farmers were, patterns found on pottery remains associated with the Lockport site were similar to those associated with agricultural groups that lived in northern and central Minnesota.[6] The agricultural pursuits along the Red River appear to have been abandoned during the fifteenth century at the height of a global climate change called the "Little Ice Age." This phenomenon, which affected temperatures around the world for over 400 years, resulted in short, cool summers which made farming in Manitoba impossible.[7] By the time Europeans arrived on the Canadian plains, agriculture was not part of the daily lives of the aboriginal residents. However, as Flynn and Simms noted, "local Aboriginal peoples were

Blackfoot men, Frank Tried to Flay and George Left Hand, sowing seed by hand, circa 1880s. A village of tipis is visible in the background. (Courtesy of the Glenbow Archives/NA-127-1.)

both familiar with the cultivation of numerous plants by their neighbours and trading partners along the Missouri River, its tributaries, and other rivers in the vicinity and [...] they cultivated fields of their own at various times and in various locations over a 400-year period prior to contact with Europeans."[8]

The written record supports the assertion made by Flynn and Simms that the Aboriginal groups on the Canadian plains were familiar with plant cultivation. In 1733 Pierre de La Vérendrye and his sons travelled overland from New France to initiate a French fur trade on the western plains. In reporting the establishment of a fort along the Red River, La Vérendrye noted that "the Cree chief intended to remain with the elders of his people near the French fort all the summer, and that he was even going to raise wheat, seed of which had been supplied him by the Sieur de la Vérendrye."[9] In 1734, La Vérendrye travelled south with a group of Cree on their annual excursion to purchase corn from the "Ouchipouennes."[10] Of the Ouchipouennes, La Vérendrye noted that fields of corn beans, peas, oats and other grains were raised by men of the community for sale to neighbouring groups.[11] The agricultural tradition of the Dakota is extensive. Anthropologist Bryce Little writes that the practice of agriculture greatly predates the arrival of Europeans and was such an integral part of the society that the Dakota name for the month of June translates as "the moon when the seedpods of the Indian turnip mature."[12] Little later asserts that the shift by the Dakota peoples in the late nineteenth century from a fur trade economy to agriculture "was more a case of re-employment of a known practice rather than any result of white-acculturation."[13] This tradition followed the Dakota onto the Canadian plains, and as late as 1951, a distant variety of "Indian Corn" was noted in the possession of Dakota peoples residing in Canada.[14] Even among the Blackfoot, the group who at contact seemed most distant from farming activities, evidence of an agricultural tradition can be found. In 1879, the significant chiefs of the Peigan, Blood and Blackfoot signed a statement that asserts that their ancestors were tillers of the soil.[15]

With this agricultural tradition amongst the Aboriginal peoples of the plains, one may ask why the only full-time practitioners at the time of European contact were the peoples of the Missouri River valley? Two possibly linked suggestions have been put forth. The most significant suggestion is the influence of the climatic phenomenon known as the Little Ice Age. A worldwide event, the Little Ice Age was a period of slightly reduced global mean temperature that lasted from approximately 1350 to 1850. According to geographer Jean Grove, the impact of this climatic episode upon agricultural pursuits in the Northern Hemisphere was dramatic. Throughout the Scandinavian countries, it is estimated that almost half of the medieval farms were abandoned during the period because the upper limit of the altitude above sea level at which cultivation could be practiced was lowered by over 150 metres.[16] The Canadian plains are believed to have been affected in a similar manner.

Historian James MacGregor suggested an alternative theory regarding the discontinuation of agricultural pursuits. He asserted that bison populations rose dramatically due to some environmental occurrence. "When this happened, they (the First Nations peoples of the northern plains) became nomadic buffalo hunters and abandoned their agricultural way of life."[17] The suggestion by MacGregor would seem to support the scientific data of the Little Ice Age impact upon arable land of western Canada. Bison are well adapted to living in cooler environments and would have definitely been an "easier" food source when compared to the numerous agricultural failures which would have occurred with the Little Ice Age climatic impact.

By the mid-nineteenth century, the Little Ice Age influence was diminishing, and the pursuit of an agriculturally based economy, though difficult, became again a possibility. Concurrently, bison populations began to decline due to increased hunting and other human activities in the plains region. The thoughts of the Aboriginal residents again turned to agriculture.

In the Qu'Appelle valley in 1857, Charles Pratt, a missionary of Cree-Assiniboine descent, commented to James Hector, a geologist on the Palliser Expedition, that the Cree in the area were growing concerned about the scarcity of buffalo: they were "anxious to try agriculture ... (and) would make a start on it if they only had spades, hoes and ploughs."[18] Two years later, Hector was told of a similar request by the Stoney people living near Howse Pass in what is now Alberta.[19] Also in 1859, a scientist exploring the prairies on behalf of the Hudson's Bay Company, Henry Youle Hind, noted a letter written from Chief Peguis to the governor of the Company. The greater part of the letter discussed concerns regarding the fulfillment of the promises outlined in the 1817 treaty between Peguis and Lord Selkirk. Peguis's main areas of concern were the size of the growing settlement and his own desire to be furnished with "mechanics and implements to help our families in forming settlements."[20]

Just over a decade later, as the bison populations continued to decline, the plains peoples expressed a heightened interest in establishing a modern agricultural lifestyle. The 1871 statements to Governor Archibald made by Chief Sweetgrass and other Cree leaders all emphasized the desperate desire to pursue subsistence farming.[21] Another Cree Chief who had an interest in agriculture at that time was Ahtahkakoop. Living along the shores of Fir Lake, Ahtahkakoop and his band made their first attempts at cultivating the soil in 1872.[22] In 1874, realizing that his people needed training in agricultural practices, Ahtahkakoop sought the assistance of John Hines, a missionary who had come to the west seeking to teach farming to Indian peoples. The band, along with Hines and his assistants, relocated to Sandy Lake in 1875 and that year cleared enough land to produce 180 bushels of wheat and barley.[23] The interest these leaders and other plains residents showed in agriculture significantly influenced the approach and desire of the First Nations people to negotiate the treaties with the government.

The early 1870s were an extremely difficult time for the First Nations peoples of the plains: bison populations were in freefall, causing hunger amongst all groups. An outbreak of smallpox in 1870 struck the prairies, causing a great number of deaths amongst the resident groups, particularly the Blackfoot. The year 1870 also witnessed what would be the last large-scale battle between First Nations groups on the Canadian plains when a group of Cree and Saulteaux attacked the Blood and Peigan camped near what is now Lethbridge, Alberta. Ever increasing numbers of Europeans were arriving in the West, starting farms and businesses. In 1870 the government of Canada acquired the North-West Territories from the Hudson's Bay Company and established the province of Manitoba with no consultation with the First Nations residents. Stories of the bitter encounters between the Aboriginal residents and the military were arriving regularly from the United States. It was in this atmosphere that treaties One through Seven were conceived and negotiated.

For several decades, the standard historical reference on this period as well as the treaty negotiations was George F. Stanley's *The Birth of Western Canada*, published in 1936. In this work, Stanley devoted significant attention to what he describes as "The Indian Problem." The problem, according to Stanley, was the inability of the First Nations peoples to understand or adapt to the changes the superior white society was bringing to western Canada.[24] He suggested that during the fur trade period, the plains peoples had come to view white society, personified by the Hudson's Bay Company, as "representative of a superior civilization and the embodiment of fair dealing."[25] The growing incursion of less scrupulous traders and of white settlers during the 19th century left the 'hapless" First Nations peoples confused about how to deal with the evolving situation.[26] When these situations were combined with the previously noted concerns of the 1870s, Stanley claimed that it became the task of the government to calm the "excited spirits" by determining "a policy which would ensure a continuance of these peaceful relations, convince the Indians of the government's good faith and assist them over the difficult transition from savagery to civilization."[27] The government, according to Stanley, chose a benevolent approach that would extinguish First Nations title to the land while establishing the peoples upon reserves. On these reserves they could be taught agriculture and religion in relative safety from the vices of European society.

Stanley's interpretation of the treaty process was to remain virtually unchallenged until the 1970s. When the reappraisals came, they focused on two themes: What role or influence did the First Nations peoples have in the drafting of the treaties, and did they have a clear understanding of treaty process and the documents they were signing? Noel Dyck was one of the first to grapple with these questions. In his 1970 MA thesis, "The Administration of Federal Indian Aid in the North-West Territories, 1879-1885," he concluded that the downfall of the initial reserve agricultural program was primarily a result of the government's drive for economy. Dyck also posed the question of First Nations involvement in the treaty process,[28] suggesting that the First Nations population was interested in pursing the treaty process, though their concept of the purpose of the treaty was different than that of the government.[29] While he is not sure whether the Aboriginal peoples were clear about the full ramifications of the treaties, his posing of the question paved the way for others to explore the issue.[30] John Taylor deliberated the question of the First Nations role in his 1975 paper "Canada's Northwest Indian Policy in the 1870s: Traditional Premises and Necessary Innovations," in which he challenged the concept promoted by Stanley that the government was completely responsible for the "wise" and "benevolent" treaties. Rather, he found that for Treaties 1, 2, and 3, many of the important treaty terms, such as agricultural aid, were not in the original government treaty drafts, but were added as "outside promises" after negotiation with the First Nations representatives.[31] The involvement of First Nations peoples in the treaty process described by Taylor, later referred to as "active ... agents" by J. R. Miller, has been supported by several authors.[32]

The debate regarding the understanding the Aboriginal treaty signatories had of the meaning and intent of the treaties has been more pronounced. An important aspect of this discussion has been to move past Stanley's notion of non-comprehension by the Native peoples and focus on the impact of cultural differences upon the interpretation of treaty terms and the corresponding process. John Taylor, in a 1981 article, noted a number of statements made by treaty signatories that seemed to indicate differing interpretations of the concept of "surrender" and the signing over of subsurface mineral rights. He felt, however, that the First Nations interpretive evidence is generally inadequate and composed of too many conflicting views to provide an effective challenge to the treaty text.[33] Alternatively, John Tobias believed that the Plains Cree leaders had a clear understanding of what they wished to gain from the treaty process and followed a strategy of negotiation based upon the tactics they had successfully used for two centuries in the fur trade.[34] These efforts were for naught, found Tobias, as the government used political, legal and physical forces to eliminate the Cree interpretations of the treaties, with the goal of obtaining complete control over the Cree peoples. In a similar vein, J. R Miller found the entire process doomed to difficulty, as the First Nations peoples believed they were establishing a treaty of friendship, assistance and mutual land usage while the government viewed the treaties as surrender of all Aboriginal title to the prairies.[35] It was under this cloud of misunderstanding that the First Nations peoples began their full-scale pursuit of an agrarian lifestyle.

The historical literature on the immediate post-treaty agricultural endeavors of the plains reserve residents follows a similar pattern to that on the treaty process. Again, George Stanley provided the benchmark analysis in his *The Birth of Western Canada*, which was to stand for several decades. His interpretation was not positive. Stanley asserted that the childlike and nomadic nature of the First Nations people was not conducive to an agricultural lifestyle. Their desire for the good old days of savage self-reliance, suggested Stanley, caused most of the reserve residents to be despondent and resentful towards the government's policy.[36] As he poignantly stated, "as long as the herds of bison tramped the prairies and the antelope sped across the plains, they were loath to abandon the thrilling life of the chase for the tedious existence of agriculture."[37] Therefore, according to Stanley, in spite of the tenacious efforts of the church and the government officials, the inability of the reserve residents to adapt to the new lifestyle caused the early failure of agricultural programs.

The 1970s witnessed the beginnings of a significant re-evaluation of Stanley's view that First Nations peoples were unsuited for farming. One of the first to present a revised view of Aboriginal farmers was Noel Dyck. In his aforementioned 1970 thesis, Dyck concluded that "the greatest obstacle in the way of the reserve agricultural program was the government's willingness to place considerations of economy above all else."[38] Sadly, Dyck went on, "Indians who were fed so little that they remained in a constant state of hunger could not become self-sufficient farmers."[39] He further explored this idea in "An Opportunity Lost: The Initiative of the Reserve Agricultural Programme in the Prairie West," where he examined why the Canadian government between 1880 and 1885 pursued policies which undermined the reserve resident's attempts to establish an agrarian economy. He began by outlining how many of the problems could be traced back to the differing impetuses for both groups to enter into the treaty negotiations. According to Dyck, the government was merely interested in gaining control of the West as frugally and expediently as possible[40]; alternatively, First Nations people viewed the treaties as the beginning of a long-term alliance.[41] As a result, Dyck asserted, the government reluctantly began to fulfill the treaty requirements to assist in establishing reserve agriculture only after being forced to pay for massive amounts of relief supplies to feed the prairie Indians in 1879.[42] The remainder of the article summarized the various means by which, in Dyck's view, the government mismanaged the administration of the agricultural program and thwarted the real interest the reserve residents had in pursuing an agricultural lifestyle. To Dyck, displaying power and control over the Aboriginal peoples was the ostensible goal of the government policies: the government missed the fact that the Cree were agitating for agricultural assistance—not because of the change in lifestyle prompted by reserve life. If these agricultural needs had been met, the "agitating" would not have occurred.[43] The result of the government's focus on control was that "the farming conducted on prairie reserves after 1885 was no longer the achievement of Indians who were seeking to become self-sufficient members of a new society; instead, it comprised the carefully supervised activities of a people who had become the involuntary wards of the government."[44]

One author who became synonymous with this reinterpretation was Sarah Carter. Her 1983 article titled "Agriculture and Agitation on the Oak River Dakota Reserve, 1875-1895" was her first contribution to the new approach of historical writing in this area. In this work, Carter repeatedly challenged Stanley's conclusions with examples of hard work, farming experimentation, and crop successes by the residents of the Oak River reserve. But in spite of the interest and aptitude First Nations people displayed during the first decade of agricultural experimentation, by the mid 1890s the reserve was no longer producing wheat crops. The reason for the downfall of the agricultural program, Carter found, was a combination of poor environmental conditions and repressive government policies.[45]

Carter further explored one of these repressive government policies in her 1989 article "Two Acres and a Cow: 'Peasant' Farming for the Indians of the Northwest, 1889-97," where she examined the implementation and impact of the "peasant" farming policy introduced by Commissioner Hayter Reed in 1889. Under this policy, reserve agriculturalists were forced to abandon the use of mechanical equipment and revert to the use of simple hand tools to plant and harvest their crops. The implementation of the policy, Carter discovered, "had a stupefying effect on Indian farming, nipping reserve agricultural development in the bud."[46] Throughout the article, Carter suggested that the policy was implemented to break down tribal unity and promote individualism as well as to reduce the amount of land the band could effectively put to crop and, in doing so, create "surplus" lands which could be surrendered and sold.[47] As well, the policy prevented the reserve farmers from competing with white farmers for the limited markets. Reed defended his policy by suggesting that in order to become "civilized" the reserve residents needed to start with the basic tools of agriculture and "progress" to the use of machinery. He suggested that the use of machinery would interrupt the steps necessary to advance a civilization and would cause the First Nations peoples to become lazy. Despite numerous reports from reserve agents and farm instructors about the detrimental effects the policy was having upon the agricultural program, Reed persisted in pursuing the policy because of political pressures from white settlers, widespread naivety regarding western Canadian agriculture, and his driving belief that "Indians were incapable of understanding these concepts, and could not operate farms as business enterprises."[48]

In 1990, Sarah Carter released *Lost Harvests*, which remains the most significant work published in the area of First Nations agriculture. The book examined the agricultural development on prairie reserves from 1874 until World War I, with particular emphasis upon the reserves within the boundaries of Treaty 4. Thematically, the book concentrates upon an in-depth exploration of the hypothesis she explored in her earlier articles, namely that repressive government policies and actions were responsible for undermining the earnest efforts of reserve residents to establish commercial farming operations. The initial reserve agricultural policy, which Carter labels the Home Farm Experiment, was implemented in 1879. The program involved establishing 'home farms' on numerous reserves, within which government farm instructors could teach the reserve residents farming methods via example. Problems with weather, slow-maturing seed and lack of markets that affected all western farmers, according to Carter, were compounded by the agent positions being staffed by ill-trained patronage appointees from Eastern Canada, who knew nothing about prairie farming nor the First Nations peoples whom they were to instruct.[49] Overshadowing the aforementioned problems, says Carter, was the "work for rations" policy through which the reserve residents were expected to meet levels of both work and food production before government rations would be issued. The unrealistic expectations of this policy led to starvation and discontent on the reserves, as well as doubts and feelings of mistrust by both the government officials and the reserve residents about the dedication to fulfilling the treaty promises.

The "Home Farm Experiment" was phased out in the mid-1880s and replaced by the 'Peasant' farming policy discussed earlier. Also at this time, the government implemented a permit system by which the agents assumed control for selling First Nations crops and a pass system that restricted reserve residents' movement off the reserves—repressive actions which according to Carter "placed restraints above and beyond those shared with other farmers in the West."[50] The government introduced also a policy of severalty onto the reserves. This policy of subdividing the reserve into individual farms, according to Carter, was implemented by Commissioner Hayter Reed because of his belief that the best way to undermine the tribal system was via individual farmers building self-reliance on their own land plots.[51] More importantly, "severalty would confine the Indians within circumscribed boundaries and their 'surplus' land could be defined and sold."[52]

Carter also explored a secondary theme of First Nations protest against the restrictive policies as they were implemented during the last quarter of the nineteenth century. She found that the reserve residents quickly came to believe they had been misled by treaty commissioners about the potential of developing an agrarian economy. Numerous examples of letters and comments of protest to government officials and missionaries are noted, protests which were largely ignored or dismissed as being spawned by laziness, incompetence and the inclination of Aboriginal peoples to complain. Despite these rebuffs, Carter wrote, "At no time, however, did Indians adopt a policy of passive submission, disinterest, or apathy. The tradition of protest continued."[53]

Not surprisingly, the results of her study are similar to the conclusions she drew in her earlier articles. As she notes in her conclusion, "histories written until very recently, obscure or overlook the Indians' positive response to agriculture in earlier years. Equally obscured and forgotten has been the role of Canadian government policy in restricting and undermining reserve agriculture."[54]

While Sarah Carter has been the dominant author in the field of First Nations agriculture, others have been active. J. R. Miller's *Skyscrapers Hide the Heavens* (1989) looked at agriculture as part of his survey of relations between the Native and non-Native populations in Canada from 1600 to present. Unfortunately, the broad scope of the work permitted only a few pages within a chapter entitled "The Policy of the Bible and the Plough," devoted to agriculture. The chapter title refers to Miller's idea that the federal government, through the use of coercive powers, forced the reserve residents to take up agriculture as part of their assimilation mandate.[55] Force was necessary because not all prairie Indians were enthused about adopting the whiteman's ways.[56] Contrary to Dyck, who saw the government acting out of the need to control people, Miller followed Carter's idea that most policies were driven by one political expediency: keeping the white voters happy.[57]

While extremely restricted in his look at the subject, a number of positives emerge from Miller's work. Notably, he accounts for variability. As mentioned earlier, Miller noted that not all First Nations peoples wanted to become farmers and indicated that poor climate was a contributing factor to the failure of the agricultural policy.[58] Like Dyck, Miller mentioned one reserve where success was gained, but he did not explore why this success was garnered in that location. Miller also provided no specific dates, only references such as "late 1880s"; having dates would be helpful in comparing the farming experiences with concurrent events so as to evaluate why success came or not. In this case a broad, thin look at agricultural policies paints the entire prairies with one brush.

The approach taken by Miller is almost duplicated in Helen Buckley's *From Wooden Ploughs to Welfare*. Like Miller's, Buckley's book is a survey of government policy; however, she focused only upon policies and interactions in the Prairie Provinces. This more limited scope allows a somewhat more in-depth look at some specifics of western reserve agriculture.

Unlike Miller, and more similar to Dyck, Buckley found that the government's desire for control was at the root of failure[59]; specifically, she suggested that the real holder of power within the government structure was the local agent. Referring to successful agriculturalists, Buckley asserts that "their success was, in essence, another aspect of control, for the grants or loans needed to make money out of farming were available to a select few, handpicked by the agent."[60] This last comment provided a possible answer to the hanging questions of both Dyck and Miller regarding the cases of sporadic agrarian success. Similarly, Buckley, like Miller, suggested climate as a possible cause of limited success down to the late 1890s. Illness and unrefined technologies are also suggested as limiting factors that affected not only Aboriginal farmers, but also the previously unmentioned early white agriculturalists in the region.[61]

Buckley provided interesting insight into the reserve agriculture. Most important was the dedication to bringing out individualism: faceless, unified-in-action government officials were not pitted against equally faceless, unified-in-action Aboriginal peoples. Buckley noted that some agents acted as individuals; she also explored the mindset of Superintendent Reed so as to explain why the peasant agriculture policy was implemented.[62] As well, she acknowledged that some First Nations gave up agriculture in the 1890s.[63] Unfortunately, the work is still a survey piece and, like Miller's, glosses over many dates and subtle changes in policy.

Another author who has contributed to the scholarship regarding Aboriginal agriculture is Peter Elias. In his book *The Dakota of the Canadian Northwest*, Elias chronicled the challenging experiences of various bands of Dakota during their efforts to establish a land and economic base on the Canadian prairies. Elias, like most of the other authors, found that the early successes of the Dakota agricultural initiatives were hampered by a harsh environment and repressive government interference. Unlike Miller and Carter, and similar to Dyck, Elias suggested that the government interference was based upon a desire to control all Indian matters on the prairies via physical presence, repressive policies, and the "coercive power of the law."[64] Though these governmental intrusions, Elias found, did serve to limit the agricultural potential of the farming Dakota bands, farmers continued to experience success and were able to develop moderately sized, small-profit agricultural operations that were perpetuated until the time of writing.[65] Any cultural or economic success, he concluded, came "when the Dakota were independent to act within the general framework of Canadian law."[66]

A comparison of the pieces by Dyck, Miller, Elias and Buckley to the works by Carter highlight a number of subtle, yet important, differences in approach and findings. Unlike the other authors, Carter seldom referred to First Nations agriculture with terms like "successful," preferring to describe the reserve agricultural pursuits with terms like "accomplished" or "improved." To utilize the term "success" would undermine her primary theory that "government polices made it virtually impossible for reserve agriculture to succeed."[67] In addressing why the government pursued these policies, Carter did not present a clear statement. She did note that some policies regarding the use of reserve land were influenced by white settlers, similar to the political expediency argument expounded by Miller[68]; she also discussed the Department of Indian Affairs (DIA) negotiations with the militia and the NWMP to help enforce Indian policy[69]; as well, Carter examined the deal struck between the DIA and Battleford merchants that prohibited the reserve residents from selling grain in the local market.[70]

These latter two points suggest a conspiracy against the First Nations peoples, but Carter did not tie these ideas together in any stated conclusion. This lack of a stated conclusion is seen in other situations in the book: a lot of information, with specific details, is presented, but readers are often left to their own devices to make connections other than to farmer persistence and government repression. Beyond these differences in focus and approach, the works of these five authors provided a viable alternative to the Stanley interpretation of early reserve agriculture; most notable is the repudiation of Stanley's notion that reserve residents were not interested in farming, thus accounting for their lack of success. Miller discussed the idea of the First Nations as being "active agents," aggressive and interested in securing the benefits of the whiteman's world.[71] Elias described the long history of successful agricultural pursuits the Dakota had previous to coming to Canada, which they were anxious to perpetuate if the government had assisted with adequate land and equipment.[72] Dyck, more directly, stated that "there is evidence not only of the willingness of prairie Indians to embark upon an agricultural way of live, but also of their continuing concern from the time of negotiation of the treaties in the mid 1870s to prepare for this eventuality."[73] Similarly, Buckley explained, "setbacks were due not to want of character or training, as many believe to this day, but to the economic and climatic conditions that made it a high risk enterprise for Indians and settlers alike."[74] Sarah Carter dedicated much of her introduction to renouncing the concept of Aboriginal lack of interest and exploring ideas similar to those mentioned by the other authors. Consensus also exists amongst the revisionist authors that the frugality of the federal government in the area of Aboriginal affairs contributed to the limited growth of reserve agriculture. Buckley asserted, in reference to treaty negotiations, that "the terms were set with a view to minimizing obligations in the light of commitments already made to the construction of the railway and other costly enterprises."[75] Noel Dyck was even more forthright, stating that "the drive for economy in Indian administration systematically retarded agricultural development."[76] This frugality in fulfilling the treaty requirements, according to the authors, facilitated the failure of reserve agriculture.[77]

It is interesting to note that the experiences of the First Nations residents of the Canadian prairies were mirrored by those living on the American plains. In "Talking with the Plow: Agricultural Policy and Indian Farming in the Canadian and U.S. Prairies," Rebecca Bateman compared the experiences of the Cheyenne and Arapaho peoples of Oklahoma with those of reserve residents of western Canada. In both areas Bateman found that the respective governments enacted policies and procedures, such as the non-use of labour-saving equipment and severalty, with the joint goals of creating excess lands to sell to white settlers and to create "the eventual cultural disappearance of Native people at any rate, rendering any permanent administration of their affairs ultimately unnecessary.[78]

A similar cross-border comparison was authored by Hana Samek in *The Blackfoot Confederacy 1880-1920*. In this work, Samek looked at the similarities and differences of the Blackfoot peoples who reside on both sides of the forty-ninth parallel, suggesting that the Canadian system of reserve administration, when compared to its counterpart in the United States during the same period, had a number of advantages.[79] However, none of these advantages made much of a difference when both administrations launched badly conceived and badly managed agricultural programs on reserves which were unsuitable to grain farming. As a result, "many Blackfeet simply gave up on farming" and the subsequent reserve allotments and surrenders "further impeded the development of a reservation economy."[80]

Thomas Wessel made similar findings. In "Agriculture on the Reservation: The Case of the Blackfeet, 1885-1935," Wessel noted that the Blackfoot people of Montana were victims of repressive rations policies and enforced agricultural projects which were inappropriate to their environmental circumstances. As a result, "instead of independent agricultural communities, the government created pockets of rural poverty physically fractionalized and politically factionalized."[81]

In 1987, R. Douglas Hurt published *Indian Agriculture in America*. In many ways his work, a survey of the agricultural experiences of American reserve residents from treaty into the twentieth century, is similar to Carter's *Lost Harvests*; the examples of repressive government policies implemented by naive and often indifferent officials are also similar. The biggest difference between Hurt's work and that of Carter and the other recent authors is his Stanleyesque assertion that "the difficulty of cultural change ... was most significant in the failure of the old nomadic and hunting tribes to adopt a whiteman's agricultural way of life."[82] Beyond this difference, Hurt concluded "severalty, cultural resistance, and the western environment, together with federal leasing and heirship policies and inadequate agricultural support, placed the Indians, not on the white man's road to self-sufficiency and civilization, but on the road to peonage."[83] Hurt's findings are similar to those suggested by Wessel and Bateman, and would indicate that the immediate post-treaty agricultural experiences for most of the First Nations peoples in North America were, unfortunately, similar.

Historians have been rather remiss in appraising the aboriginal farming activities that occurred on the plains after the turn of the century. The plethora of books and articles that examine the experiences of immigrant agriculturalists during the early part of the twentieth century rarely mention Native people's endeavours and, if they do, only as a cursory note. Even amongst the earlier-mentioned works of Buckley, Miller and Carter, the evaluation of post-1900 agriculture is limited.

Within this scant history, two themes dominate the discussion. All the authors find that the government's primary focus during the period was to encourage the reserve residents to surrender land from their reserves so that the growing number of white settlers could make use of these properties.[84] Both Miner and Carter follow their earlier explorations of repressive government policies by noting the lengths to which the government went to encourage land surrenders. These actions included changing the Indian Act in 1906 and 1911 to make the process easier.[85] Under these amendments, the government was able to: release to the reserve residents up to 50 percent of the land sale monies in cash, a tempting situation for a cash-poor society; see reserve land expropriated for the use of land development companies and municipalities; and remove reserve residents from reserves near communities with over 8,000 residents.

The second theme to emerge involves the First Nations' resistance to the increasing government incursion. James Dempsey stated that during the period after the turn of the century, "government domination had reached its peak and resistance was at a low ebb."[86] Miller's comments would support this assertion as he suggested that cases of successful resistance were few as the government would use "tools of compulsion,"[87] specifically, changes to the Indian Act which would impede or eliminate the reserve residents' ability to challenge the Department's desires. Carter's views on this issue are similar: "Indian resistance to surrender was generally pronounced and adamant to begin with but was generally broken down through a variety of tactics."[88]

The period also featured the establishment of the File Hills Colony in Saskatchewan. Created in 1901, the colony was the brainchild of W. M. Graham, then an Indian agent at the Qu'Appelle agency; it was a farming settlement composed of select graduates of the local residential schools. These young men and women were brought together, expected to marry, set up modern and successful farms on pre-selected plots within the colony, and live according to the Euro-Canadian ideals they had been taught in school. Based upon these objectives, E. Brian Titley in his article "W. M. Graham: Indian Agent Extraordinaire" found that the colony was "undoubtedly a success."[89] In "Demonstrating Success: The File Hills Farm Colony," Sarah Carter agreed

that, from an agricultural perspective, the colony was successful; however, in perpetuating her earlier-mentioned themes of repressive government policies and Aboriginal resistance, she noted numerous activities that were forbidden in the colony.[90] She also provided examples of resistance amongst the colonists to these suppressive rules, specifically those involving the continuance of traditional ceremonies.[91] However, government reaction to opposition was the same here as on the other reserves, and Carter found that objections and grievances were ignored or else met with changed policies so as to secure government success.[92] An alternate point of view regarding the File Hills Colony is expressed by Eleanor Brass in *I Walk in Two Worlds*. As one of the first children born on the File Hills Colony, Brass's 1987 autobiography offers a rare glimpse at a First Nations perspective of the impact of these policies. Generally, Brass reflects positively upon the File Hills Colony, providing numerous examples of the agricultural and economic successes her family and neighbours enjoyed. Her view of the repressive and paternal administration by Graham and other government officials is very matter-of-fact and denotes no sense of grievance; for example, in referring to the earliest days of the colony, she comments that "Mr. Graham made his own plans which were felt to be quite strict at times. A few beginners could not stand up to these rules and soon left for other parts."[93]

Another early twentieth-century reserve agricultural program that has been accorded some investigation is the Greater Production Campaign (GPC), a federal government agricultural program launched in the early 1918, ostensibly to increase food production across the nation for the good of the war effort.

Governor General Earl Grey inspects Cree men threshing on the File Hills Colony, Peepeekisis reserve, Saskatchewan, 1906. (Courtesy of the Glenbow Archives/NA-3454-18.)

However, several authors have suggested that the program implemented on the western Canadian reserves was problematic.

In *Lost Harvests*, Sarah Carter devoted two pages to the GPC, with particular attention to the repressive aspects of the policy. She found that the project "was plagued by problems of mismanagement and the financial returns were not impressive. The experiment was soon phased out."[94] Miller, holding a similar view to Carter's, briefly described the "ill-starred" "Greater Production" scheme through which Ottawa could "help themselves" to reserve lands for the good of the war effort.[95] James Dempsey is of similar mind, stating that the GPC "was an indication of how easily the government could override Native rights by simply amending the Indian Act."[96]

Three authors have a slightly different interpretation of the GPC. E. Brian Titley devoted a number of pages to the campaign in his book *A Narrow Vision*, which examined the public career of Duncan Campbell Scott, long-time head of the Department of Indian Affairs. Although a good portion of the discussion is dedicated to the strained relationship between Scott and William Graham, the individual appointed commissioner for the campaign in 1918, Titley also provided a good summary of the program: while expressing concern over the "gradual erosion of Indian control of their reserve lands,"[97] he concluded that these extraordinary intrusions were understandable in a time of war and justified by the economic success of the campaign.

In *Canadian Indian Policy During the Inter-War Years 1918-1939*, J. L. Taylor considered the contemporary arguments both for and against the campaign. In his appraisal, Taylor dismissed the arguments made by detractors of the program as narrowly focussed and politically motivated, believing that "it is difficult to establish criteria for success in connection with a project like Greater Production."[98] He also states that the GPC did not result in permanent loss of land by the reserve residents.[99]

The first scholarly review of any of the primary documents associated with the GPC program was authored by anthropologist A. D. Fisher. In the article, published in volume 4, number 1 of the 1974 *Western Canadian Journal of Anthropology*, Fisher examined the views and concerns expressed by R. N. Wilson, in his 1921 memorandum "Our Betrayed Wards." R. N. Wilson had been hired by the residents of the Blood Reserve in Alberta in 1920 to assist them in challenging the federal government to cancel the program. In his critique, Fisher stated that Wilson was correct in asserting that the implementation of the GPC on the Blood Reserve resulted in the destruction of the reserve's agricultural base. While much of the blame can be attached to the new government policies, Fisher cautioned, there were several circumstances that aggravated the situation. He suggested that minister of the Interior, Arthur Meighen, while a capable man, was too busy and too far removed from Alberta to know the impact of the Act. As well, overzealous agents at the local level and poor weather conditions also contributed to the industry's collapse.

Following the discussion of these earlier works, which merely summarized the program, the present author attempted a more detailed analysis of the Greater Production Campaign in "'Better than a Few Squirrels': The Greater Production Campaign of the First Nations Reserves of the Canadian Prairies." In this study I examined the political situation that allowed the program to come into place and the methods by which W. M. Graham was able to gain, and use, significant coercive power over the lives of the reserve residents in the implementation of the GPC. However, although Graham produced numerous accounts of the tremendous benefits the First Nations farmers were reaping from the program, I found that the GPC "served largely to advance the career of W. M. Graham and to pay for departmental expenses" and provided few benefits beyond what the government was already supposed to be doing to promote reserve agriculture.[100] In spite of the power Graham came to wield, I suggest that the reserve residents did offer significant resistance to the program, the most important being the struggle launched by the Blood Reserve: not only did the Blood people take the significant step of hiring a legal advisor—they were also able to bring their concerns to national attention, force a meeting with the minister of Indian Affairs, and "prompt the Department

Binding oats on St. Paul's Anglican mission farm, Blood reserve, 1912. (Courtesy of the Glenbow Archives/ NA-769-10.)

to take the unusual step of changing the department's policies for the benefit of the reserve residents."[101]

The sum of scholarship on the GPC, excepting to some degree the work by this author, is a short list of publications that essentially survey the subject, leaving many details of the campaign unexplored. Such is the case for most aspects of First Nations agricultural history: most studies of agricultural pursuits are considered within the framework of a larger study of government policies. While agriculture is often a central feature of the policy analysis related to western reserves, as is the case in the works of Dyck, Buckley and Miller, the level of exploration of the subject is somewhat minimal as the agricultural policies are studied in conjunction with other socioeconomic initiatives of the government. Only Sarah Carter explored the topic in any depth. However, her focus upon the dual themes of government repression and First Nations resistance, while important avenues of approach, do somewhat limit the scope of her studies. Aside from agents, senior bureaucrats and other government officials, what other individuals could have induced success or failure? Did any reserve residents influence failure?[102] These questions need to be explored in greater detail. As well, why do discussions of agriculture within the published literature essentially all end in 1920? In their defence, the authors might offer that the government was no longer interested in promoting agriculture—hence the selling and leasing of reserve land after 1896.[103] However, new reserve agriculture policies continued to be introduced, and agriculture is still a significant resource base on most of the prairie reserves.

The modern period needs research. The field of historical study of First Nations agriculture has grown greatly during the past 20 years. From the general acceptance of Stanley's concept of the First Nations as an uninterested group who could neither comprehend nor adapt to agriculture, the field has blossomed to include a number of works which identify the long association and interest the Aboriginal peoples of the prairies have had with agriculture, and the numerous obstacles they have had to battle in an attempt to practice an agrarian lifestyle. However, in spite of this new literature, the history of the reserve agriculture traditions and the importance of agriculture within the economic and social spheres of the reserve residents are still not truly appreciated, particularly by the government. Consider the recently completed Royal Commission on Aboriginal Peoples. After several years of research, in November of 1996 the commission released a final report containing over 4,000 pages of background information and recommendations on the various aspects of the past relationships between Aboriginal peoples, non-Aboriginal peoples and the levels of government as well as recommendations on how to redefine these relationships. A significant portion of the report deals with land issues, yet agriculture is accorded little more than passing comment within the historical background of the paper. Obviously more work is still needed to enhance our understanding of the significance of agriculture to the lives of First Nations peoples. A new field of historical study has been broken.

NOTES

This article first appeared in *Prairie Forum* 28, no. 1 (2003): 99-115.

1   The First Nations groups to be studied are the ancestors and descendants of those groups which signed Treaties 1, 2, 4, 6 and 7 with the Canadian government between 1871 and 1877.
2   W. L. Morton, "A Century of Plain and Parkland," *Alberta Historical Review* 17 (Spring 1969): 1.
3   John Hawkes, *Story of Saskatchewan*, Vol. 1. (Regina: S. J. Clarke Publishing Co., 1924), 80.
4   Brian Fagan, *People of the Earth* (Boston: Scott, Foresman and Company, 1989), 347.
5   *First Farmers in the Red River Valley* (Winnipeg: Manitoba Culture, Heritage and Citizenship, n.d.), 2.
6   Catherine Flynn and Leigh Syms, "Manitoba's First Farmers," *Manitoba History* 31 (1996): 7.
7   *First Farmers*, 11.

8 Flynn and Syms, "Manitoba's First Farmers," 10.
9 Lawrence Burpee, *Journals and Letters of Pierre Gaultier de Varennes de la Verendrye and His Sons* (Toronto: The Champlain Society, 1927), 106.
10 Burpee, *Journals,* 119. The Ouchipouennes are likely the Mandan, a well-established community of Aboriginal agriculturalists who resided along the Missouri River in what is now North Dakota. It is also possible that the Ouchipouennes is a reference to the Hidatsa or Arikara who were also sedentary agricultural groups living along the Missouri River.
11 Ibid.
12 Bryce Penneyer Little, "People of the Red Path: An Ethnohistory of the Dakota Fur Trade, 1760-1851" (PhD dissertation, University of Pennsylvania, 1984), 46.
13 Ibid., 95.
14 James Howard, *The Canadian Sioux* (Lincoln: University of Nebraska Press, 1984), 5.
15 United States, Department of the Interior, *Annual Report,* 1879, pt 3, page 80 as in Oscar Lewis, *Effects of White Contact Upon Blackfoot Culture* (Washington: Smithsonian Institute, 1942), 8.
16 Jean Grove, *The Little Ice Age* (New York: Routledge, 1990), 414.
17 James MacGregor, *Behold the Shining Mountains* (Edmonton: Applied Arts Products, 1954), 159.
18 Irene M. Spry, *The Palliser Expedition* (Toronto: The Macmillan Company, 1963), 60.
19 Ibid., 240.
20 Henry Youle Hind, *Narrative of The Canadian Red River Exploring Expedition of 1857 and of the Assiniboine and Saskatchewan Exploring Expeditions of 1858,* Part 2 (Edmonton: M. G. Hurtig Ltd., 1971), 175.
21 Archives of Manitoba (AM), MG 12, File 272: Chief Sweetgrass to Canadian government (with attachments from other Cree Chiefs), 14 April 1871.
22 Deanna Christensen, *Ahtahkakoop* (Shell Lake, Saskatchewan: Ahtahkakoop Publishing, 2000), 153.
23 Ibid., 204.
24 George F. Stanley, *The Birth of Western Canada* (1936; Toronto: University of Toronto Press, 1973), 194.
25 Ibid., 197.
26 Ibid., 198.
27 Ibid., 204, 215.
28 Noel Dyck, "The Administration of Federal Indian Aid in the North-West Territories, 1879-1885" (MA thesis, University of Saskatchewan, 1970), 81.
29 Ibid., 12.
30 Ibid., 13.
31 John Leonard Taylor, "Canada's Northwest Indian Policy in the 1870s: Traditional Premises and Necessary Innovations," in Richard Price (ed.), *The Spirit of the Alberta Indian Treaties* (Montreal: The Institute for Research on Public Policy, 1980), 5.
32 J. R. Miller, *Skyscrapers Hide the Heavens* (Toronto: University of Toronto Press, 1989), 163. In addition to Miller, others who have written on the active involvement include J. E. Foster, "The Saulteaux and the Numbered Treaties: An Aboriginal Rights Position?," in *The spirit of the Alberta Indian Treaties,* Hugh Dempsey, *Treaty Research Report: Treaty Seven* (Ottawa, Department of Indian and Northern Affairs, 1987);

Sarah Carter, *Lost Harvests* (Montreal: McGill-Queen's University Press, 1989); John Taylor, *Treaty Research Reports for Treaties 4 and 6* (Ottawa, Government of Canada, Indian and Northern Affairs, 1985); and Arthur J. Ray, Jim Miller, and Frank Tough, *Bounty and Benevolence: A History of Saskatchewan Treaties* (Montreal and Kingston: McGill-Queen's University Press, 2000).

33 John Taylor, "Two Views on the Meaning of Treaties Six and Seven," in Richard Price (ed.), *The Spirit of the Alberta Indian Treaties* (Montreal: Institute for Research on Public Policy, 1981), 44-45.

34 John Tobias, "Canada's Subjugation of the Plains Cree, 1879-1885," *Canadian Historical Review* 64, no 4 (1983), 521-22.

35 Miller, *Skyscrapers*, 168-69.

36 Stanley, *The Birth of Western Canada*, 217.

37 Ibid., 218.

38 Dyck, "Administration," 81.

39 Ibid., 81.

40 Noel Dyck, "An Opportunity Lost: The Initiative of the Reserve Agriculture Programme in the Prairie West," in F. Laurie Barron and James B. Waldram (eds.), *1885 and After: Native Society in Transition* (Regina: Canadian Plains Research Center, 1986), 122.

41 Ibid., 123.

42 Ibid., 124.

43 Ibid., 130-33.

44 Ibid., 133.

45 Sarah Carter, "Agriculture and Agitation on the Oak River Dakota Reserve, 1875-1895," *Manitoba History* 6 (Fall 1983): 8.

46 Sarah Carter, "Two Acres and a Cow: 'Peasant' Farming for the Indians of the Northwest, 1889-97," *Canadian Historical Review* 70, 1 (1989): 28.

47 Ibid., 30, 31.

48 Ibid., 40.

49 Carter, *Lost Harvests*, 84-86, 98-99.

50 Ibid., 158.

51 Ibid., 193.

52 Ibid., 236.

53 Ibid., 255.

54 Ibid., 258.

55 Miller, *Skyscrapers*, 189.

56 Ibid., 164.

57 Ibid., 201.

58 Ibid., 199.

59 Helen Buckley, *From Wooden Ploughs to Welfare* (Montreal: McGill-Queen's University Press, 1992), 43, 56.

60 Ibid., 54.

61 Ibid., 51.

62 Ibid., 53.

63 Ibid., 54.

64 Peter Douglas Elias, *The Dakota of the Canadian Northwest* (Winnipeg: University of Manitoba Press, 1988), 83.

65   Ibid., 82, 108, 128, 164.

66   Ibid., 224.

67   Carter, *Lost Harvests*, ix.

68   Ibid., 185-86.

69   Ibid., 150-51.

70   Ibid., 188.

71   Miller, *Skyscrapers*, ix.

72   Elias, *Dakota*, xvi.

73   Dyck, "Opportunity," 121.

74   Buckley, *Wooden*, 52. 75.

75   Ibid., 35.

76   Dyck, "Opportunity," 127.

77   Buckley, *Wooden*, 52; Miller, *Skyscrapers*, 162, 200; Dyck, "Opportunity," 122, 153; Carter, *Lost Harvests*, 51, Carter does not actually state that the limited government spending inhibited farming, but does provide numerous cause and effect statements which imply this assertion. For example, on page 63 she comments on the government's non-desire to spend money on animals and farming implements. On page 65, she notes that the inability of the band to take in the crop was due to lack of equipment and draught animals.

78   Rebecca Bateman, "Talking with the Plow: Agricultural Policy and Indian Farming in the Canadian and U.S. Prairies," *The Canadian Journal of Native Studies* 16, no. 2 (1996): 219-25.

79   Hana Samek, *The Blackfoot Confederacy, 1880-1920* (Albuquerque: University of New Mexico Press, 1987), 180.

80   Ibid.

81   Thomas Wessel, "Agriculture on the Reservations: The Case of the Blackfeet, 1885-1935," *Journal of the West* 23, no 4 (1979): 17.

82   R. Douglas Hurt, Indian *Agriculture in America* (University of Kansas Press, 1987), 152.

83   Ibid., 153.

84   Buckley, *Wooden*, 56; Miller, *Skyscrapers*, 202; Carter, *Lost Harvest*, 237; E. Brian Titley, *A Narrow Vision* (Vancouver, UBC Press 1986), 22; James Dempsey, *Warriors of the King: Prairie Indians in World War I* (Regina, Canadian Plains Research Center, 1999), 15.

85   Miller, *Skyscrapers*, 202; Carter, *Lost Harvests*, 244-45.

86   Dempsey, *Warriors*, 15.

87   Miller, *Skyscrapers*, 202.

88   Carter, *Lost Harvests*, 247.

89   E. Brian Titley, "W.M. Graham: Indian Agent Extraordinaire," *Prairie Forum* 8, no 1 (Spring 1983): 28.

90   Sarah Carter, "Demonstrating Success: The File Hills Farm Colony," *Prairie Forum* 16, no 2 (Fall 1991): 165.

91   Ibid., 169-71.

92   Ibid., 171.

93   Eleanor Brass, *I Walk in Two Worlds* (Calgary: Glenbow Museum, 1987), 11.

94   Carter, *Lost Harvests*, 251.

95   Miller, *Skyscrapers*, 203.

96 Dempsey, *Warriors*, 74.
97 Titley, *Narrow Vision*, 41.
98 John Leonard Taylor, *Canadian Indian Policy During the Inter-War Years 1918-1939* (Ottawa: Indian and Northern Affairs Canada, 1984), 20.
99 Ibid., 25.
100 Bruce Dawson, "'Better than A Few Squirrels': The Greater Production Campaign on the First Nations Reserves of the Canadian Prairies," in Patrick Douaud and Bruce Dawson (eds.), *Plain Speaking: Essays on Aboriginal Peoples and the Prairie* (Regina: Canadian Plains Research Center, 2002), 21.
101 Ibid.
102 In this assertion I am not promoting the "untrainable" argument asserted by Stanley. Rather, did any First Nations individual, through mistake or malice, damage agricultural aspirations at the local or regional level?
103 The three authors (Dyck is excused, for his study period ends in 1885) allude to this idea within their respective works.

Farming

5. "The Settlers' Grand Difficulty": Haying in the Economy of the Red River Settlement

Barry Kaye

When Lord Selkirk founded a colony on Red River in 1812, his intention was to establish a commercial farming settlement but, from the onset, the colonial economy was based only partially on crops and domestic animals. Repeated crop failures compelled all settlers at Red River, whatever their origins and agricultural skills, to turn continually to the game resources of the plains and the fisheries of lake and stream for a large part of their food supply. From its commencement, the colony was linked by necessity to the nomadic economy that was the basis of the fur trade, a link that the colony, even at its optimum stage of development, was not able to break. Hunting, fishing, and fowling, in addition to agriculture, were the supports of the colony, just as they were of the fur trade.

It was the continuing uncertainty of both agriculture and the plains buffalo hunt that first created and later sustained this basic dichotomy in the Red River economy prior to 1870. This dichotomy also reflected the disparate peoples and cultures that made up the settlement. Among Red River's population were colonists, primarily Highland Scots, but including also numbers of French Canadians, British mixed-bloods and Orkneymen, whose prime concerns were their crops and animals. By 1830, however, these colonists were out-numbered by peoples who showed little predilection for agriculture, and lived chiefly from hunting, fishing and trading in furs. Predominant among this latter group were the French-speaking mixed-bloods of St. Boniface and St. Francois-Xavier, but some French Canadian freemen, British mixed-bloods and many of the full-blooded Indians at Red River lived in a similar manner. One distinguishing characteristic of this group was its heavy dependence on "wild" resources, or on what Palliser called the "natural productions" of the land.[1] In the colony's early years, would-be farmers at Red River were also dependent

Church and Mission School at the Upper Settlement, built by the Rev. J. West.

Working with scythes, circa 1821-23, about two miles below the Forks on the west bank of the Red River. In the background are the church and mission school founded by Reverend John West of the Church of England in 1820. (Courtesy of the Glenbow Archives/NA3421-2.)

on the same "wild" resources, for not until the late 1820s was anything like a viable agriculture beginning to emerge. Until then, tillage endured only as a partner of the nomadic economy that underlay the fur trade. The most obvious expression of this partnership was the autumn migration of most of the colonial population to Pembina, sixty miles to the south, where they lived by hunting during the winter months.

Increased agricultural productivity after 1827, and the acquisition of domestic animals from the United States, enabled potential farmers in the Red River population to discontinue their yearly trip to Pembina and, to a great extent, freed them from dependence on the buffalo herds. However, the settlement of large numbers of French-speaking Métis at the colony after 1823 ensured the perpetuation of the essential dichotomy of the Red River economy. As a result, by 1830, one of the most striking internal divisions within the settlement was that between areas where agriculture was the main support of the economy and areas where hunting and fishing supplied most of the staples of life.

The sharpness of the distinction between the farmers and the hunters at Red River should not be overemphasized, however. It was blurred by the fact that some dominantly hunting peoples, despite their general preference for the free, wandering life of the plains, carried on small-scale agriculture. Another obscuring factor was that most Red River farmers continued to exploit many of the freely available "wild" resources after 1830. The farmer-colonists supplemented the produce of their river-lot farms with sugar from riverside maple groves, with a variety of wild fruits gathered at appropriate seasons, and with fish, especially whitefish, which they caught in Lakes Winnipeg and Manitoba. Some of them also took part in the spring and autumn goose hunts at Shoal Lake. The main concern here, however, is with one aspect of the farmers' dependence on the natural environment: their dependence on the plains as a source of winter fodder for their livestock.

During the colony's difficult early years, the few domesticated animals at Red River suffered from a general lack of adequate husbandry. Harvests were often meagre and provided little or nothing to spare for animal feed. Haymaking was a casual, hit-and-miss affair, not the well-organized and regulated activity it became later. As a result, winter feed was often in short supply, while sheltering stables were inadequate or absent.

The settlers' initial experience with livestock provided plenty of evidence that Red River farmers would have a difficult struggle as they tried to find time and energy to gather, transport and put aside sufficient quantities of winter fodder to feed the increasing livestock population. Prior to 1827, they clearly failed to meet the challenge this problem presented. The journalist at Fort Garry made several references to cattle and horses dying for want of feed in 1826, the year of the great flood.[2] By late April of that year, winter feed was in such short supply that the settlers had to resort to desperate measures to prevent their animals from starving. Some farmers fed their cattle from dwindling grain stocks, thus raising apprehensions of seed scarcity at spring planting.[3] Others without grain felled trees on their lots to enable cattle to feed on the branches, "an alternative," the journalist noted, "that is only had recourse to, in the last extremity, where wood is scarce.[4]

The same journalist was convinced that these difficulties were due largely to the indolence of settlers in failing to lay in a sufficient supply of hay the previous summer. There may be some truth in this, but inexperience more than indolence was probably the cause. Many of the settlers were rearing livestock for the first time and lacked familiarity with the environment. Only experience could bring knowledge of the best haying spots and of the quantities of winter feed needed to bring the various types of livestock through a Red River winter. Moreover, many settlers came to own livestock before they had erected winter shelters for them. As late as 1830, only 233 stables were listed in the Red River census.

In spite of these early setbacks, livestock were gradually integrated into the Red River economy after 1827. By the 1830s, many Red River farmers had a horse for riding in summer and for sledge-pulling in winter, plus a pair of oxen for ploughing and carting hay and wood. Many also had a small herd of cattle and a number of pigs, sheep and poultry. Yet, of all the cultural groups at Red River, it was the Highland Scots of the Lower Settlement who were most interested in raising livestock. At the other extreme, the numerous Métis valued their buffalo runners but most of them had little interest in pastoral activities. Livestock numbers, as recorded in various censuses from 1827 to 1856, are shown in Table 1.

Livestock	1827	1833	1840	1846	1849	1856
Cattle	635	2,572	4,166	3,842	3,917	6,609
Sheep	—	—	1,897	4,223	3,096	2,245
Pigs	225	2,033	2,149	3,800	1,565	4,929
Horses	237	492	1,292	2,360	2,085	2,681
Oxen	247	1,219	1,749	2,681	2,097	3,006

Table 1. Number of livestock reported for the Red River Settlement in various censuses, 1827-1856. Source: *Red River Censuses*, 1827-1856.

The greater productivity of Red River agriculture after 1827 did not result in a reduced dependency on the natural vegetation of the plains for livestock feed. The production of enough winter feed from the plains to sustain their animals through the long and often severe winters continued to be the greatest problem for livestock farmers. During the open season, livestock were turned out to graze on the unimproved land of the settlers' lots, normally two-thirds or more of the whole or, especially with oxen and horses, out on the open prairies beyond. Pigs were unhusbanded ordinarily and left to live off the country, finding sustenance from the mast and roots of the timbered river fringes or from the prairies beyond. Most of the horses at Red River were of the small Spanish type and these hardy animals normally spent the winter outdoors. The remaining livestock, however, had to be sheltered and fed during five or six months in winter.

Some animal feed resulted from the settlement's arable farming; in good crop years a little barley and oats were used as feed, potato surpluses were sometimes used to fatten pigs and sheep, and the straw of the threshed grain was occasionally fed to oxen. But the colonists were almost entirely dependent on the wild hay that they could bring in from the plains to feed their livestock in winter. The prairie grasses found in abundance throughout the Red River

Valley offered an excellent opportunity for hay making and provided the main support for livestock farming at Red River. As a result, there was little incentive to introduce tame hay. Occasional experiments were conducted with the cultivation of red clover, white clover and timothy but none met with any great success and, prior to 1870, no kinds of tame grasses were grown for hay making. It was wild hay from the plains that provided the main support for animal husbandry at Red River. The *Nor'Wester* could report in 1869 that "our prairies are covered with nutritious grasses which are the only food for our stock both summer and winter."[5]

The growth of the colony's population brought forth a number of regulations and restrictions affecting hay gathering and animal grazing. These were intended to reduce squabbling by giving each settler an equal and fair chance to mow sufficient hay to carry him over the winter. Within the "settlement belt," or inner two miles, each colonist was able to mow the hay and graze his animals on his own two-mile-long lot as and when he wished. Beyond the two-mile limit, settlers had the exclusive right to cut hay on the outer two miles immediately at the rear of their holding. This important right was recognized by the Hudson's Bay Company and was known as the "hay-privilege." Some light timber was cut in the outer two miles and by 1870 the outer two miles of several parishes contained considerable ploughed land, but it was valued chiefly for the hay it provided.[6]

Not all settlers at Red River had a "hay-privilege" behind their farm, however. In the parishes of St. James, St. Boniface and St. Vital, many did not enjoy this privilege. In these parishes, part of the outer two miles was cut off by the junction of the Red and Assiniboine Rivers, and by the junction of the River Seine with the Red. In St. James, the lots of some settlers abutted on the "hay-privilege" of St. John's parish and partly on the Hudson's Bay Company's reserve land at Upper Fort Garry.[7] As a result, in 1860 the residents of St. James petitioned the Council of Assiniboia to give them the same haying privileges enjoyed by other colonists.[8] On the east side of the Red, where the river lots were squeezed between the main river and the River Seine, much of the land beyond was the property of the Roman Catholic church. Nor was the "hay-privilege" a feature of the new settlements established after the early 1850s along the Assiniboine west of St. Francois-Xavier.

Beyond the "hay-privilege," all colonists had equal rights to the hay and timber of what was called "the common."[9] Access to this "common ground for hay-cutting," however, was regulated by the Council of Assiniboia. The usual date on which the plains beyond the "hay-privilege" were thrown open to farmers was July 20.[10] One settler recalled that "before the day fixed for the beginning of hay-cutting each year, the best hay meadows were spied out, and each man had planned where he was to cut hay."[11] On the designated day, the farmer took his scythe, hand rakes and carts, and set off for the lower stretches of the plains where the slowly evaporating water produced an especially rich growth of grass. Sloughs, marshes and other wet spots within easy reach of the colony were the favoured places. With experience, the colonists eventually became familiar with those sections of the plains which would best reward their hard labours. The "Big Swamp" to the west of the Lower Settlement, Long Lake beyond Baie St. Paul on the Assiniboine, the Grosse Isle, northwest of Upper Fort Garry, and the "Weedy Hills" (location unknown) are mentioned in the records as favoured haying areas. The marshy Netley Creek area, on the edge of the Red River delta, was another important haying location. The Hind Expedition observed large numbers of haystacks there in September 1857.[12] In dry years, not all settlers could make sufficient hay in the vicinity of the colony, so some haymakers were forced to travel farther afield. At such times, the wet margins of lakes, such as the marshy Long Lake area and the marshlands along the southern edge of Lake Manitoba, attracted more settlers.

At these and other unnamed locations, the farmers set up their tents and commenced to cut down the grasses on or soon after July 20. To preempt his chosen hay land, the mower cut a circle around it with his scythe, and "each man's ownership of the area he had marked out was always respected."[13] Once the grass was cut, it was cocked and made into large haystacks. The stacks were frequently fenced and covered with tree branches in an attempt to protect them from fire, wind and wandering animals.

While most of the occupied land within the "settlement belt" was held in severalty, two areas within the colony, both close to the Forks, were held in common and to these a number of settlers had equal right. The two areas were the Point Douglas and St. Boniface Commons or Reserves. This right, probably first granted by Lord Selkirk in 1817 and later confirmed by the Council of Assiniboia, was given to the residents with small holdings on Point Douglas and in St. Boniface, on the narrow neck of land at the junction of the Seine and the Red. Access to grazing and haying land was difficult for settlers in both these areas, so they were allowed to graze their animals and cut hay on the common, which evidently represented the pastoral centre for the grazing of the 'commonable' animals. In 1864, Point Douglas residents protested to the Council of Assiniboia about encroachments that settlers on the Assiniboine River had made on what they called their "special reserve." They claimed that they alone had the right to graze animals and make hay on the reserve.[14] The commons also may have functioned to some extent as market and social centres. In many ways, they resembled the village greens characteristic of Western European villages or, in Canada, the commons that were established during the seventeenth and eighteenth centuries by French settlers along the lower St. Lawrence River.

Despite the labour expended on haying, the regulations of the Council of Assiniboia, and the seemingly endless unenclosed plains beyond the colony, the acquisition of sufficient winter fodder was the farmers' "grand difficulty."[15] Red River farmers found it far from easy to gather enough hay to keep their animals fed and in good condition until spring, by which time it was invariably in short supply. As a result, by the end of winter Red River cattle often looked "more like death than life."[16] According to Palliser, many cattle and horses were "lost every winter from the people not laying in a sufficient stock of hay."[17] But losses were especially severe when a long, very cold winter, following a dry summer, had rendered hay scarce. Livestock mortality rates were very high during the drought years 1846-48. As George Simpson observed in 1847, "the blight that destroyed the grain and potato crops of last year greatly injured the hay, which is found to be of unsound quality; and that, together with the length and severity of the winter, occasioned a great mortality among the cattle and sheep...."[18] Writing in the following year, Andrew McDermot noted that "in consequence of the most dreadful winter that was ever known in Red River at least one-sixth part of all the cattle was starved to death."[19] The unusually severe winter of 1856-57, which was followed by a very backward spring, caused the deaths of nearly one-seventh of all the colony's cattle.[20] The census of 1856 had reported 6,609 cattle, which would suggest that more than 900 cattle perished during that particular winter. The colony's sheep also fared badly during winter. Hind, for example, noted that 184 sheep were lost during the winter of 1855-56.[21]

The dry years of the 1860s also reduced the pastures and the numbers and condition of Red River livestock. The droughts of 1863, 1864 and 1865 caused extreme shortages of winter fodder, resulted in an acute scarcity of hay by springtime, and had the usual effect of reducing the weight and condition of the stock. In spring 1864, Samuel Taylor, a farmer from the lower part of St. Andrew's parish, recorded in his journal, "a great many cattle dying for want of something to eat," and later added that "hay is scarce all over this year some had not one straw long ago."[22] The following March, the same writer noted that "there is people down from up amongst the French and Scotch, in great need of hay."[23] The situation was just as bad in the late winter and spring of 1866; "there is a general complaint all over for hay, and none to get to buy anywhere," wrote Taylor.[24] Hay continued to be scarce in 1867; Taylor noted in March and April that "bad times begining [*sic*] for hay ... no warm weather yet and hay little ... a great many people out of hay.[25] These conditions of scarcity inevitably raised mortality rates amongst all kinds of livestock, as they had in the dry years of the 1840s.

The not infrequent heavy winter livestock losses suggest that some Red River farmers were keeping too many animals for the amounts of winter feed they could reasonably expect to gather, especially in a dry year. Yet livestock raising at Red River was on a small scale. Even among those settlers most concerned with livestock, the herds and flocks were not large. However, despite the modest livestock numbers, the amount of prairie hay that had to be cut, hauled and put up for the winter months was considerable. Hay, an extremely bulky product, was normally measured by the cartload. There are no precise figures on the number of loads required to winter cattle or sheep at Red River but, if Palliser is correct, then it took five loads to winter an ox and ten loads to winter a horse.[26] We get some indication of total hay requirements from *Nor'Wester* reports on prairie fire losses. In October 1860 the paper reported that:

> We are sorry to say, that the fires on the east bank of the Red River have done a great deal of injury to the people of the middle district. Mr. William Bunn, we are told, has lost all his hay—say 140 loads, Mr. Angus Matheson, part—say 45 loads; Mr. John Gunn about the same quantity; and Mr. George Munroe 30 loads.[27]

It is clear that one of the hardest and most time-consuming jobs facing many Red River farmers each year was the harvest of well over one hundred cartloads of hay on "the common" during the haying season.

The major disadvantage of almost total dependence on the plains' wild hay resources was fluctuation in grassland growth from year to year. One settler estimated that three or four tons of hay could be made from an acre of prairie.[28] The actual yield in any one year was determined largely by rainfall conditions, and the colony's graziers were particularly vulnerable to the effects of drought. In dry years, not only was the growth of prairie grasses less rank, but the annual prairie fires ran farthest and most often, thus further reducing already scanty hay supplies. The *Nor'Wester's* description of the fire situation in August 1864 is equally applicable to many other years:

> Destructive fires have raged over the plains this month. Along the Assiniboine and Red River they caused considerable loss, burning much of the hay belonging to settlers on both sides of these rivers, and in some instances, we believe, pan of the crops. The loss of a winter's hay is a very serious one at any time, but it is particularly so this year, when it was with the greatest difficulty that farmers succeeded in getting enough to keep their stock alive till spring.[29]

Haymakers tried to protect their precious stocks from destructive fires by surrounding them "with a ploughed or burned ring at least eight feet wide situated about twenty yards from the stacks."[30] Such precautions undoubtedly reduced losses but it was a rare year when disastrous fires failed to claim all or most of some settler's hay. Settlers unfortunate enough to lose their supplies had to rely on the generosity of neighbours to see them through the winter.[31]

The quantity of hay each settler could harvest was severely limited by the shortage of hired labour. A few Indians from the Indian Settlement of St. Peter's sometimes hired themselves out at this busy time,[32] but haying was essentially a family affair in which both women and youngsters regularly took part.[33] The amount of hay that could be put away for winter was also limited by the slow hard methods employed in making it. A few mowing machines were introduced into the Red River Settlement from the United States during the late 1850s and 1860s, but throughout most of the colony prior to 1870 hand labour continued to dominate harvesting (see Figure 1).

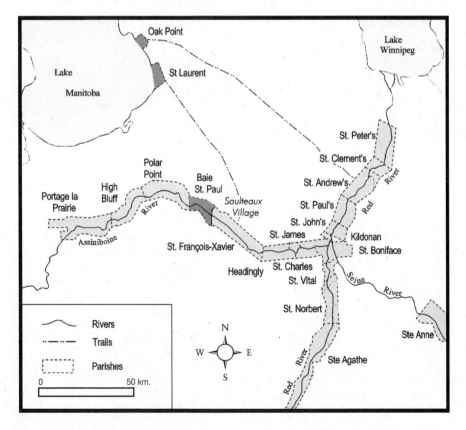

Figure 1. The Red River settlement and surrounding area, circa 1870. (Courtesy of the Canadian Plains Research Center.)

Cutting the prairie grasses beyond the limits of the "hay-privilege" was carried out usually during late July and early August, though a little inferior hay was made later once the crops were harvested.[34] However, annual climatic variations frequently caused modifications in the haying routine. In dry years, when hay seemed likely to be scarce, anxious farmers commenced mowing before the official date laid down by the Council of Assiniboia, even though they ran the risk of sacrificing their haying rights.[35] In late July 1863, the *Nor'Wester* learned that some settlers had begun "to put hay before the legal date, and that Mr. Mactavish [Governor of Assiniboia] gave orders to have the hay seized and sold for the benefit of the public."[36] And when a wet spring and early summer made movement on the plains difficult, the commencement of haying was put off for a few weeks to give the prairies more time to dry out.[37]

No matter when haying commenced, the farmers' problems were further complicated by the fact that barley, the first crop to come to maturity at Red River, invariably ripened at about the same time as the prairie grasses were ready for cutting, thereby forcing them to do two arduous tasks within a short monthly period.[38] Haymaking, therefore, was a time of strenuous activity for the Red River farmer, an activity "which every year half breaks both his back and his heart."[39] The hay harvest also coincided with the mosquito season and, despite the obvious risk of prairie fires, bonfires were often lit in the hope that the smoke would keep spiteful insects at bay.

Once their hay was stacked, the settlers had to make numerous journeys hauling it by cart or sledge to the riverside farms. The amount of travelling increased during drought years, for lack of water on the plains forced haymakers to get their drinking water from the Red or the Assiniboine. Taylor, in August 1863, recorded that "people had to cart out water to hay making and reaping."[40] The *Nor'Wester*, one year later, wrote that

> Despite some smart showers in June and July there is barely enough herbage on the plains to feed the cattle, and owing to the entire absence of water there they cannot go far enough to pick the best of it. They have all to march back once or twice a day to drink at the river.[41]

Haying activities were curtailed as the harvesting of wheat, the colony's main crop, got underway but were taken up again in the fall and continued throughout the winter, for the hay was rarely put away before the onset of cold weather. Despite the threat from annual fall fires, hay was left out on the plains, from whence it was hauled home when needed. The *Nor'Wester* informed its readers that during the last three weeks of September 1861, "strings of carts, sometimes a quarter of a mile in length, might be seen traversing over bleak and level prairies with their loads."[42] Agricola, a correspondent on agricultural matters in the *Nor'Wester,* indicated that the winter hauling of hay was:

> a business which of itself occupies the greater part of the time of the sturdiest members of the family for most of the winter, and sometimes compels a weak-handed family to send out little boys to work for which none but hardy men are really quite fitted. Twenty four and thirty hours are nothing unusual for a single trip over bleak and sometimes trackless plains.[43]

The Taylor journal confirms that the hauling of hay (and timber) was an almost daily activity of Red River livestock farmers during the winter. The sledge was the usual mode of winter hay transport but the cart continued in use when snow was scarce or absent.

Prairie hay was a "free" resource available to all and the vast majority of settlers with livestock took advantage of this to meet their own requirements. Yet hay did have some commercial value at Red River. Farmers who had failed to put up sufficient hay to last the winter, or had lost hay through fire, created most of the demand and tried to make up their deficiencies through purchase as hay ran out.[44] The Hudson's Bay Company also occasionally bought hay to feed its livestock at Lower Fort Garry.[45] Information on hay prices is slight but, during the 1860s, the market price in normal times seems to have ranged from two shillings and sixpence to six shillings per cartload. As fodder supplies decreased towards spring the price of hay invariably rose sharply and, at that season, sometimes brought as high as twenty shillings for a cartload weighing 800 pounds.[46]

There were years when severe drought and the consequent insufficiency of hay caused the normal annual haying routine at Red River to be modified. This occurred in 1847 and 1864 when, in response to the drought hazard, farmers made hay at distances of sixty miles or more from the colony. Rather than hauling home their hay, farmers with substantial herds wintered their cattle at the source of supply. At such distant locations rough sheds were hastily constructed to shelter cattle and they were kept there until the spring thaw. This practice was known as "out-wintering" and favoured locations for it were the Long Lake area and the southern margins of Lake Manitoba.[47]

The Red River Settlement's haying economy did not long survive the creation of the Province of Manitoba in 1870 and the disintegration of the old order that followed the commencement of permanent agricultural colonization on the prairies beyond the Red and the Assiniboine riverfront settlements. The commons of Point Douglas and St. Boniface were destroyed by the expansion of the new urban centre of Winnipeg. The ploughing up of the "hay-privilege," which had been underway since about 1860, gathered pace in the years after 1870. During the same decade, the common land beyond the "hay-privilege" was divided up into quarter-section farms by Dominion land surveyors. However, the many poorly-drained areas in the Red River lowland, some of which had formed the favoured hay grounds of the Selkirk colonists for forty years were avoided by the incoming settlers and, no doubt, continued to contribute to feeding the livestock of old settlers and new.

NOTES

This article first appeared in *Prairie Forum* 9, no. 1 (1984): 1-11. The author wished to thank the Hudson's Bay Company for granting him permission to consult and quote from the Company's archives.

1   John Palliser, *Further Papers Relative to the Exploration by• the Expedition under Captain Palliser* (London: G. E. Eyre and William Spottiswoode, 1860), 38.
2   Fort Garry post journal, 30 January, 4 and 19 April 1826, Hudson's Bay Company Archives, Winnipeg (hereafter H.B.C.), B235/a/7, fos. 20, 28 and 31.
3   Journal of Reverend D. T. Jones, 29 April, 1826, Church Missionary Society Archives, Microfilm copies, Provincial Archives of Manitoba, (hereafter C.M.S.), Reel A77, p. 197.
4   Fort Garry post journal, 31 April 1826, H.B.C., B235/'a/7, fo. 31.
5   *Nor'Wester* (Red River Settlement), 3 April 1869.
6   The extension of cultivation into the "hay-privilege" is discussed in Barry Kaye, "Some Aspects of the Historical Geography of the Red River Settlement from 1812 to 1870" (Master's Thesis, University of Manitoba, 1967), 255-261.
7   *New Nation*, 6 May 1870: Archer Martin, *The Hudson's Bay Company's Land Tenures* (London: W. Clowes and Sons Ltd., 1898), 213.
8   E. H. Oliver (ed.), *The Canadian North West: Its Early Development and Legislative Records*, 2 vols. (Ottawa: Government Printing Bureau, 1914), 1: 456-457.
9   *Nor'Wester*, 28 July 1860.
10   Until the early 1860s, the date of the commencement of haymaking on "the common" was fixed at July 20. However, beginning in 1862, the Council of Assiniboine started to vary the opening date in order to take into account haying prospects and conditions on the plains. Initially, at least, the change aroused considerable opposition among Red River farmers. See Nor'Wester, 14 May and 23 July 1862.
11   W. J. Healy, *Women of Red River* (Winnipeg: Women's Canadian Club, 1923), 150.
12   Henry Youle Hind, *Narrative of the Canadian Red River Exploring Expedition of 1857 and of the Assiniboine Exploring Expedition of 1858*, 2 vols. (London: Longman, Green, Long-man, and Roberts, 1860), 1: 123.

13   Healy, *Women of Red River*, 150.
14   Oliver, *Canadian North West*, 1: 541.
15   *Nor'Wester*, 10 May 1864.
16   Ibid.
17   Palliser, *Further Papers*, 55.
18   George Simpson to Governor and Committee, 1 July 1847, H.B.C., A12 3, fo. 424.
19   Andrew McDermot to G. M. Cary, 24 July 1848, G. M. Cary papers, Provincial Archives of Manitoba, Winnipeg.
20   Simpson to Governor and Committee, 30 June 1857, H.B.C.; A12/8, fo. 497.
21   Hind, *Narrative*, 1: 229.
22   Samuel Taylor journal, March and April 1864, Provincial Archives of Manitoba, Winnipeg.
23   Ibid., March 1865.
24   Ibid., March 1866.
25   Ibid., March and April 1867.
26   Palliser, *Further Papers*, 55. A cartload of hay probably weighed about 800 pounds.
27   *Nor'Wester*, 29 October 1860.
28   Canada, Journal of the House of Commons, Vol. X, 1876, Appendix, Evidence of J. Sutherland.
29   *Nor'Wester*, 18 August 1864.
30   J. J. Hargrave, *Red River* (Altona, Manitoba: Friesen Printers, 1977, first published 1871), 178.
31   Healy, *Women of Red River*, 150; R. G. MacBeth, *The Selkirk Settlers in Real Life* (Toronto: W. Briggs, 1897), 47.
32   Journal of Reverend John Smithurst, 5 July 1841, C.M.S., Reel A96.
33   In its issue of 14 May 1862, the *Nor'Wester* claimed that only ten percent of the farmers at Red River, consisting of men of "large means," were able to hire labour for haymaking and harvesting. The rest had to rely wholly on family labour.
34   *Nor'Wester*, 14 May 1862.
35   Taylor journal, July 1864.
36   *Nor'Wester*, 22 July 1863.
37   Ibid., 28 July and 14 September 1861; Alexander Ross to James Ross, 8 September 1856, Alexander Ross family papers, Provincial Archives of Manitoba, Winnipeg.
38   *Nor'Wester*, 14 May 1862; Oliver, *Canadian North West*, 1: 510.
39   *Nor'Wester*, 10 May 1864.
40   Taylor journal, 23 August 1863.
41   *Nor' Wester*, 18 August 1864.
42   Ibid., 1 October 1861.
43   Ibid., 10 May 1864.
44   Taylor journal, February 1866.
45   Journal of Reverend J. Smithhurst, 10 December 1846, C.M.S., Reel A96.
46   Hargrave. *Red River*, 178.
47   *Nor'Wester*, 18 August, 2 November and 6 December 1864.

6.   Estimates of Farm-Making Costs in Saskatchewan, 1882–1914

Lyle Dick

How much money did a settler on the Western Canadian prairie need to start farming? An important prerequisite to the study of agricultural economics in the settlement period is an understanding of the costs of farm-making. The question, moreover, has important historiographical implications. If it could be demonstrated that most settlers were able to begin with a negligible amount of capital, greater credence might be given to the hypothesis that economic democratization did occur on the prairie frontier. If, on the other hand, a substantial investment was required, the popular notion of the penniless "sodbuster" would be shown to be little more than a romantic myth. Recently, economists Robert Ankli and Robert M. Litt have estimated that prairie farm-making costs in 1900 were $1,000.[1] Their paper introduces a heretofore neglected area of Canadian economic history. Yet farm-making costs were not fixed for all settlers, but could be affected by a multiplicity of variables, including the settler's cultural background, the type of farming practised, family size, relationships with neighbours and the topographical character of his land. There is a demonstrable need for detailed analysis on the micro-level to differentiate the costs of farm-making among different settlement groups.

This paper examines farm-making costs in two communities in southeastern Saskatchewan—the Anglo-Saxon Ontarian homesteaders around the village of Abernethy, and a group of German-speaking peasant settlers at the neighbouring community of Neudorf. To provide quantitative comparisons, all 461 Department of Interior homestead files relating to six townships—three at Abernethy and three at Neudorf—were examined. Most of these files contain the Application for Patent, a form which required the homesteader to state, among other things, the extent and value of his improvements, including his dwelling and outbuildings, and acreage broken, cultivated, and fenced. Since this form was filled out after the settler had performed his requisite duties under the Dominion Lands Act, the values recorded in the files do not reflect the initial expenditures so much as the accumulated improvements made in the first three or more years. For the Abernethy area, the mean year of application was 1894, for Neudorf 1901. In addition to the quantitative data, qualitative sources such as pioneer diaries, early newspapers, colonization journals, and immigration pamphlets have been consulted to provide a larger con-text in which to treat the Abernethy findings.

In approaching the question of farm-making costs, it would be useful first to determine the extent of financial resources available to the settler. The absence of concrete data in this instance requires that an approximation of resources be reconstructed through individual case examples.

Most settlers in the townships surrounding Abernethy came from Ontario. Arriving between 1882 and 1905, they formed part of the preponderant Ontarian migration to Western Canada in the first three decades after Confederation. In his studies of Peel County after 1850, David Gagan has shown how the prevailing Ontarian inheritance system forced surplus sons off the farm, many of whom joined the massive migration to the prairies.[2] Gagan's figures respecting the average debt per farm suggest that little surplus was available to the second and third sons of most Ontarian farmers. The sons might have worked for a few years before taking up a homestead, but a demographic analysis of Abernethy homesteaders shows that this group contained a large proportion of very young men—40 per cent were 24 years or younger, 23 per cent, 21 or younger. Given the then current scales of pay in Ontario,[3] one thousand dollars seems a very large amount for men with only two or three years in the work force.

Wages were higher in Western Canada than in the East, although the maximum wage obtainable there for experienced farm hands was $35 per month, including board.[4] Threshing labourers were paid at a higher rate, but the threshing season lasted only two months, and the thresher was often obliged, if he could find work for the winter, to hire on at a very modest salary. Agricultural historian John Thompson has argued that most harvest hands would have experienced great difficulty in raising even $600,[5] which was the figure cited by Saskatchewan premier Walter Scott in 1906, as the minimum required to establish a homestead.

Similar inferences may be drawn from the case studies of individual settlers. After graduating from The Ontario Agricultural College, W. R. Motherwell, the third son of a Lanark County farmer, came west in the spring of 1881. He did not immediately file for a homestead, but worked for two summers and harvest seasons in Manitoba before settling in the Pheasant Plains area of the District of Assiniboia in 1883. It is difficult to determine how much money Motherwell brought to his homesteading venture, but the possibility of financial assistance from his impoverished parents seems remote. It is equally improbable that he saved large sums during two expensive years at college. Taking into consideration the current farm wages, the most plausible range of capital that Motherwell might have earned and saved in the two years following his graduation is $400-$600.

Testimonials of other settlers in the Qu'Appelle region are also indicative of the amount of liquid capital which many newcomers brought. John Burton, aged 21, homesteaded two miles north of the present village of Abernethy in 1882. A native of Bruce County, Ontario, Burton related that he had only $65 when he arrived.[6] Samuel Copithorn who took up land north-west of Balcarres, emigrated from Toronto with a year's earnings of $200 in his pocket.[7] Both settlers eventually became successful farmers. Indeed, the pages of contemporary agricultural journals and immigration pamphlets are filled with dozens of similar testimonials *vis-à-vis* meagre starting capital.[8]

A reading of the local newspapers from the settlement period confirms that many settlers brought only a few hundred dollars to their farming ventures. In April 1889, the *Regina Leader* reported the arrival of 53 Canadians and 22 settlers of other nationalities. The total value of their effects was $6,000, and cash, $12,000, giving a net figure of $240 per settler as a starting capital.[9] In another report the *Leader* noted that 50 Galician families established themselves at Batoche with a capital investment as low as $40.[10] In 1885, German immigrants established the colony of New Toulcha north of Balgonie. Of 13 farmers surveyed by the *Leader*, none had possessed more than $250 at the outset. Each German had shared ownership of a team of oxen, harness, plough and wagon with another settler.[11] It could be misleading to treat these testimonials and journalistic accounts as representative of the majority, but there is ample evidence to support the view, commonly held in the period that "energy, experience, judgement, and enterprise,"[12] were probably more important to the establishing of a successful farm than a large initial investment.

If a large starting capital was beyond the reach of many settlers, it would be useful to explore alternative forms of financing that might have been available. Under the terms of the Dominion Lands Act, a settler was prohibited from mortgaging homestead lands prior to the issuing of a patent.[13] While the minimum "proving-up" period was three years, the majority of settlers took even longer to fulfill their homestead duties.[14] The homesteader might negotiate a bank loan, but with little collateral it seems unlikely that banks would have advanced him more than a few hundred dollars.

Land purchased from corporate interests was mortgageable, but required a down payment. c.p.r. land could be acquired with an initial payment of one-sixth or one-tenth of the purchase price, followed by five equal annual installments. In the period before 1900, c.p.r. land commonly sold at prices ranging from $2.50 to $5.00 per acre,[15] depending on its distance from the railroad and grain handling facilities. In these instances, a settler was obliged to pay out at least $66.67 initially for a quarter section purchased on time, but often this payment was a much larger sum. As the sections available for free homestead finally began to fill out after 1900, c.p.r. land prices jumped dramatically. After 1900, therefore, purchased land tended to be prohibitively expensive for the newcomer of limited means.

It is important to determine how cheaply a farm could be established by tabulating the individual costs involved. One of the first, and most essential, expenditures encountered by the prairie settler was that of shelter. Ankli and Litt have estimated the cost of housing to have been $200 to $300 for most settlers. They base this conclusion on two sources: an article on early housing written from a series of pioneer questionnaires distributed by the Saskatchewan Archives Board in 1955; and James M. Minifie's *Homesteader*,[16] in which Minifie cites the valuation of $400 placed by his father on his homestead Application for Patent form. Yet, a quantitative investigation of homestead applications suggests a lower figure. While in the Abernethy district, 106 settlers, largely of Anglo-Saxon Ontarian origin, assessed their houses an average value of $253, at Neudorf, 147 Germans valued their dwellings at an average of only $168.

The criteria employed in these valuations may have varied from one applicant to the next; it is not clear, for example, to what extent the applicants took into consideration the cost of materials, the labour cost or the overall market value. However, the paltry values recorded by a large proportion of the settlers suggest that capital expenditures for housing were often very small indeed. In the two study areas, 46 per cent of the homesteaders valued their houses at $100 or less and 17 per cent recorded values of $50 or less. On the other hand, some settlers placed very high values of $1,000 or more on their houses, but it should be remembered that these valuations were made at the time of application, several years after the entry. Even these relatively prosperous settlers frequently put up cheaper temporary structures at the outset, that served for a year or two before the building of a more substantial residence. Consequently, the values found in the homestead files are probably higher than the actual initial investment.

In the Neudorf area, the log dwellings listed in the homestead files were preceded by even more primitive initial shelters. A local resident has described the building of these habitations:

… The first thing they would do was dig a hole in the ground
and bank it up with sods, put a few poles and hay on top, make
a big oven of clay in the middle of the house, no floor, and live
in it until they built a log house.[17]

Such mean lodgings were common among the poorer European peasant
immigrants, but even some of the Anglo-Saxons built similar shelters. An
English family in the Primitive Methodist Colony northeast of Abernethy
"spent their first winter in a dugout in the east side of a hill. The roof was
covered with green poles, dirt, and hay."[18]

Savings on housing costs stemmed principally from the availability of
indigenous building materials. Sod was obtained simply by ploughing the
prairie turf. Logs could be collected at little or no capital cost, particularly in
the early period, when settlement tended to cling to the parkland belt and
the woodland-prairie margin. In these areas the settler's actual investment
was limited to the purchase of nails and tools, lumber for doors, door frames,
windows, and occasionally, flooring and roofing. Poorer settlers typically opted
for a sod roof and a dirt floor. One farmer in Manitoba in the early 1880s suggested that $60-$75 was required for pine flooring, window and door frames.[19]
Another, writing in 1895, estimated the cost of lumber finishing for all his farm
buildings to have been $110.[20] A 1902 immigration pamphlet set the lumber
cost of doors and windows for a log house at $50.[21] At Neudorf valuations of
homestead dwellings ranged as low as $5, suggesting that a settler could build
a shelter with an almost negligible investment as long as he was prepared to
forego all amenities and comforts for the first year or two.

A poorer pioneering family outside of their sod dwelling.

Most Abernethy and Neudorf settlers possessed an initial advantage in choosing lands in close proximity to wooded areas. As the range of homestead lands expanded into the true prairie of the Palliser triangle after 1900, many settlers were obliged to purchase more of their building materials. Even there, however, a habitable shanty could be put up with a relatively modest outlay. Willem de Gelder, a Dutch immigrant who homesteaded north of Morse, Saskatchewan in 1910, built a 10' × 12' lumber and shiplap shanty for $100 including extra labour costs.[22]

This last example illustrates that the cost of shelter was also related to the size of family that was to be housed. Western Canadian settlement was characterized by a large proportion of bachelor homesteaders and in the Abernethy area, more than one-third of the homestead applicants were unmarried at the time of application. Since some of the applicants had married in the interval between entry and application, an even larger portion was single at the outset. Bachelors tended to select their lands adjacent to the homesteads of their friends with whom they would share accommodation during the first year or two. In addition to providing needed company on the lonely frontier, these arrangements were of economic importance. Settlers could save money by sharing the expenses of food and shelter, and by pooling their equipment and labour.

The Dominion Lands Act permitted a settler to postpone establishing a residence on his homestead provided he lived within a radius of two miles from his quarter section after the initial entry.[23] He was still obligated to build a "habitable" dwelling and live in it for three months prior to application, but this provision allowed him to live with neighbours for up to 33 of the 36 month "proving-up" period. That many took advantage of the clause is evident from a reading of individual homestead files. Shared living arrangements were common not only among young bachelors, but also among married homesteaders who had left their wives in Ontario for one or two years until they were in a position to bring them out to a finished home.

The settler also required a shelter for his livestock. In the Abernethy area study, only one of the homesteaders reported the existence of a barn, although roughly half of the Anglo-Saxons and 75 per cent of German settlers stated that they possessed stables. The average value reported for the Germans' stables was $75 compared with $139 for the Anglo-Saxons. As was the case with the house values, these assessments for stables probably reflect the market value, and are higher than the actual capital investment. Most respondents stated that their stables were made from logs. This obviously represented a great saving in material costs. Granaries and other outbuildings were also principally constructed of logs. Abernethy applicants valued their granaries and outbuildings at an average of $79; Neudorf homesteaders appraised their outbuildings at $58. Overall, if the structural improvements on homesteads in the study areas are broken down by quartile, a wide variation in assessed values is revealed (see Table 1).

Table 1. VALUE OF HOMESTEAD STRUCTURAL IMPROVEMENTS BY QUARTILE, ABERNETHY AND NEUDORF DISTRICTS, 1882-1917 IN CURRENT DOLLARS*

	I	II	III	IV
Value of Dwelling	$5-75	$75-150	$150-250	$250-1500 (n: 303)
Value of Stable	5-30	30-50	50-100	100-725 (n: 147)
Value of Granaries & Out-buildings	2- 20	20- 50	50- 75	75- 400 (n: 81)
Value of Fencing	5- 20	20- 50	50-100	100- 500 (n: 111)
TOTAL	$17-145	$145-300	$300-525	$525- 3125

*In terms of constant dollars, house values increased overall after 1900. The mean valuation for pre-1900 dwellings was $201 (1900 dollars) compared with a mean of $258 between 1900 and 1914. If valuations are grouped by five intervals, however, the trend is less clear.

Interval	Number of Observations	Mean Valuation
1886-89	53	$209
1890-94	40	354
1895-99	106	139
1900-04	43	228
1905-09	40	296
1910-14	14	238

The low average recorded between 1895 and 1899 reflects the preponderance of poorer settlers at Neudorf in this period (89 of 106). Apart from this period, house values averaged $271 before 1900, and actually dropped slightly after the turn of the century.

Beyond the structural costs, establishing a residence entailed expenditures for sundry household items, including a stove, furniture, bedding, and kitchenware. Estimates for these costs vary from one pioneer account to the next, a fact which may be indicative of the diversity of needs and resources of different shelters. With respect to stoves, it appears that the Ankli-Litt estimate of $40 is higher than the essential minimum. In 1891 the *Qu'Appelle Vidette* carried an advertisement for "cheap patent stoves" ranging in price from $16 to $80.[24] Isaac Cowie's pamphlet *The Edmonton Country* published in 1901, reported that cooking stoves in that locality could be purchased for $23 to $26.50.[25] Georgina Binnie-Clark, who settled north of Fort Qu'Appelle in 1905, later wrote that she had purchased a second-hand stove for $15.[26]

Waiting out the winter. A young bachelor in his sod house huddles over a typical early stove. (Courtesy of the Saskatchewan Archives Board/R-A2321.)

Other necessary chattels could be purchased relatively cheaply. Cowie's pamphlet recorded the price of hardwood chairs to be from 55 cents to $1; tables, $3 or more; and bedsteads, from $4 up.[27] Alternatively, all of these articles could be handmade. Tables and chairs might be constructed with the available supply of timber and beds were commonly made by sewing a tick and filling it with straw. Under the most frugal circumstances, settlers were still obliged to purchase tools, such as axes, saws and nails; lamps, bedding and kitchen utensils. Contemporary immigration pamphlets suggest that the costs of furnishing a log house could range between $20 and $75.[28] The minimum tools required by a settler building his own log house would seem to include a spade, cross-cut saw, hammer, chisel, brace and bits, planes, an auger, axe, and some nails. In 1902, these items cost between $11.40 and $21.70 at Edmonton.[29]

No settler could avoid the necessary expenditures for provisions, although estimates of food costs diverge greatly from one contemporary account to the next. In 1882 John Macoun wrote that a settler with a family of five would spend about $250 on provisions during the first year.[30] A year later, however, a British settler in Manitoba estimated his cash requirements for groceries to be only $20.[31] Obviously the outlay for groceries varied with the size of family and the kind of lifestyle the settler pursued. Two accounts—an immigration pamphlet published in 1902[32] and a Dutch homesteader's tabulation written in 1910[33]—place the cost of provisions for a bachelor settler at about $90 to $120.

This capital requirement would tend to increase in the case of a larger family, although poorer peasant settlers could and did make do with less. An immigration pamphlet published in 1882 stated that the cost of provisions for one family of five Mennonites at a subsistence level had been $93.[34] Their diet consisted almost solely of flour, pork and beans. Alternatively a settler could spend much more. In 1907, for example, Georgina Binnie-Clark spent $245 for groceries, flour, meat, repairs and veterinary fees, although in her view, the "degree of necessity" was a much smaller amount.[35]

Fencing was a requirement for farmers who raised livestock but the most striking aspect of farm fencing in the Abernethy district in the homesteading period was its absence. Of 461 applicants, only 111 reported the presence of fencing on their homesteads. Among this group, there were wide variations in acreage fenced. The average extent was 44 acres at Neudorf and 63 acres at Abernethy but the majority enclosed 35 acres or less and more than 20 per cent fenced in areas of 20 acres or less. Settlers also reported widely divergent expenditures for fencing. One appraisal stated a cost of only 19 cents per acre; others valued homestead fencing at up to $4 per acre. Overall, the mean cost for Abernethy settlers was about $1.25 per acre, a figure which is confirmed by most estimates in the period.[36] Given that most settlers began farming with no more livestock than a yoke of oxen or team of horses, a sufficient initial expenditure for fencing in the 1880s and 1890s was $10 to $15 for a small pasture of 10 acres.

Water was an essential requirement for all settlers. Of the 19 Abernethy settlers who reported wells on their homestead applications, 14 recorded values of $50 or less and the average valuation was $23.50. Others who were less fortunate were obliged to abandon their homesteads altogether for lack of water. Still, the variability in these cases is so great that they cannot reasonably be averaged with the costs incurred by those individuals who did strike water. Assuming that the settler did not need to hire a professional well-digger, he could avail himself of a government-owned digging apparatus for a nominal fee. One homesteader at Morse, Saskatchewan, wrote that he spent 75 cents per foot to dig a well in 1912.[37] Another source for 1912 placed the cost of well-digging at $1 a foot for the first 50 feet. Thereafter the cost increased steadily until it reached $2 per foot at the depth of 100 feet.[38] The writer observed that in wooded areas, water could generally be found at a depth of 20 to 40 feet. On the open prairie, despite the presence of occasional springs, a well digger would usually be obliged to go to a much greater depth. Assuming that a settler struck water at 20 feet, as several Abernethy settlers appear to have done, the cost of digging a well, in 1900 dollars, was about $14. To crib his well the Morse homesteader estimated that he needed 300 feet of lumber, costing $9.[39] If this estimate is converted into 1900 dollars, the overall cost of constructing a well of 20 to 40 feet was $21 to $35.

Livestock constituted another farm-making cost that was highly variable. Professor Ankli has estimated the cost of horses to have been $75-$100 at the turn of the century, and bases this figure on the 1901 Census, which gives the average value of horses in Manitoba and the North-West Territories as $96 and $62 respectively. Since the census average includes horses of all sizes, the figures are misleading. The basic operations of breaking sod and pulling implements required heavy draught horses which were more expensive than the average. In 1881 a settler in Manitoba estimated the cost of a team of horses and harness to have been $325.[40] Walter Elkington, who settled north of Fort Qu'Appelle in 1891, wrote that teams of horses ranged between £30 and £60, or $150 to $300.[41] In 1912 a Dutch homesteader at Morse, Saskatchewan, purchased a team of large sorrel horses for $450.[42] The following year, Boam's *The Prairie Provinces of Canada* included an estimate of $360 for a team of "good horses."[43]

A farmer might alternatively purchase oxen to pull his implements. While a team of oxen sometimes cost as much as $200 or more,[44] most estimates in the 1880-1900 period range between $100 and $130.[45] Costs varied according to the age and quality of the animals, and the time of year in which they were purchased. Oxen sold in the spring, for example, would command a higher price than those sold in the fall.[46] The usual practice was to purchase a yoke of oxen for the purpose of prairie breaking, and to exchange these for more expensive horses a year or two later.

One farm-making cost that few settlers could avoid was the requisite outlay for a wagon. Wagons were essential to the hauling of farm products to market centres and supplies from the town back to the homestead. They were used in the collection of wood for building and fuel, and were an essential means of transport to social and religious activities. Settlers could, and often did, begin farming with Red River carts worth $10,[47] but once they had begun to transport large quantities of farm produce, wood, and hay, the purchase of a more serviceable wagon was mandatory. Four sources for the 1880s give $80 as the standard price of a farm wagon,[48] a figure which does not seem to have changed for the entire period leading up to the First World War. These estimates pertain to the entire unit, including a four-wheel chassis, and removable wooden box. For winter travel, settlers were also obliged to purchase a set of sleigh runners for the wagon box. One Manitoba resident wrote in 1883 that the price of a pair of runners was then $30.[49]

Early Abernethy and Neudorf area settlers would often travel to agricultural fairs in CPR mainline towns to purchase or trade livestock. An 1898 livestock market in Qu'Appelle is pictured here. (Courtesy of the Saskatchewan Archives Board/R-A9937.)

Expenditures for farm implements could exceed $1000, or several thousand dollars, if a settler purchased a threshing machine. For purposes of initial farm making, however, many settlers' requirements were much more modest. In most accounts, a small investment of $40 was sufficient to purchase a prairie breaking plough, a stubble plough, and a harrow.[50] These simple implements permitted the breaking and preparation of several acres for crop. Cultivation was of necessity very crude as the prohibitive cost of sophisticated implements required that the poorer farmers broadcast seed, stook, harvest and thresh their crops by hand, using such primitive tools as cradles and flails.[51] Professor Ankli's estimate of $330 as the cost of implements for "substantial settlers" does not seem unreasonable, although it is difficult to accept an overall estimate without some indication as to the size of farm for which these implements are considered sufficient. Obviously, too, a farmer's expenditure for implements was related to the kind of crops he cultivated, and the extent to which he raised livestock. If a farmer raised cattle, for example, it was incumbent that he invest in a mower and rake, for purposes of haying. Table 2 provides price quotations for implements essential to farm 160 acres.

Table 2. PRICE QUOTATIONS FOR SELECTED FARM IMPLEMENTS ON THE PRAIRIES, 1884-1915.*

	1884[52]	1889[53]	1897[54]	1909[55]	1914[56]	1915[57]
Breaking Plough (Walking)	22	22	20	22	20	24
Stubble Plough (Walking)	16	20	18	22	14.5	24
Iron Harrows	18	20	15	20	18.5	20
Drill Seeder	55	65	90	88	85	100
Mower	80	63	55	52	56	52
Binder	290	150	155	150	145	150
Horse Rake	35	20	28	33	33	33
Sleigh Runners	30	28	25	37	35	38
Total in Current $	546	388	406	424	407	461
Total in 1900 $	494	371	477	369	310	324

*Price quotations were deflated using the JI wholesale Price Index, after Mitchell, 1868-1925 (Urquhart and Buckley, Historical Statistics of Canada, p.291). Since each set of price quotations relates to a different location and documentary source, allowance should be made for variations in freight rates and retail mark-up. An effort was made to ensure that quotations pertain to the same kinds of implements, but it must be recognized that technological changes had modified some of these implements by the end of the period.

Most settlers did not incur expenses for seed in the first year of settlement, as few were able to prepare any of their newly-broken land in time to put in a crop. In the study areas of Abernethy and Neudorf, only 45 or about 10 per cent of the 461 homestead applicants reported that they had placed one or more acres under cultivation in the first year. Sixty-two per cent reported no cropping, and the remaining twenty-eight per cent did not answer the question. It should be noted that before 1900 comparatively few settlers began farming on improved land. Purchasers of CPR and colonization company lands were similarly obliged to break virgin prairie prior to cultivation. Most sources suggest that 1½ to 2 bushels of seed grain was sufficient to sow one acre of wheat. At $1 per bushel, seed sufficient to sow 10 acres could be purchased for $15 to $20, but even this small expense could be deferred if the settler obtained a seed grain advance from the Dominion government.

If the essential tasks of wood gathering and house building limited the amount of acreage cropped in the first year, these factors also served to curtail the extent of breaking for most settlers. Anglo-Saxon settlers at Abernethy broke an average of 15 acres in the first year, compared with an average of 8 acres broken among the German settlers at Neudorf. If a farmer elected to contract out his breaking he usually paid from $1.50 to $3.00 an acre for this work,[58] and an additional $2.00 if he also wished to have the land backset in the fall.[59] Ankli is correct in his statement that by 1900 steam plowing offered a cheaper alternative to breaking with horses or oxen. Yet the continuing debate in the agricultural press in the early 1900s suggests that steam plowing was slow to supersede horses and oxen as the preferred method.[60] The relatively small extent of land clearing in the Abernethy area suggests that the majority of settlers performed their own breaking with mould-board plows.

A factor which must not be overlooked in assessing farm-making costs was the role of cooperative enterprise among settlers. While the principal mode of land disposal was that of individual "free" homesteads, as opposed to group settlement, settlers tended to select homesteads close to their friends, to permit the pooling of implements, livestock and labour. Pooled labour was common not only among Continental European immigrants, but in the Anglo-Saxon communities as well, as the memoir of an English settler near Fort Qu'Appelle indicates:

> The people in Canada are very good in helping one another; if a man wants a new house or stable put up, he gets the material ready and goes around among his neighbours and asks them to come and help him on a certain day; this is called a "bee" and it is not only done for buildings, but for ploughing, sowing, reaping, or if a man has had a misfortune and is behind in his work, the neighbours go in together and give him a day's work.[61]

Similarly, expensive implements, which were beyond the means of many farmers, were sometimes loaned by more prosperous settlers in exchange for a day's labour from the borrower.[62] Alternatively, some settlers found that they could still save money by contracting out various farming operations. During the first five years, for example, W. R. Motherwell hired a neighbour to do his binding.[63] Settlers also defrayed capital costs by purchasing implements and livestock in partnership with their neighbours. John Teece, who homesteaded near Abernethy in 1883, held a one-fourth interest in a team of horses, wagon and plow during the first four years. Thereafter, he possessed his own yoke of oxen, wagon and plow, but purchased a one-fourth interest in other implements. Shared ownership does not seem to have impeded Mr. Teece's progress as a farmer; in 1913 he reported that he was in possession of nine quarter sections of land and $10,000 worth of livestock.[64] Another way in which partnerships operated was to cultivate crops "on shares."[65] One Manitoba settler related that farmers commonly worked out arrangements

A well-attended barn-raising bee at Herb Keith's farm in the Cottonwood district north of Pense, Saskatchewan, 1912. (Courtesy of the Saskatchewan Archives Board/R-B12988.)

whereby one settler would perform the cultivation duties for two farms while his partner continued to work on the railway to support both homesteads.[66]

Further evidence that cooperative enterprise played a major role in the farm-making process lied in the fact that fewer than 25 per cent of the German homesteaders recorded owning cattle in the first year of farming, and only ten per cent stated that they owned horses in the same year. At Abernethy, approximately one-third of the homesteaders reported that they had owned cattle in the first year, compared with fewer than one-fourth reporting the initial ownership of horses. It is possible that some settlers misread the question on the form, and therefore failed to answer it completely. Such a small percentage of respondents, however, leaves little doubt that a substantial portion of this homesteading population did not own oxen or horses at the time of entry.

It is important to recognize that farm-making was usually a gradual process in which homestead development was phased in over a period of several years. Settlers usually made only a partial beginning in the first year, and gradually improved their holdings over the "proving-up" period of three years or more. In the first year a typical pattern was to select land early in the spring and put up a tent or very rudimentary shelter. No field work could be performed until after the spring run off, but once it was dry enough to plough, the settler

The German settlers in the Neudorf district, such as those pictured here at an area wedding, not only shared language and customs, they often worked out cooperative arrangements in order to operate and maintain their farms.

would break five to fifteen acres. If his land was stony, rocks would have to be picked, and the rate of breaking would be slowed. Working industriously the settler could harrow a few acres in time to plant small amounts of wheat, oats, and root crops sufficient to provide him with flour, livestock feed, and vegetables for the winter. Late June and July was customarily devoted to haying. Homesteaders who settled within reasonable proximity to sloughs would be able to cut enough hay to meet their own requirements. Any surplus could generally be sold in the towns or to other farmers. During the harvest months of August to October, the new settler would often find employment with threshing crews harvesting in the developed farm communities in Manitoba and along the CPR main line. For the duration of the winter, young settlers frequently hired on with other farmers, or found work with the CPR.[67] Those who stayed on their homesteads tended to such tasks as fencing. Commonly, they worked in the bush most of the winter, cutting wood for their own needs, and marketing surplus cords in town to support themselves.[68] This income might be supplemented by sales of milk, eggs and other livestock products. Generally, farming in the first year of settlement was of a subsistence nature, and was carried out chiefly to offset costs while the homesteader prepared his farm for future profit-oriented agriculture. Only after the passage of several years, and the accumulation of the necessary buildings, tools, prepared acreage, and other improvements, did many settlers begin to treat their farming as a full-time occupation.

In his report on the Canadian North-West in 1904, James Mavor observed that "the amount of capital necessary for the establishment of a colonist varies with the district, with the kind of cultivation he intends to adopt, and with the standard of comfort of the colonist himself."[69] For the present study three ranges of farm-making costs have been tabulated: the minimum, the average, and the substantial. The minimum range assumes that the settler possessed few resources other than his own labour, or, if married, the labour of his family. If a bachelor, such a settler lived in a sod, log, or dugout shanty, and possibly shared accommodation with another homesteader or two. For this tabulation, the range of expenditures in the lower quartile of homestead improvements, i.e., $12 to $75, had been assumed to represent the range of structural costs incurred by the modest settler. He also shared ownership or borrowed the requisite implements—a breaking plow and harrow, and a yoke of oxen. For water, the modest settler might dig a shallow well for $21-$35, or he might also haul water from nearby sloughs or streams, and spend nothing. Without horses, his fencing requirements were negligible, although Abernethy homesteaders in the lower quartile of structural costs spent between $5 and $20 in fencing small pastures of 5 to 15 acres. The principal expenditures in this minimum range of expenditures were for provisions, which might vary between $50 and $90, depending on the size of the family and the settler's ability to supplement his grocery purchases with game and fish from nearby woods and waters. For peasant settlers with larger families, higher provisioning costs were obviated by the general practice of hiring children out to Anglo-Saxon settlers. Overall, farm-making costs for the modest settler have been calculated at between $242 and $436.

The young daughter of eastern European peasant immigrants at work feeding chickens. (Courtesy of the Saskatchewan Archives Board/R-A20815.)

The second range of farm-making costs represents the probable amounts that the average settler spent to set up a homestead independently in the first year. It must be admitted that the calculation of average, in the sense of typical, costs, is a somewhat tenuous undertaking. The data, such as it exists, permits some informed guesses. If the homesteaders' own assessments of value of farm improvements can be assumed to reflect their actual expenditures, then the average settler's building costs would fall within the interquartile range of homestead improvements, i.e., the range between the twenty-fifth and seventy-fifth percentiles of the surveyed population. This range implies an expenditure of $145 to $525 for a house, stable, granaries and fencing. Since well construction cost 70 cents per foot for the first fifty feet, the average settler might spend $21 to $35 for digging and cribbing. As noted earlier, most accounts suggest that provisions for a single homesteader were in the range of $90 to $120. Small hand tools suitable for the purposes of an owner of 160 acres were said to cost up to $40 ($31 in 1900 dollars). The average settler needed comparatively few implements in the first year: two ploughs, costing $26-$34, and a mower and rake, worth another $60-$74. Since he did little or no cultivation in the first year he had no need either of a drill seeder or of seed grain. It is assumed, however, that he needed to purchase his own wagon, worth $80. Overall, the expenditures of the average settler fall in the range of $590-$1193.

An established "average settler" proudly poses with his family, animals, buggy and home.

The "substantial settler" may have contracted out his threshing to a crew and rig such as that pictured here.

The substantial settler may be defined as a settler who began farming with all the necessary accoutrements in the first year. The upper quartile in the distribution of homestead structural improvements may be taken to represent this relatively prosperous settler's investment in farm buildings, that is $570-$2625. Since he owned a larger house (the upper quartile in terms of square footage ranged from 480 square feet up) he also probably purchased another heating stove to supplement his cooking stove, for a total expenditure of $28-$46. Similarly, he purchased or brought more furniture, say $44-$68.

We assume that the substantial settler purchased a full range of farm implements for about $300-$500, and a wagon for $70-$82. He was unlikely to have purchased his own threshing machine immediately, but, like most other farmers, contracted out his threshing. Assuming that he began cultivation in the first year, he would spend $305-$517 in hiring out the breaking of 100 acres. To work the larger acreage the substantial settler needed at least two teams of horses, costing $338-$696, and two harnesses, at $38 each. We might also expect him to have purchased perhaps 6 dairy cattle to provide an interim income prior to the fall harvest. These cattle would cost an additional $23-$48 each, or $138-$288 in total. To provide seed for 100 acres of wheat, he would spend $65-$151, depending on the current price. In all, the substantial settler would have spent about $2093-$5873 (see Table 3).

Table 3. ESTIMATED FARM-MAKING COSTS FOR THREE CATEGORIES OF SETTLERS,
1882-1914 (CONVERTED TO 1900 DOLLARS)*

	Number of Observations	Minimum Capital Requirements	Average Homesteader's Costs	Substantial Settler's Costs	
Entry Fee	(N/A)	8–12	8–12	8–12	
Dwelling	(303)	5–75	75–250	250–1500	
Stable	(147)	5–30	30–100	100–725	
Granaries	(81)	2–20	20- 75	75- 400	
Well	(19)	21–35	21–35	42–70	(2 wells)
Fencing	(111)	5–20	20–100	100–500	
Provisions	(3)	90–130	90–130	130–217	
Stove	(4)	14–23	14–23	28–46	(2 stoves)
Furnishings	(3)	22–34	22–34	44–68	
Implements	(6)	13–17	94–131	310–494	
		(2 plows ½ int.)	(2 plows, 1 mower & rake)	(full complement)	
Small Tools	(3)	14–31	14–31	14- 31	
Livestock:					
—yoke of oxen	(6)	47–77n (½ int.)	93–153	——	
—team of horses	(4)	——		338–696	(2 teams)
—dairy cattle	(3)	——	——	138–288	(6 cattle)
Harness					
—for oxen	(2)	10–19 (½ int.)	19–37	——	
—for horses	(1)	——	——	76–76	
Breaking					
—100 acres	(3)	——	——	305–517	
Seed					
—150 bushels	(N/A)	——	——	65–151	
Wagon	(7)	35–41 (½ int.)	70–82	70–82	
Total		$291–564	$590–1193	$2093–5873	

*Due to the great variability in reported values of structural improvements, estimates for the costs of dwellings, stables, granaries, wells, and fencing have been left in current dollars.

To conclude, the costs of prairie farm-making between 1882 and 1914 varied greatly among settlers of different economic and cultural origins. Costs also varied according to a host of individual variables, including the size of the settler's family, the availability of indigenous building materials, water, and wild game, his working relationships with his neighbours, and opportunities for co-operative ownership of implements. A crucial variable was the timing of settlement. In the 1880s incoming settlers usually possessed a good choice of free grant and CPR lands with easy access to timber, water, and unoccupied pasture lands. After 1900 the price of purchased lands increased rapidly and the selection of homestead lands of high quality diminished.

Despite these differences, it is apparent that a prairie settler could get started with a small investment of about $300-$550, albeit in a rudimentary way. If the settler made a more independent beginning, he would spend $600-$1200. Alternatively, the settler with a large accumulated capital could spend $2100 or more to begin farming on a substantial scale. These figures tend to support Ankli and Litt in their tabulation of farm-making costs for substantial settlers, but suggest a somewhat lower limit to cost estimates for average settlers. Where this paper departs significantly from Ankli and Litt, is in its conclusion that a settler could, if necessary, begin farming with a very small investment.[70]

The low minimum capital requirement implies a degree of economic democracy in the prairie settlement experience. Throughout the period 1883-1914 the standard wages for experienced farm hands remained fairly constant in current dollars, including board. Even in the context of inflation at the end of the period, a hired man usually earned enough to start farming on free grant land with a year's accumulated wages. At the same time it must be admitted that homesteaders who arrived early and claimed the best lands were in a comparatively better position than later arrivals to turn a small investment into a small farm. Farmers who arrived with substantial capital also had a head start in bringing their farms into market-oriented production. Poorer settlers often spent an inordinate amount of time establishing themselves. The homesteader who began with little was more vulnerable and had fewer options than his more wealthy counterparts. Any single economic disaster, such as the loss of his buildings to a prairie fire, or the failure to find water on his land, could bring an early end to his homesteading venture. These qualifications notwithstanding, homesteading was open to most settlers in the earliest period. A tentative conclusion would seem to be that a large capital investment was not essential for the establishment of a successful farming operation.

NOTES

This article first appeared in *Prairie Forum* 6, no. 2 (1981): 183-201. This paper was originally presented to the Eleventh Conference on Quantitative Methods in Canadian Economic History, Queen's University, Kingston, February 27-28, 1981. The author wished to thank Professors Michael Percy, Robert Ankli, James Richtik and Gerald Friesen for their comments.

1   Robert E. Ankli and Robert M. Litt, "The Growth of Prairie Agriculture: Economic Considerations," in Donald H. Akenson, ed., *Canadian Papers in Rural History* (Gananoque: Langdale Press, 1978), vol. 1, pp. 35-64.

2   David Gagan, "Land Population and Social Change: The 'Critical Years' in Rural Canada West," *Canadian Historical Review*, LIX (1978), pp. 295-318; and "The Indivisibility of Land in Nineteenth-Century Ontario," *Journal of Economic History*, XXXVI (1976), pp. 126-141. Gagan has noted that cash inheritances averaged $450 for sons who were not the principal heirs of Ontario farmers. One child in five received nothing at all. Gagan, "The Indivisibility of Land," p. 136.

3   J. F. Snell, "The Cost of Living in Canada in 1870," *Social History*, XII (1979), p. 189. Wage rates for Ontario farm labourers for four localities ranged between $12 and $15 per month, with board, and averaged $13.50. One Qu'Appelle area settler from Toronto described a year's earnings in Ontario as $200. See the testimonial of Samuel Copithorn, *Qu'Appelle Vidette*, March 2, 1893. Even the more arduous tasks of threshing by cradle brought only about $1.50 per day, or about $40 per month during harvest season, and summer wage rates generally ranged between $1 and $1.50 in the early 1870s in Ontario. Richard Pomfret, "The Mechanization of Reaping in Nineteenth Century Ontario: A Case Study of the Pace and Causes of the Diffusion of Embodied Technical Change," *Journal of Economic History*, XXXVI (1976), p. 414.

4   Saskatchewan Archives Board, Regina (hereafter SABR), Local Histories File "Qu'Appelle" Annual Report of A. J. Baker, Immigration Agent, Qu'Appelle to Dominion Minister of Agriculture, 31 December 1888.

5   John Thompson, "Bringing in the Sheaves: The Harvest Excursionists, 1890-1929," *Canadian Historical Review*, LIX (1978), p. 488.

6   Letter of John Burton to the Editor, *Qu'Appelle Vidette*, March 2, 1889.

7   "Started on $200," *Qu'Appelle Vidette*, March 3, 1889.

8   See for example, "Experience and Opinions of Settlers," *Manitoba Official Handbook*, pp. 40-44.

9   *Regina Leader*, Supplement, April 9, 1889.

10 *Leader*, May 19, 1898.

11 "Successful Colonization in the German Colony of New Toulcha," *Leader*, December 4, 1888, p. 8.

12 *Western World*, March, 1891, p. 63.

13 An Act Respecting Public Lands," *Statutes of Canada*, 49 Vic., 1886, Chapter 54, S. 52, p. 832.

14 The average proving-up period at Abernethy was more than 5 years.

15 James B. Hedges, *Building the Canadian West* (New York: Russel & Russel, 1939), pp. 67-68.

16 James M. Minifie, *Homesteader* (Toronto: Macmillan, 1972), p. 140.

17 Annie I. Yule, *Grit and Growth: The Story of Grenfell* (Wolseley: n.p. 1967), p. 27.

18 Ibid., p. 76.

19 *Letters from a Young Emigrant in Manitoba* (London: Kegan Paul, 1883), p. 97.

20 W. M. Elkington, *Five Years in Canada* (London: Whittaker & Co., 1895), p. 135.

21 "Western Canada: Manitoba, Alberta, Assiniboia, Saskatchewan, and New Ontario," 1902, p. 68.

22 Willem de Gelder, *A Dutch Homesteader on the Prairies,* translated by Herman Ganzevoort (Toronto: University of Toronto Press, 1973), pp. 27-29.

23 "An Act Respecting Public Lands," *Revised Statutes of Canada,* 49 Vict. 1886, Chapter 54, S. 38, s.s. 6 & 7, pp. 830-31.

24 *Qu'Appelle Vidette,* November 5, 1891, p. 4.

25 Isaac Cowie, *The Agricultural and Mineral Resources of the Edmonton Country* (Western Canada Immigrant Association, 1901), p. 51.

26 Georgina Binnie-Clark, *Wheat and Women* (Toronto: Bell and Cockburn, 1915), p. 60.

27 Cowie, *The Edmonton Country,* p. 51.

28 The Star Almanac 1893, "Hints to Intending Settlers," (Montreal: Hugh Graham, 1893), p. 255; *Letters from a Young Emigrant in Manitoba* (London: Kegan Paul, French & Co., 1883) pp. 96-98; John Macoun, *Manitoba and the Great North-West* (Guelph: The World Publishing Co. 1882), "Advice to Immigrants," pp. 637-39.

29 Isaac Cowie, *Edmonton Country,* p. 51.

30 John Macoun, *Manitoba and the Great North-West,* pp. 637-39.

31 *Letters from a Young Emigrant,* p. 130.

32 "Western Canada," 1902, p. 68.

33 Willem de Gelder, A *Dutch Homesteader,* p. 88.

34 Canada. Department of Agriculture. "The Province of Manitoba and North West Territory, Information of Intending Immigrants," 5th ed. (Ottawa, 1881), pp. 14-19. Reprinted in Kevin H. Burley, ed., *The Development of Canada's Staples, 1867-1939* (Toronto: McClelland and Stewart, 1971), pp. 37, 38.

35 Georgina Binnie-Clark, *Wheat and Women,* p. 350.

36 Ibid., p. 166. Georgina Binnie-Clark has provided a cost breakdown for a two strand barbed wire fence in 1907:

8 rolls wire	$24.00
660 pickets	$19.80
Labour	$20.00
	$63.80 per mile

Since two miles of fencing was required to enclose a quarter section, Binnie-Clark would spend $127.60 or $0.80 per acre.

37 Willem de Gelder, *A Dutch Homesteader,* p. 35.

38 Leo Thwaite, *Alberta: An Account of Its Wealth and Progress* (London: George Routledge & Sons, 1912), p. 88.

39 Willem de Gelder, *A Dutch Homesteader,* p. 40.

40 *Letters from a Young Emigrant,* pp. 86-88.

41 Elkington, *Five Years in Canada,* p. 135.

42 Willem de Gelder, *A Dutch Homesteader*, p. 72.
43 See for example, John Macoun's *Manitoba and the Great Northwest*, p. 637; *Letters from a Young Emigrant in Manitoba*, pp. 96-98; Isaac Cowie, *Edmonton Country*, pp. 18-19; *Toronto Globe*, October 15, 1898, "Hints to Settlers," quoted in Annie Yule, *Grit and Growth*, p. 67.
44 Henry J. Boam, comp., *The Prairie Provinces of Canada*, edited by Ashley G. Brown (London: Sells Ltd., 1914), pp. 388-89.
45 Janice Action comp. *Lemberg Local History* (Lemberg: n.p. 1972), p. 164.
46 Willem de Gelder, *op. cit.*, pp. 36-37
47 "The Province of Manitoba and North West Territory Information for Intending Immigrants," in Kevin H. Burley, ed., *The Development of Canada's Staples, 1867-1939*, p. 37.
48 All of the following sources, cite $80 as the price of a wagon in the 1880s: (1) Macoun, *Manitoba and the Great Northwest*, p. 637; (2) *Farming and Ranching in the Canadian Northwest, the Guide Book for Settlers* (Ottawa: King's Printer, 1886); (3) "The Province of Manitoba and North West Territory: Information for Intending Immigrants," p. 37; and (4) R. Goodridge, *A Year in Manitoba* (London: W.R. Chambers, 1882), pp. 107-108. Cowie gives $75 as the price of a wagon in 1897 and Boam's *The Prairie Provinces*, published in 1914, states that farm wagons at that time could be purchased for $70. Second hand wagons, of course, could be purchased more cheaply. In 1907 Georgina Binnie-Clarke purchased a used wagon for $45. *Wheat and Women*, p. 37.
49 *Letters from a Young Emigrant in Manitoba*, p. 130.
50 See, for example, *Western Canada: Manitoba, Alberta, Assiniboia, Saskatchewan, and New Ontario* (pamphlet, 1902), p. 68; *The Manitoba Official Handbook* (Liverpool: n.p., 1892), pp. 44, 45.
51 Saskatchewan Archives Board, Saskatoon (SABS), Pioneer Questionnaire No. 6, George A. Hartwell, Pheasant Forks, District of Assiniboia, 1882, p. 5.
52 Canada. Department of Agriculture, Annual Report, *Sessional Papers* 48 Vic. (1885), No. 8, Table B, p. 150. "Table showing the Price of Agricultural Implements as Sold at Different places in the United States and Canada, during the season of 1884." Prices of implements at Brandon, Manitoba were selected for this paper.
53 Canada. Department of Agriculture Indian Head Dominion Experimental Farm Ledger Book, 1888-89. The figures quoted in the ledger book represent actual expenditures for farm implements. Comparable prices for 1889 farm implements are revealed in an Oak Lake settler's letter to the *Manitoba Colonist*, No. 42, November 1889, p. 14.
54 Isaac Cowie, The *Edmonton Country*, p. 51.
55 Saskatchewan Archives Board, Saskatoon, (SABS), Department of Agriculture Records. Agricultural Machinery Administration Series II Catalogues Group A. 20 "Massey-Harris." Box 1, "No. 1 Retail Price List 1909 Manitoba Branch." Massey Harris Company Ltd., Winnipeg, Manitoba.
56 SABS, Royal Commission, "to inquire into the conditions surrounding the sale in Saskatchewan of farm machinery (etc.) ..." Exhibits. File A(v), Cockshutt Plow Company, Box 1. "Retail Price List 'B'," 1914.
57 SABS, Department of Agriculture Records. Agricultural Machinery Administration, II Catalogues. Group A. 20. Massey Harris Box 1. "No 2 X Retail Price List, 1915-1916." (effective May 1, 1915) Regina Branch, Massey Harris.

58 Provincial Archives of Manitoba, Diary of Claude H. Manners, Moosomin, 1883.
59 Isaac Cowie, *The Edmonton Country*, p. 19, gives $1.95 as the cost of backsetting in 1901.
60 See "Wheat: Cost of Production," *Canadian Thresherman*, November 1905, p. 3. According to this article steam plowing entailed a cost of 57.5 cents per acre. "Using the Traction Engine on the Plow," *Nor'-West Farmer*, March 5, 1902, p. 193.
61 W. M. Elkington, *op. cit.*, p. 64.
62 Letter of William Compton, Opawaka, Manitoba, to the Editor, *Western World*, August 25, 1890, printed in the September 1890 issue.
63 "A Farm Started in '82," *Nor'-West Farmer*, November 20, 1928.
64 Letter of John Teece to J. Cromie, December 27, 1928. SABR, Saskatchewan Department of Agriculture, Statistics Branch, file re: General Publicity, 1914.
65 J. M. Bonnor, "Early History of Blackwood District," (unpublished local history manuscript), Blackwood, Saskatchewan, 1963. See also SABS, Homestead File 71859.
66 *Letters from a Young Emigrant in Manitoba*, pp. 85-86.
67 SABS, Department of Interior Homestead Files.
68 See, for example, *Letters from a Young Emigrant* (1883), and Elkington, *Five Years in Canada* (1895).
69 James Mavor, *Report to the Board of Trade on the North West of Canada, With Special Reference to their Wheat Production for Export* (London: King's Printer, 1902), p. 26.
70 Statistical surveys of prairie farmers conducted in the 1920s and 1930s indicate that a large proportion of settlers began with relatively little capital. In 1930-31, the Saskatchewan College of Agriculture carried a series of surveys in areas representative of the different soil groups in the province. Overall, these surveys showed that farmers who settled before 1900 had begun with an average capital of $357. Evidently, the costs of farm-making increased quickly after 1900, as farmers who arrived in the 1900-05 period reported an average initial worth of $1249. Starting capital increased rapidly thereafter to a high of $5000 for those who commenced farming after World War I. Further evidence for the low capital outlay thesis is found in a 1926 study of 360 Manitoba farmers. Fully 60.1 per cent of the surveyed group reported having started farming with less than $500. Sources: R. W. Murchie, *Agricultural Progress on the Prairie Frontier*, Canadian Frontiers of Settlement Series V (Toronto: Macmillan, 1936), pp. 72-73; and R. W. Murchie and H. C. Grant, *Unused Lands of Manitoba* (Winnipeg: Department of Agriculture and Immigration, 1926), pp. 71-72.

7.   Adapting to the Frontier Environment:
     Mixed and Dryland Farming near Pincher Creek, 1895-1914

Warren M. Elofson

There is a need for closer investigation of both the frontier and the natural environments in the history of western Canadian agriculture.[1] This article represents one attempt to work towards that objective. Its focus is the farming community which formed around the southern Alberta town of Pincher Creek between about 1895 and World War I. In this community, soil, terrain, and weather patterns change dramatically as one moves from one area to another, revealing the divergent farming methods that have been necessary to accommodate them.[2] Consequently, the process of adaptation to the environment can be traced rather effectively. In the earliest years of settlement, however, specific frontier conditions dictated that the ability of agriculturalists to modify techniques would be significantly more limited than in later periods. It took some time before a sophisticated compromise was made with the complex characteristics of the natural environment. The central argument here is that by 1914 the farming community near Pincher Creek had taken some crucial steps in that direction.

The geography and climate in southern Alberta change a great deal as one moves east to west. The east is predominantly level, devoid of indigenous forests and, for agricultural purposes, suffers from a deficiency of moisture particularly in the Palliser Triangle, the great semi-arid belt, the heart of which runs from central Saskatchewan to within a few miles of Lethbridge. By contrast the western edge of the province has uneven terrain and is largely covered with trees, and, as the outer edge of the Rocky Mountains whose lofty peaks catch the rain and snow from moisture-laden clouds moving inland from the Pacific Ocean, it usually obtains ample precipitation. The area best suited to arable farming lies between the two extremes. From the Lethbridge district west the country is still relatively flat and treeless and normally receives a little more rainfall than the more easterly regions. As Paul Voisey has demonstrated, this area was ideal for dryland grain production.[3] It quickly became apparent to farmers that wheat, which needs more sunshine and a longer frost-free season than say oats and barley, the predominant crops in the parklands to the north, would be their staple.

Pincher Creek is located on the western edge of the wheat belt at the point where it suddenly rises into the foothills of the Rockies. On the lower elevations immediately surrounding the town and to the east is found some of the best prairie farmland in the province. The large open tracts are level enough for widespread cultivation, the growing season and the hours of sunshine are usually sufficient for ripening wheat and drought is less common than in most Alberta dryland farming districts. The area beginning a few miles more or less straight north, south and west of Pincher Creek is generally referred to as the foothills. It is less amenable to cultivation. Besides being extremely uneven, it is crisscrossed with the streams and winding tributary rivers which carry the runoff from the higher mountains to the Old Man River before, very near Pincher Creek, it starts its journey inland. In this region the land outside the valleys can be used only for grazing as it is too rough for farm implements and the soils are too thin for anything but the hardiest of highland grasses which grow in the open spaces between the spruce, pine and poplar forests. While the valley bottoms and some of the gentler hillsides that rise out of them are relatively fertile, the process of working them with horse and plow was comparatively painful for the first farmers because the dark loam soils were heavy and the sod was very thick after centuries of growth and decay of tall native grasses. While rainfall in the foothills region is substantially greater than on the prairie, the hours of sunshine and the growing season tend to be consistently too short to ripen cash crops, including wheat.[4] Therefore, arable land is utilized much as it is in the mixed farming regions of central and north-central Alberta. It is limited largely to forage hay, which thrives on moisture, and oats or oats mixed with barley, which are often harvested early as feed grains or cut and stacked for roughage in years when the ripening process is interrupted by autumn frosts.

The first era of heavy population growth in the vicinity of Pincher Creek was the nearly two decades that began in the late nineties. As this period opened, Pincher Creek was a mere village of about 150 people clustered around two log churches, a one-room log school house and an assortment of crudely hewn stores and tiny businesses.[5] Along the streams and rivers of the foothills were several British ranches that had as yet managed to keep their large holdings more or less free from invasion by homesteaders. Settlement was sparse and the cattle and horses of the ranches could wander over great

Main Street, Pincher Creek, Alberta, circa 1912. (Courtesy of the Glenbow Archives/NA-1995-3.)

distances, unrestricted by man-made fences or buildings. By 1914, however, the district had become a fully developed agricultural hinterland with a farming community approximating its current size.[6] The change was reflected most clearly in the town of Pincher Creek whose growth responded to the demand for goods and services from a burgeoning rural society. The town's population had increased about tenfold and its downtown core, composed of two- and three-story buildings, included a department store, hotel, blacksmith shop, implement warehouse, furniture store, a weekly newspaper, a creamery, and the "largest merchandising business in Southern Alberta."[7]

There were two major waves of settlement in the district before the war. The first occurred in the decade following 1895 and the second began as the first subsided. Interestingly, the homesteaders who composed the initial wave were drawn mainly to the foothills country with all its evident agricultural limitations, rather than to the prairie. Like thousands of frontiersmen who migrated to the West in the late nineteenth century, they brought with them a great array of farm animals. Migrating farmers travelled with numerous animals for two reasons. One was cultural: many had been raised on mixed farms in eastern Canada, Europe or the United States and had been schooled to believe that they should occupy themselves in the feeding and caring for livestock as well as in plant husbandry.[8] The second factor was the conditions of the frontier itself. When they first established their homesteads where none had been before, they would be unable to purchase many essentials of life because grocery stores, dairies, butcher shops and bakeries would be inaccessible. The only way to ensure that they had the necessary ingredients for existence and a balanced diet, therefore, was to take the fundamentals with them. That meant that along with seeds for their first crops and beasts of burden and implements to plow and harvest, they required seeds for gardens to grow their own vegetables and corn, and animals upon which they could depend for such basics as pork, beef, poultry, milk, butter and eggs.

In the Pincher Creek area the first farmers had limited freedom in choosing homesteads because they brought the nucleus of future livestock herds and flocks with them. They had to take up holdings in areas where livestock's basic requirements could be readily met. This pretty well precluded tracts that were exclusively prairie because of several critical deficiencies. The most glaring was probably natural shelter. Cattle, horses, pigs, and chickens need at least some protection from the elements. But the pioneer farmers' priorities, apart from the regular farming routine, had to be building houses for themselves, clearing and breaking future arable acres, and erecting fences to keep the animals out of gardens, haystacks and the first crops. Initially, most frontiersmen had no time to construct even the crudest of shelters for livestock, so it was essential that there be forests, hills and valleys to protect them from the intense heat of the summer sun as well as the biting cold winds of winter.

It was also crucial in their minds that their animals find an abundance of natural sources of drinking water in the foothills in a plethora of streams, rivers, sloughs and lakes.[9] Moreover, they must have been attracted to the natural supplies of winter roughage. The prairie was endowed with a short hardy grass which was adequate in the summer months,[10] but was often of insufficient length or volume to be cut, raked and stacked for hay. In the foothills on the other hand, the tall grasses growing in the valleys were ideal.[11] Thus, in an era when the demands on their labours were great, they could not only harvest essential crops of hay but they could also keep the amounts required to the minimum. Eventually, of course, disastrously cold and precipitous winters demonstrated the dangers of heavy reliance on winter pastures. However, in the period when they were getting established few settlers abstained from the practice.[12]

When a new agricultural region opens, farming for profit is a dubious undertaking. Major markets are not available because domestic populations are sparse, and transportation systems and facilities for handling and storing farm goods for shipment to distant markets are usually either nonexistent or grossly inadequate.[13] In the late nineteenth and very early twentieth centuries, furthermore, the process of breaking land with horses or oxen and single-bottom plow was both strenuous and excruciatingly slow. In the best conditions the average pioneer could bring no more than about twenty acres into production each year.[14] So the central philosophy when mixed farming areas were first developed was to concentrate on securing the feed supply of the livestock. As more land was broken, domestic grasses such as timothy and brome were planted for both pasture and hay, and oats almost always became the first grain crop because of its value both as roughage and as a feed grain.[15] The farmers were determined to produce these crops in sufficient quantities to ensure the survival of their livestock, and therefore of themselves and their families.

In the first few years after the turn of the century, however, the farmers in the foothills began marketing their surplus produce for two reasons. First, they had become productive enough to create them—both directly, from increased acreages of cultivated land, and indirectly, from the growing number of healthy animals that this land could support; and secondly, because merchants had come to town to supply all the facilities of trade.[16] At first the local community itself provided a market as Pincher Creek's second frontier to the east filled with grain farmers who needed to purchase milk, eggs, poultry and beef. Soon, however, demand also came from the lumber frontier of the Kootenays and the mining frontier of the Crowsnest Pass. In 1904, the local newspaper boasted that

> The farmers will have every facility for getting rid of their crops this year. The market to the west of us in the Kootnay [*sic*] is increasing rapidly year by year, while the new coal towns springing up in Alberta, and which can be supplied by us on terms that no other districts can compete with, will require a large amount of produce and supplies. Not only hay and grain, but vegetables, eggs, [meat] and butter. Our merchants now ship all the eggs they can procure in the country and all the butter also, over the local demand, to the Crow's Nest Pass towns.[17]

In the decade and a half after 1900, growing demand from the coal and lumber camps influenced the mixed farmers around Pincher Creek to take a far more intensive approach. This was reflected in their efforts to adapt to the natural conditions around them. We look now at the major environmental pressures that affected farming in the western foothills.

Significantly, in the earliest part of the settlement era the frontier encouraged crudeness in mixed farming. Generally, the first agriculturalists in a new land fall well short of the highest standards and the most sophisticated farming methods of their time. While they are clearing and breaking land and constructing shelters for themselves homesteaders cannot give much attention to the refinement of technique. Near Pincher Creek this situation was particularly hard on the animals. Cattle, horses, sheep and hogs had to be left a great deal to fend for themselves. While they were fenced out of gardens and the first few acres of hay and then grain crops, they were rarely confined to any particular spaces. They roamed more or less freely, mixed with the animals of neighbouring homesteaders, bred at will and gave birth in the wilds with a minimum of human intervention. Because of this, there was a very high death loss. Diseases including black leg, pink eye and the mange could not be treated, predators including wolves and coyotes could not be kept away, and, ultimately, despite all the natural buffers, the effects of blizzards, extreme temperatures and deep snow in some winters could not be ameliorated sufficiently to keep a high enough percentage of the animals alive.[18] As the animals were harmed by exposure to the elements, so quite obviously was the quantity of foodstuffs derived from them. For the farmers to make the best of emerging market demand, therefore, change was necessary. This began once they were well established. Gradually they brought enough land under cultivation to ensure sufficient feed first to supply their own animals, then for themselves and those working in the lumber and coal camps. Secondly, they began to construct networks of barbed wire fences, corrals, proper barns and other buildings so that their herds and flocks could be accessed, fed, sheltered and protected year round.

Of all branches of mixed farming, that which responded most dramatically to the challenge of frontier circumstances and subjection to the elements was probably beef. Before the foothills had been densely populated, many farmers had not only allowed their herds to roam but had also increased their numbers rather unrealistically, presumably because of the seemingly unlimited extent of the open range. Early on some realized this error and worked towards change.[19] It was not until early in the twentieth century, however, that most settlers were able to construct the facilities needed to exercise the requisite control. Then much of the foothills, and the formerly closed leases of the great ranches, were divided up into a multiplicity of small holdings, each fenced off and separated from the rest. This is illustrated in the following discussion of the problems confronting cattlemen attempting in 1904 to initiate a united approach to a well-known infectious bovine hide disease[20]:

> In a range country it is possible for ranchers to take prompt
> and effectual methods for prevention and cure [of the mange].
> Where the range is not intersected with fences it is easy to
> handle large herds of cattle, run them through dips, and hold
> them until every head in that particular district has been dipped.

But in ... Pincher Creek country ... the building of public dips ... would be almost useless. For one thing there ... [is] not a large enough vacant space of land, unappropriated, which could be used for herding any large bunch of cattle; and for another thing, the cattle ... [are] now held by so many people in small bunches, that it would be almost impossible to see to the treatment of all infected animals.[21]

By this stage the range had "vanished and cattle herds" had been "cut down to what the owned land could support."[22] The movement to improve was a characteristic of operations large and small. One of the original great ranches, the Walrond, provided an example for the rest to follow as it relied increasingly on deeded land, moved away from pure ranching and became essentially a huge mixed or "stock" farm. In a typical year after the turn of the century its hands kept close enough control of the herds to brand "nearly 2300 calves" and to ship "nearly 1500 cattle" to market. They also threshed some "2000 bushels of oats" and enough roughage to winter all their calves and bulls even though they lost "375 tons of hay by fire."[23]

The farmers who raised beef cattle began not only to care for their animals more tenaciously, they also took steps to raise the quality of their herds. Initially, most of the cattle had originated south of the border where indiscriminate breeding on the open range allowed characteristics of some of the poorest animals to spread widely.[24] "It is a great pity that so much inferior stock has been brought in from the United States during the past season by settlers taking up their residence here" the Department of the Interior recounted in 1895.[25] In the earliest period the Canadian range was as open and uncontrolled as the American and very little upgrading occurred.[26] However, efforts by at least a small core of individual farmers to raise the quality of the stock soon became apparent. From 1898 a Cattle Breeders Association operated in the North-West Territories and after 1905 in Alberta. Its members imported purebred cattle from Great Britain, eastern Canada and elsewhere.[27] In the *Pincher Creek Echo* purebred Hereford, Angus and Shorthorn herd sires and brood cows were advertised for sale regularly from 1903. While it would be unrealistic to conclude that the farmers were able to establish herds of the finest quality before the war, it is apparent that significant steps were taken to ameliorate some of the worst manifestations of the open range. By 1914 the cattle in the pastures of the foothills exuded more characteristics of the fleshier, higher yielding British varieties than they had earlier and fewer of those of the taller, thinner, far less meaty Texas Longhorn and other mixed types that had infiltrated the herds.[28]

One development which demonstrates the concern for improvement in beef production is the decision by a number of mixed farmers in the Pincher Creek area to finish their cattle on dry feed after the grazing season. Prior to the early twentieth century, participation by any but the largest cattlemen in the fat beef trade as a whole had been made most difficult by frontier conditions which dictated that a vast number of small herds were scattered over huge areas and largely uncontrolled and inaccessible for much of the time[29]:

> Under the present system buyers travel laboriously through the country picking up a few cattle here and a few there. Every little band or every head is bargained for separately, and, as the ultimate sale of these animals is by weight, it may be concluded that in sheer self-defence the buyer is compelled to purchase on the safe side.[30]

When supplying markets either locally or in the mining towns or lumber camps, such conditions were tolerable because of lack of competition. Breaking into the huge European markets, necessary for the industry to reach its potential, was another matter altogether. It required acceptance from much more discriminating palates and ideally the shipment of carload lots of some eighteen to twenty at a time in order to reduce the transportation cost per head.[31] It also demanded that the animals be covered with an extra layer of hard fat so that after the long trek to Britain there should still be significant amounts of good meat on their bones.[32] The only way the cattlemen could get into this trade therefore was to separate from their herds all the animals that were close to being ready for slaughter after they came off the pastures in the autumn, and push enough good feed to them over a 90- to 120-day period not just to keep them healthy but to fatten them properly. In the years that immediately preceded the war a number of the farmers in the foothills began to do this. In reality the transition was not incredibly difficult once they realized the need to stop relying on winter pastures and to keep all their cattle close to feed and shelter during the time of the year when the elements could be the least forgiving. It was rather a natural next step entailing the expansion of existing facilities including corrals and both feeding and watering systems and daily attention to the animals throughout the winter. At first the usual practice was to feed roughage (hay and green feed) alone. Eventually, some farmers realized that incorporating substantial amounts of grain into their rations went further in producing the well-marbled meat the Europeans demanded.[33] Interest in grain finishing in 1913 prompted the government farm at Claresholm (seventy kilometers north of Pincher Creek) to begin experimenting to develop the best possible system to fatten cattle while sheltering them from the cold winds of winter.[34]

World War I was a temporary impediment to a new approach in Pincher Creek and elsewhere because it brought such a great demand for all grains. In 1913 the price had already begun to soar and even the government experimenters concluded that farmers would make more money by selling marketable inventories rather than feeding them, though the price of beef at 7.75¢ per pound dressed was unusually high.[35] However, finishing remained an alternative for using up oats, barley and frost-damaged (and therefore price-reduced) wheat.

The laying of the foundations of the grain-fed beef trade in Pincher Creek prior to 1914 is evidence of development in the cattle business. The practice of properly concluding the fattening process with rations that consisted of a high percentage of oats and barley grew slowly and became general after the war. Then western beef packing led by the Burns and Canada Packers plants in Calgary and facilitated by refrigerated railway cars carrying dressed instead of live meat to the East and overseas would further the process of refinement in the industry as a whole.[36]

Measured by livestock numbers, beef appears to have been the largest industry that the cattlemen in the foothills districts of Pincher Creek engaged in prior to 1914. It was not the only one. Many farmers also took up dairying. These must have lived within a few miles of town since milk is a highly perishable commodity and had to be delivered without delay by horse and wagon over relatively poor country roads or trails. In many cases the farmers were responding to the need to acquire capital to keep them going during the long periods between the seasonal beef cattle sales. By 1908 there was a private creamery operating in the town and there were enough dairy farmers for it to advertise regular daily pickup routes. These ran through the good cattle country "from Mountain Mill ... from Fishburn gathering cream from Spring Ridge and Dry Fork and ... from Twin Butte via Utopia and Dutch Flat."[37] By this time the cream separator had made an appearance,[38] enabling the farmers to sell their cream for the manufacture of butter and cheese while keeping the skim milk for their own consumption and as a feed for their pigs.[39] While this brought some important streamlining to the business at a formative time, dairy production was not more efficiently conducted during any part of the study period than beef. Here too, the most fundamental flaw was the low quality of the livestock. The great majority of the cows were the progeny of the range either in Canada or the United States (or both). They were typical frontier crossbreeds often euphemistically referred to as "dual purpose" but in reality a somewhat less than totally satisfactory compromise of beef and dairy qualities. In 1908 the Alberta Department of Agriculture reported: "In order to attain the highest success in dairying it would be neces-sary to introduce dairy cattle ... if we want a real dairy herd we must replace the beef bulls with dairy ones."[40]

Most farmers actually milked hardy Shorthorn crosses which produced milk, but not as efficiently as Holsteins or Guernseys, and also beef, but not with the high carcass yields obtainable from Hereford or Angus varieties. Prior to the war few farmers had the buildings to keep and feed their cows indoors during the winter. Therefore, they did not milk year round. When the grazing season ended, they simply ceased production until spring and attempted to maintain their cows out-of-doors as inexpensively as possible. At a convention for butter and cheese makers in 1908, dairy farmers were urged to shelter and feed their animals properly over the winter so that they could milk them "nine or ten months, instead of five, as was common now."[41] A report the next year attributed a 40 percent drop in cream production in the spring and summer of 1907 to the previous winter when cattle had been exposed to horrendous climatic conditions.[42]

If dairying standards and practices on the frontier were crude, it is also evident that the movement toward improvement began in this period. In 1914 the Department of Agriculture reported what it considered crucial steps forward in the quality of the livestock. From that report it is also unmistakable that much remained to be done:

> During the year there has been a phenomenal demand for milch cows and in the neighborhood of 10,000 were brought into the province. Many of them were fairly good cows, but there was also a proportion of genuine culls amongst them. All were sold at high prices—prices which were only justified by the high returns obtainable in the cities for milk and cream.[43]

Most dairy farmers appear to have continued to produce only seasonally and to feed their cows in the winter just enough for survival. The majority ostensibly were milking indoors by the end of the study period but few if any had built special parlours for that purpose, using instead regular stock barns which their cows shared with horses, pigs and other animals.[44] In that circumstance cows usually feel some stress and fail to lactate to their maximum. In future the upgrading of stock and the building of special parlours where cattle could be milked and fed year round would enable farmers to become much more efficient. Then many would withdraw from other forms of animal husbandry to double and triple the size of their dairy herds. As that happened the dairy industry around Pincher Creek would begin to approximate its modern form.

Youth butchering a pig on the A. E. Cox ranch, Pincher Creek, Alberta, circa 1906-08. (Courtesy of the Glenbow Archives/NA-2001-15.)

Another traditional enterprise which many Pincher Creek settlers undertook in these years was pork. While most mixed farmers tended to run a few hogs from the earliest years, they were restrained from getting into the business seriously by specific frontier conditions,[45] particularly the inadequacy of shelter. Hogs lack a heavy coat of hair like cattle or horses to protect them from the elements and therefore were particularly vulnerable in the period before the settlers were able to construct sties. "The feeding of hogs around a straw stack during the winter is a business to which is attached the minimum of profit" it was lamented in 1899.[46] Hogs were often allowed to roam with the cattle and it was well into the twentieth century before most could even give birth indoors. Their first man-made housing, moreover, almost always consisted of a corner of the regular stock barn where artificial heating was unavailable and where the diseases which passed easily from one species to the other were a constant threat.[47] Farmers were prepared to take the chance on a few head primarily for their own consumption, but they were unwilling to risk the enormous death losses which might have ensued had they expanded their operations.

Apart from the poor living conditions for the animals the frontier hog industry around Pincher Creek was also hampered by a dearth of facilities for marketing. In 1903 a private packing plant was started in the town but it remained small and local and had no markets over and above the Crowsnest Pass and Kootenay regions upon which to build a dynamic trade.[48] At that time the province as a whole was still forced to import 75 percent of its cured pork.[49] Our "most energetic farmers have been compelled to go out of the hog industry," a special commission reported in 1908, because during a portion of most years there is "no market at all" and producers are "not receiving a living price."[50]

The hog industry, like the others associated with animal husbandry around Pincher Creek, also suffered during this fledgling stage from severe deficiencies in the livestock itself. Here the difficulty was probably not indiscriminate mixing on the frontier as pigs, even when left on their own, lack the mobility to roam over great distances and therefore to mate with every variety and type of their species imaginable. Initially, inbreeding must have been a factor in an isolated country where good outside stock was difficult to procure. However, the main dilemma was that most of the hogs first brought into the area including the Duroc, Jersey, Poland China, and Chester White were simply the wrong breeds. They were lard hogs that had thrived in the United States on unlimited supplies of corn and had been used primarily as a means of producing lard for populations without ready access to vegetable oils. In the Canadian West there were no supplies of corn and the market both locally and in the East and overseas, demanded a pork or "bacon hog longer in the side."[51]

The movement to more suitable breeds was gradual but did eventually bring results. The Department of Agriculture and the Canadian Pacific Railway sponsored importation from the East and Great Britain of purebreds including Berkshires, Tamsworths and Yorkshires which were (and are) desired for their red meat.[52] In 1903 the Territorial Swine Breeders Association emerged and through its short life span helped its members to do the same.[53] By 1908 the Pincher Creek Agricultural Exhibition and Race Meet was encouraging higher standards by awarding prizes to owners of purebred bacon sows both over and under seven months.[54] What really brought positive results, however, was the eventual growth in the export market just prior to the war when the

United States duty on Canadian pork was removed and access granted to Chicago and Seattle. This not only prompted a greater sense of urgency among western farmers about attaining better bloodlines but did much to inspire a general upgrading of housing and handling facilities on the farm and of purchasing, packing and transportation systems outside. In 1913 a government representative noted that

> The number of hogs raised in the province has increased by leaps and bounds and the high prices that have been obtained for some years has made this line of work a most profitable one. ... The result was a large export of hogs at remunerative prices. Large shipments were also made to Eastern Canadian markets particularly Toronto and Montreal. It is very gratifying to note that the quality of these hogs attracted universal attention, demonstrating to the amazement of many the possibilities of Alberta stock and feeding stuffs. It certainly has caused the Ontario feeder to sit up and take notice to see Alberta hogs invading his special preserve.[55]

In Pincher Creek the numbers of animals per farm at this point must have risen well beyond the demands of self-sufficiency. The average herd was likely increased to around fifteen head and fed and farrowed indoors for a considerable period on rations of oats, frozen barley and wheat, skim milk and the surplus leaves, stocks and roots of garden vegetables. By World War I the hog industry around the town had not produced any of the modern sanitized hog "factories" one sees in the province today, but it had become a permanent part of the mixed farming culture, and it had raised its standards considerably beyond those which had marked it as a frontier venture.

The one other major enterprise taken up by mixed farmers in the Pincher Creek districts was horses. From the beginning horses were suited to frontier agriculture because they can look after themselves under the most trying climatic conditions. In winter they grow a very thick coat of hair and can withstand the coldest temperatures. Moreover, they are adept at rummaging through deep coverings of snow to get at the grasses below. So winter losses among rangeland horses were lower than in the cattle herds. There was considerable demand for stock from Great Britain and places like South Africa where British troops needed mounts to control the Empire.[56] One interesting aspect of the horse business, however, is that the major demand came from the agricultural frontier itself, as settlers pouring into the West required beasts of burden for transportation and for sowing, reaping and harvesting their crops.[57]

Haying with horses in the Pincher Creek area (likely on J. de Lotbiniere-Harwood's ranch, 8 miles north of the community), circa 1900-03. (Courtesy of the Glenbow Archives/NA-2382-8.)

This business also suffered from low standards when supply depended on stock which accompanied the homesteaders and before fences, corrals and stables made proper breed selection possible. The quality of the first Pincher Creek animals was reflected in the comments of an early pioneer who recalled that the "horse rancher was not very particular as to what kind of horse he had so long as they had four legs and the usual complement of eyes, ears and other organs."[58] In 1906 stallions were registered by owners across the province, according to the Horse Breeders' Ordnance, to be made available at a fee for breeding privately owned mares. Of the 491 that were registered only 162, including eighty-one Clydes, thirty-five Percherons, ten Shires, ten Standardbreds and ten Thoroughbreds, were pure.[59]

Here too, change for the better came before World War I. Dealers began to import certified purebreds from the East and overseas.[60] Pincher Creek farmers bought up this stock and, as they took control of the open range, they started producing a better commodity.[61] Progress was handicapped by the settlers' relative shortage of cash and the impossibility of wiping out very quickly the aftereffects of the frontier. The numbers marketed from the Pincher Creek area peaked around 1906 when sales involving 683 were recorded. But by 1913 the number had dropped to 444.[62] The explanation points to several factors. Just before the war, the construction industry which had utilized a great deal of horse power, fell victim to recession as railway building and

The first steam outfit being brought into Pincher Creek, circa 1899-1900. (Courtesy of the Glenbow Archives/NA-1853-1.)

municipal construction declined. Furthermore the steam tractor was used more and more for cartage work in the cities and towns and for trade within the province.[63] Even more importantly, the period of expansion in the foothills had ended as the best lands had been settled and farmers had begun increasingly to pour into the wheat regions on the dryland prairie. As will be seen, the natural environment as a whole in Pincher Creek's second frontier lent itself to exploitation less by horse and plow and more by tractors and a host of new implements for sowing and harvesting.

The final branch of mixed farming near Pincher Creek large enough to merit mention was poultry. Its process of adaptation to frontier conditions was similar to the hog industry. The earliest farmers needed chickens for eggs and meat and many brought a few to their new home. However, initially the birds did poorly because of the elements and because, largely unattended, they fell prey to predators including both wolves and coyotes. Also, insufficient mixing of bloodlines was troublesome in the days when outside stock was hard to attain. When a provincial breeding plant was built in Edmonton in 1908 "several hatches were taken of eggs being secured in the province." A large percentage of infertile eggs was forthcoming. This condition applied "to practically all the eggs received from various points." The farmers were advised that they must take "greater care ... in selecting only strong, healthy well-developed birds from which to breed" and warned against "inbreeding."[64]

Elsie Honeyman's son feeding chickens on the family farm near Pincher Creek, date unknown. (Courtesy of the Glenbow Archives/NA-2045-8.)

In 1906 the Alberta government, noting the "suitability of this Province for raising poultry," spoke of the "great need ... for some effort on the part of the farmers ... to supply the demand for poultry and eggs which at present has to be met from points outside."[65]

Improvement inevitably came once farmers were in a position to deal with some of the most difficult environmental obstacles. It was relatively simple for them to acquire first-rate stock once the provincial breeding station appeared and to protect it from both the climate and predators, as a hundred or more chickens can be housed quite comfortably in coops no larger than nine feet by twelve feet. Demand in the town of Pincher Creek and in the Crowsnest Pass and Kootenay regions eventually stimulated expansion which drew comparisons with that in the pork sector.[66] By 1914 most mixed farmers kept chickens, protected them from predators, fed them indoors during the seasons of severe weather and marketed what produce they did not need for their own sustenance. There is no evidence that any farmers close to Pincher Creek specialized in poultry alone or developed the large-scale chicken farms one sees in the province today.

The picture which emerges of the typical mixed farmer to the west, north and south of the town of Pincher Creek by 1914, is that of a small producer working first with his 160-acre homestead, then perhaps as he is joined by a wife, expanding to twice that amount of land.[67] On this operation he usually ran a hundred or so chickens, a basic small herd of perhaps forty beef cattle, four to six milk cows, some fifteen hogs and, while demand was still strong, twenty or so grade mares which were bred for a fee to the stallions of farmers who specialized in keeping sires.[68] In most cases the quality of the livestock had improved as sufficient capital had been gathered to purchase better breeding stock and as the requisite facilities had been built and expanded to control it properly and protect it from weather, wild beasts and disease. What this amounted to was intensive farming on a scale which paled in comparison to the great ranches which had mushroomed into existence in the 1880s and early 1890s, but which enabled the farmer to more closely supervise his fields and stock than had been possible in the past. This change was dictated in part by government policy which ultimately had come down strongly in support of the 160-acre homestead for the masses instead of the one hundred thousand acre closed lease to the select few.[69] However, it was also decided by climatic and other environmental conditions which demanded the same detailed refinement of practices that had taken place over centuries in the East and Europe. In 1914 the process of improvement was far from complete. It had, however, begun and farmers had plainly shown "that it is possible to produce larger yields from smaller areas by a little more care and attention."[70]

As the mixed farmers in the foothills to the west, north and south of Pincher Creek were beginning to take the first steps away from the raw and unrefined approach, other settlers were establishing farms on the flatter, drier, open prairie regions closely surrounding the town and to the east. The populating of the prairie, which began in earnest about 1905, proceeded more quickly and in some respects with more dramatic effects than had that of the foothills. Those who witnessed it were most impressed[71]:

> In ... [the early] days the so-called "mossback" farmer had not yet crowded out the ranchers [one pioneer later wrote]; not much garden stuff was grown; there was a little fencing for pasture and that was all. Cattle and horses roamed at will until the annual or semi-annual round-up. If anyone had then predicted that in a few short years the world-famous "Alberta Red" would be growing in the District, he would have been laughed to scorn.[72]

The wheat farmers could not be self-sufficient even when they first took up homesteads because they did not bring and could not, in the prairie environment, support a full compliment of farm animals.[73] To survive they had to seed and harvest their crops, sell them for cash and then purchase the basics from local merchants. The commercial mixed farming sector made this feasible not only because it produced many provisions that the grain farmers needed, but also because it had spurred trade and economic development for the entire area. By the time the wheat farmers arrived the community of Pincher Creek had in place elevators, grain, implement and other merchants, and railways that linked the region to national and global markets.

Because they had to purchase so many essentials of life, the prairie farmers felt greater urgency about getting into full production than had their mixed farming predecessors.[74] Many turned to mechanization in an attempt to turn all their grasslands immediately into productive wheat fields. Rather than relying on a team of horses and the traditional single-bottom plow, some hired custom operators with teams of six or eight horses and eight- or even ten-bottom plows. Others brought in steam tractors believing they could both pull larger implements and work longer hours without respite.[75] Some drylanders including W. J. Chism and W. R. Dobbin, both large wheat farmers, were using steam for both breaking and harvesting by 1904.[76] The prairie landform was ideal for this invariably unwieldy machinery. The great open spaces provided large fields for it to work and make its most cumbersome turns and the relative scarcity of trees, rivers and valleys allowed it readily to move from one field to another. "The steam plough has now passed beyond its experimental stage in Western Canada," the *Farm and Ranch Review* could hypothesize in 1906, "and is successfully used in many parts, particularly the Claresholm district."[77] Tractors on the wheat frontier facilitated mechanization in all facets of plant husbandry. Soon steam was used to power harvesters,[78] and binders,[79] and for all forms of cultivation and seeding.[80] In a surprisingly short time some steam engines were in turn replaced by the faster, less ungainly and more conveniently operated gasoline variety.[81]

Working a field, likely on the Pincher Creek area ranch of Dolphous Cyr, using a steam engine and an 8-bottom plough, circa 1900-03. Note that horse power has not been entirely replaced by machinery, as there is a two-horse team at left. (Courtesy of the Glenbow Archives/NA-2382-9.)

Mechanization, then, to which the relatively smooth and treeless prairie lent itself, allowed many farmers to get into production quickly and thus to attain vital cash incomes for survival in a geographic area inimical to both subsistence mixed farming and self-sufficiency. The relative importance of the environment in the establishment of cash wheat farming is demonstrated, as Paul Voisey has suggested, in that it diametrically clashed with conventional wisdom. From the beginning the farmers were subjected to the admonitions of mixed farming advocates including "representatives from every level of government, from corporations with interests in the west, and from agricultural experts and observers of every description," who preached the benefits of a sophisticated rotation of crops, the use of manures on the soil, and diversification into a wide range of products including eggs, milk and meat.[82] In Pincher Creek, as elsewhere on the prairie, wheat farming took hold primarily because the force of nature prevailed over the dictates of man.

This is not to contend that in conforming to their surroundings, the wheat farmers necessarily rose to the most refined techniques of more developed agricultural societies much more rapidly than their counterparts in the foothills. Indeed, during the earliest days their standards in some respects were equally crude. This is especially apparent in their attitude to their most important natural resource—the earth itself. The vast majority of these farmers appear to have been almost totally indifferent to soil conservation over the long term. Arguably, in the first years this reflected their impatience to extract every possible cent from their soil to attain the requisite cash income. It also, however, reflected a frontier mentality about land in general. A major reason why settlers come to a frontier is that by contrast with their previous home, land is abundant and cheap. This induces them to visualize it almost as a limitless commodity even before they arrive. Moreover, as they settle, they see a small and sparse population base grappling with the problems of isolation where roads, transportation facilities and communication are either lacking or in short supply. Therefore, the land looms as a great and challenging adversary. Far from fearing that it will ever run out or become exhausted, the homesteader starts to see it as a barrier to be broken down or a wild uncontrolled thing needing domestication. This naturally induces farmers to be careless with it. In the short term, the more a farmer cultivates, the more productive the land becomes. The soils of the southern prairies tend to be light and shallow and easily turned. Consequently, many of the first wheat farmers were inclined to work them relentlessly, indeed, to "mine"[83] them. Culver T. Campbell,[84] son of the great prophet of the gospel of the dryland farming method, which was really only a pseudoscientific rationalization of soil mining, gave a public lecture in the town in 1908.[85] In the misguided belief that it would bring moisture up from the depth of the earth below through capillary action, he advised deep plowing in the fall and frequent tilling thereafter, the burning of stubble and other methods to prevent the build up of organic matter, and the pulverization of the top soil.[86] This was exactly wrong for a country where getting enough precipitation could be a problem. It exposed the top soil to the air, caused it to dry out, turned it into a dust mulch, lowered organic content, reduced the levels of much needed nutrients, and promoted wind erosion. The farmers unflinchingly embraced the Campbellite system, however, because it seemed to them the fastest way to turn virgin prairie holdings for which there was no foreseeable end into the greatest possible cash rewards. When reflecting local opinion, the *Echo* consistently called for "thorough cultivation" of the surface "by disc and harrow"[87] and suggested that expansion of the wheat fields could go on forever.[88] "It will not be long before the whole Pincher Creek flat and the valley down to Fishburn will be one great wheat field," the paper announced in 1904, "and the country [will] experience a period of prosperity undreamt of even by cattlemen."[89]

If the wheat farmers in this frontier setting tended to overwork their land, they were also inclined to summer fallow only minimally, choosing instead to grow successive crops year after year.[90] This not only helped to wear down the soil but also led to weed infestations[91] and to general outbreaks of the disease known as smut.[92] In the decade following the beginning of the war the results of the prairie farmers' excesses would be little less than disastrous. Crops were lost, soils dried out and, in many areas, wind erosion took a heavy toll. Around Pincher Creek the problem was less severe than further east.[93] Even so, average crops in some years dropped to between six and a half and nine bushels per acre.[94] The farmers took many years to mend their ways. Not until the "Dirty Thirties" did the wheat producers on the prairie south of Calgary adopt the conservationist methods of strip farming, summer fallowing and less frequent and shallower cultivation which, as James Gray has demonstrated, has become accepted procedure in Alberta's dryland regions in the modern era.[95]

Nevertheless, Pincher Creek's first prairie farmers made some noteworthy advances in specific areas. The movement to mechanization described above continued, and by the end of the study period, automation had been introduced to every facet of wheat husbandry from soil cultivation to planting, sowing and harvesting. The most important developments, however, were probably those associated with establishing new strains of wheat to accommodate the climate. Next to occasional shortages of precipitation in the area, the most frequent climatic problem was probably frost. While the total seasonal hours of sunshine were normally sufficient for wheat, crops were frequently damaged in the early autumn when the thermometer dropped below the freezing mark—particularly on the western edge of the wheat belt where the land rises to the higher elevations. Almost from the time the first settler arrived on the prairie a search began for a variety of wheat that could withstand colder temperatures and/or ripen earlier.[96] In the earlier period Kansas Red, later known as Alberta Red, was imported and used on many farms near Pincher Creek[97] and the famous Marquis variety developed for the Canadian West by William Saunders and his son Sir Charles gained widespread acceptance in the 1920s.[98] The regions close to the foothills south of Calgary also came to rely heavily on winter wheat which was sown in the fall so that it could get a very early start (and thus beat the frosts) the next season. It was this which some would argue made the early development of the wheat economy possible.[99] In 1905, 141,294 bushels were reportedly grown on 4,938 acres near Pincher Creek in comparison to only 4,714 bushels of spring wheat on 222 acres. By 1914 the figures for winter wheat had risen to 173,996 bushels on 10,991 acres and for spring wheat to 59,080 bushels on 5,578 acres.[100]

Today among all the prairie districts in southern Alberta those closest to the foothills tend to rely the most heavily on winter wheat. There are three major climatic factors. One appears to be the deeper coverings of snow that settle on the land at the higher elevations and protect the plants from extreme temperatures at the coldest time of year. Another is that this area in the winter, while subject to chinooks, normally lacks long spells of temperatures above the freezing mark. Further east such spells often encourage the wheat to begin to grow and then to be devastated when the weather turns cold again. A third factor is heavier springtime precipitation which enables the tender sprouts to spring very quickly to life.

In discussing farming in general near Pincher Creek it has been necessary in this article to oversimplify the picture somewhat. Thus, mixed farming in the foothills and wheat farming on the Prairies have been treated as if they were mutually exclusive. Such often was not the case. Some farmers, particularly after they were able to expand beyond their original 160-acre homestead, owned land that lay on foothills terrain in one place and dryland prairie in another. Such farmers inevitably raised animals and grew forage and feed as well as wheat. They were molded by two frontiers and were forced to adjust to two quite distinct types of environmental pressures. But, like the mixed farmers, they were involved in both animal and plant husbandry and once they understood the varied nature of their holdings, they could utilize their skills and techniques, indeed, much of their machinery and equipment, on the plains as well as in the valleys and gentler slopes of the high country. After World War I, wheat and mixed farming would evolve at about the same pace.

By then mechanization was underway, especially on the Prairies, but it still had a long way to go. The number of steam engines on the fields was to grow in the postwar period and, along with other implements, the gasoline tractor was slowly to gain acceptance over several decades.[101] In the foothills the movement to higher quality livestock and better systems and facilities would continue in every facet from dairying, to beef, hogs and poultry. It does seem evident, however, that the entire process of development had taken a firm hold in the era preceding 1914. It was then that mixed farming and its various branches, and cash wheat farming, had stepped beyond the frontier stage in some important respects and had managed to make major adjustments to accommodate the natural environment. The basic patterns had been established and the need thereafter was to bring further refinement to existing practices but not substantially to modify the infrastructure.

NOTES

This article first appeared in *Prairie Forum* 19, no. 1 (1994): 31-50.

1   For which see Elofson, "Adapting to the Frontier Environment: The Ranching Industry in Western Canada, 1881-1914," in D. H. Akenson, ed., *Canadian Papers in Rural History*, vol. 8 (Gananoque: Langdale Press, 1992), 307-27.

2   For contemporary comment see *Pincher Creek Echo*, 12 August 1909, p. 8.

3   Paul Voisey, *Vulcan: The Making of a Prairie Community* (Toronto: University of Toronto Press, 1988), 77-127.

4   The Alberta Department of Agriculture, *Annual Report*, 1913 indicates Pincher Creek consistently received more precipitation than the southern districts to the east. It averaged 15.50 inches on the years for which complete figures are given while Lethbridge and Medicine Hat got 14.43 and 10.27 inches respectively (p. 33). See also D. C. Jones, *"We'll All Be Buried Down Here": The Prairie Dryland Disaster 1917-1926* (Calgary: Alberta Records Publication Board, 1986), 121.

5   Canada, *Census of Canada*, 1921, vol. 1, p. 200, gives the population of the town as follows: 1891—0; 1901—355; 1911—1,027.

6   The 1921 census (ibid., vol. 5, pp. 268-70), records the "rural population" of census division 2—of which Pincher Creek was about 1/7th—at 22,112, and the occupied farm area at 2,189,244 acres. In 1991 the rural population of the municipal district of Pincher Creek was 3,108 (Statistics Canada, *Census Divisions and Census Subdivisions*, 1991 census, p. 101).

7   A. L. Freebairn, *Pincher Creek Old Timer's Souvenir Album* (Pincher Creek: Pincher Creek Old Timers' Association, 1958), 5. By 1941 the town population had actually dropped to 994. By 1951 it had risen again to 1,456 (Canada, *Census of Canada*, 1951, vol. 1, pp. 5-74). It is currently 3,660 (*Census Divisions*, p. 101).

8   For contemporary comment see, "Mixed Farming the Rule," Canada, *Sessional Papers*, 1895, vol. 28, no. 13, p. 150.

9   For the importance of water reserves see D. H. Breen, *The Canadian Prairie West and the Ranching Frontier, 1874-1924* (Toronto: University of Toronto Press, 1983), 78-81, 140-44.

10   Alberta, Department of Agriculture, *Annual Report*, 1906, p. 71.

11   This was not as important for horses as it was for cattle which do not possess the common sense to rummage through the snow for the feed below.

12   See, for instance, *Pincher Creek Echo*, 4 March 1909, p. 3.

13   As John McCallum has argued in chapter 6 of *Unequal Beginnings: Agriculture and Economic Development in Quebec and Ontario until 1870* (Toronto: University of Toronto Press, 1980), agricultural production tends to produce transportation systems not vice versa.

14   That is, when clearing as well as breaking had to be done. See P. A. Russell, "Upper Canada: A Poor Man's Country? Some Statistical Evidence," in *Canadian Papers in Rural History*, vol. 3, 129-47.

15   "Crops in Alberta," Canada, *Sessional Papers*, 1895, vol. 28, no. 13, part 2, p. 150.

16   See, for instance, *"Prairie Grass to Mountain Pass": History of the Pioneers of Pincher Creek and District* (Calgary: Pincher Creek Historical Society, 1974), 7—the recollections of Frank Austin and Myra (Austin) Harshman. Regular advertisements in the *Pincher Creek Echo* demonstrate that by 1904 grocery stores capable of servicing the entire community had been established, along with blacksmiths, clothiers and implement dealers.

17   *Pincher Creek Echo*, 14 June 1904, p. 2.

18   For which see Elofson, "Adapting to the Frontier Environment," 320-27.

19   North-West Territories (NWT), Department of Agriculture, *Annual Report*, 1889, p. 54.

20   For the mange see Elofson, "Adapting to the Frontier Environment," 321-22.

21   *Pincher Creek Echo*, 24 May 1904, p. 3.

22   *"Prairie Grass to Mountain Pass,"* 7—the recollections of Frank Austin and Myra (Austin) Harshman.

23   Ibid.

24   Elofson, "Adapting to the Frontier Environment," 309.

25   Canada, *Sessional Papers*, 1895, vol. 28, no. 13, part 1, p. 26.

26   Alberta, *Annual Report*, 1908, p. 153.

27   Ibid., 1914, p. 49.

28   See Elofson, "Adapting to the Frontier Environment," 309.

29   For participation by the great ranches in the finished beef trade see ibid., 320-23.

30   NWT, *Annual Report, 1902*, p. 59.

31   Glenbow Archives, M480, 16 September 1895.

32   Elofson, "Adapting to the Frontier Environment," 324.

33   Alberta, *Annual Report*, 1908, p. 151.

34   Ibid., p. 38.

35   Ibid., 1913, p. 136; prices prior to 1906 had tended to run between 2.5 and 5 cents, NWT, *Annual Report*, 1902, p. 54.

36   L. V. Kelly, *The Range Men* (High River, AB: Willow Creek Publishing, 1988), 229.

37   *Pincher Creek Echo*, 22 May 1908, p. 1.

38   By 1903 separators were being advertised regularly in the newspaper.

39 Some operations such as the Starlight Ranch, themselves, diversified into cheese production for the local market (*Pincher Creek Echo*, 30 September 1909, p. 5).

40 Alberta, *Annual Report*, 1908, p. 92.

41 Ibid., p. 90.

42 Ibid., p. 82.

43 Ibid., 1913, p.6.

44 Ibid., 1907, p.8.

45 *Pincher Creek Echo*, 6 November 1903, p. 2.

46 NWT, *Annual Report*, 1899, p. 69.

47 See V. H. Lawrence, "Pigs," *Alberta Historical Review* 24 (1976): 9-13.

48 *Pincher Creek Echo*, 6 November 1903, p. 1.

49 Ibid.

50 Alberta, *Annual Report*, 1908, p. 31.

51 NWT, *Annual Report*, 1901, pp. 92-93.

52 Ibid., 1900, pp. 68-72.

53 Ibid., 1903, p. 193.

54 *Pincher Creek Echo*, 8 October 1908, p. 1.

55 Alberta, *Annual Report*, 1913, p. 7.

56 *Pincher Creek Echo*, 13 October 1903, p. 4.

57 Ibid., 7 June 1904, p. 2.

58 Ibid.

59 Alberta, *Annual Report*, 1906, pp. 72-73.

60 *Pincher Creek Echo*, 29 July 1909, p. 4.

61 See, for instance, Glenbow Archives, M480, 8 June 1906.

62 Alberta, *Annual Report*, 1905, p. 70; 1906, p. 77; 1908, p. 7; 1913, p. 137.

63 Ibid., 1913, p. 136.

64 Ibid., 1908, p. 128.

65 Ibid.

66 Ibid., 1913, p. 7.

67 The average family farm in Alberta rose from 288.6 acres in 1901 to 352.5 acres in 1921 (Canada, *Census of Canada*, 1921, vol. 5, p. xv).

68 These figures are rough estimates based on livestock and farm numbers reported for the Macleod, Claresholm, Lethbridge, Raymond and Crowsnest Pass area in ibid., vol. 5, p. 725. I have put the numbers for Pincher Creek above the average for the area on the assumption that farmers in the dryer regions to the east would have pulled the average down.

69 See, Breen, *The Canadian Prairie West*, 3, 45.

70 Alberta, *Annual Report*, 1913, p. 7.

71 See, for instance, Freebairn, *Pincher Creek Old Timer's Souvenir Album*, 6.

72 Glenbow Archives, Pinkham MSS., M975.

73 Of course all of them had work horses and some almost certainly kept some other animals—a cow for milk, for instance, or a few chickens for eggs. Significant numbers, however, were out of the question; see P. Voisey, "A Mix-up over Mixed Farming: the Curious History of the Agricultural Diversification Movement in a Single Crop Area of Southern Alberta," in David C. Jones and Ian MacPherson, eds., *Building Beyond the Homestead* (Calgary: University of Calgary Press, 1985), 182-83.

74 See, for instance, *Pincher Creek Echo*, 31 May 1904, p. 2.
75 Paul Voisey makes the point that the controversy over the question of the relative viability of tractors and horses was not resolved until well after World War I (see *Vulcan*, 140-44). However tractors do seem to have shown superiority in both plowing and threshing almost from the beginning.
76 *Pincher Creek Echo*, 2 August 1904, p. 4.
77 *Farm and Ranch Review*, May 1906, p. 16.
78 See *Pincher Creek Echo*, 13 September 1904, p. 3.
79 See advertisement by J. J. Scott in ibid., 25 August 1905, p. 1.
80 Ibid., 28 July 1910, p. 8.
81 Ibid., p. 1; for use of the gasoline engine in harvesting see ibid., 10 November 1910, p. 5.
82 Voisey, *Vulcan*, 77.
83 See ibid., pp. 103-6.
84 The original Campbellite was Hardy W. Campbell.
85 *Pincher Creek Echo*, 26 June 1908, p. 3.
86 Ibid., 15 April 1909, p. 1; see also, E. B. Ingles "Some Aspects of Dry-Land Agriculture in the Canadian Prairies to 1925" M.A. thesis, University of Calgary, 1973), 7-48.
87 *Pincher Creek Echo*, 30 October 1905, p. 4; see also, 15 April 1909, p. 1.
88 Ibid., 31 May 1904, p. 2.
89 Ibid., 20 September 1904, p. 2.
90 The *Echo* also warned the farmers against growing "all wheat and always wheat" (20 January 1910, p. 6).
91 *Pincher Creek Echo*, 13 July 1906, p. 4; 13 May 1909, p. 5.
92 Ibid., 5 September 1905, p. 2.
93 For which, see, David C. Jones, *Empire of Dust: Settling and Abandoning the Dry Belt* (Edmonton: University of Alberta Press, 1987).
94 Alberta, *Annual Report*, 1919, 106-8; 1918, 139-41. For figures covering a wider area in these years see, Jones, *"We'll All Be Buried,"* 27-28.
95 J. H. Gray, *Men Against the Desert* (Saskatoon: Western Producer Prairie Books, 1967).
96 *"Prairie Grass to Mountain Pass,"* 185.
97 Ibid.
98 *The Canadian Encyclopedia* (Edmonton: Hurtig, 1985), vol. 3, p. 1647.
99 *Pincher Creek Echo*, 13 September 1904, p. 1.
100 Annual figures for the years from 1905-14 are provided in the Alberta Department of Agriculture annual reports.
101 See, for instance, R. E. Ankli, H. Dan Helsberg, and J. H. Thompson, "The Adoption of the Gasoline Tractor in Western Canada," in *Canadian Papers in Rural History*, vol. 2, 9-40; R. Bruce Shepard, "Tractors and Combines in the Second Stage of Agricultural Mechanization on the Canadian Plains," *Prairie Forum* 11, no. 2 (Fall 1986): 253-71.

8. Farming Technology and Crop Area on Early Prairie Farms

Tony Ward

INTRODUCTION

Early farms on the Canadian Prairies were small, with an average cropped area in 1880 for Manitoba of just twenty-five acres. Between then and the start of the wheat boom in 1900, the typical farm in the Brandon area grew to 143 cropped acres.[1] This article documents improvements in farming equipment, techniques and crop strains that led to much of this expansion.[2] The productivity of farm labour is estimated, and the implications of that increased productivity for farm size and costs are calculated, to show the significance of technological change in agricultural equipment.

The study begins with 1880, by which time Winnipeg was connected to the United States by rail, and construction of the Canadian Pacific Railway was imminent. It extends through to the height of the wheat boom in 1910, when settlement had transformed the Prairies.

LITERATURE

There is an extensive literature on the topic of Canadian prairie settlement which identifies many factors that contributed to the slow progress of agriculture before 1900. Norrie (1975) analyzed the critical problem of inadequate rainfall, and identified the need for the development of dry farming techniques. Lewis (1981) examined the issue of the inadequate network of railway branch lines. Railway freight rates were initially very high, as Green (1986) showed, but by 1900 had fallen due to technological change in the railway sector (Green 1994). Borins (1982) showed that the length of the frost-free period and the amount of summer rainfall were important determinants of the rate and pattern of settlement.

There has been no detailed analysis of technological change in agriculture. Dick (1980) looked at the productivity of non-land inputs to agriculture, and found a rapid increase in the early twentieth century. McInnis found rapid growth in value added per worker around the turn of the century, suggesting:

> Further investigation of the substance of Canadian agricultural development will probably reinforce the plausibility of this pattern as the waning years of the century are shown to be a period of decided agricultural improvement. ... the overall rise in agricultural output per worker was more rapid in the late nineteenth than in the early twentieth century (McInnis, 1986: 757-58).

Ankli (1974) and Dick (1981) analyzed the initial cost of establishing a grain farm. In this article the only costs examined are those of the equipment and draft animals, to show how productivity increases reduced the labour and machine components of farming costs.

Not all writers have felt that the growth in farm size can be attributed to technical change. Kislev and Peterson (1982) analyzed the increase in the average size of American farms between 1930 and 1970, and determined that it could be explained entirely by changes in factor prices. They suggest (p. 575) that "we explain virtually all of the growth in the machine-labor ratio and in farm size over the 1930-70 period by changes in relative factor prices without reference to 'technological change' or 'economies of scale'." That approach, however, ignores the reasons for the relative price changes. As Table 6 of this article shows, the reduction in machinery costs is attributable to the development of improved implements that could cover greater areas of land for what was frequently the same capital cost.

Kislev and Peterson also see the reduction in labour cost per acre as due to factor price changes. Real farm wages rose over the period they examined. The fall in per acre labour costs must therefore have been due to an increase in the output per unit of labour, which can be attributed primarily to the rise in productivity resulting from improved machinery. Factor prices are important, but they are a *reflection* of factor scarcity and productivity, rather than its *cause*.

THE EARLY PRAIRIE FARM

A typical homesteader arriving on the Prairies took up at least 160 acres of land, frequently more. Farm income, however, was determined not by the total area of land, but by the area of crops grown, which depended on the amount and the productivity of labour. There was little hired labour available, so the work was done primarily by the farm family.[3] Between the start of spring and the first frost in fall, there was only a short period for the farmer to sow, ripen and harvest his grain. The early prairie farm then can be characterized by scarce labour and surplus land. Given these limitations, and the agricultural technology available, farmers could grow only a small area of crops.[4] The following sections analyze some of the ways in which the constraints on farm size and revenues relaxed between 1880 and 1910.

THE CHANGING TECHNOLOGY OF GRAIN FARMING

This section analyzes the equipment used at the two critical times of the grain farmer's year: spring and fall. The productivities of the items discussed are used later in this article to estimate potential crop areas. Activities carried out during the less critical periods of the year are considered briefly in the next section, followed by a study of changes to crop strains after that.

Spring Activities

Ploughing[5]

In 1880 the most effective plough was the chilled-steel plough, first developed about a decade earlier. This was heavy, with a high draft, and needed a team of two oxen or three horses (McKinley 1980: 9) to cover about 2.1 acres in a day's work. (This corresponds to a labour requirement of $1/2.1 = 0.476$ days per acre of grain.) A big advance in ploughing technology in the 1890s was the development of the sulky gang plough. Early sulky ploughs consisted of a single ploughshare mounted on a pair of large wheels with a seat for the driver. This reduced fatigue and increased the number of hours the driver could work, thus increasing output by about 5 percent. Broehl (1984: 201) noted that the first Gilpin sulky appeared in 1875, but a letter in the *Farmers' Advocate* (December 1882: 32) commented that no manufacturers were yet ready to sell such ploughs in Canada. Early sulkies with two side wheels were unstable since they could rock backward and forward about the axle, and therefore were not popular.

The idea of the gang plough was to attach two or more ploughshares to a single frame, thereby enabling one man to plough more land at each pass. Early implements of this type had a very high draft—twice that of a single share plough, so they needed four to six oxen or six to eight horses to pull them. Few farmers had that many draft animals, and managing such a large team was difficult, particularly on the turns at the end of each furrow, so the gang plough never became popular. The concept of the sulky gang plough was obvious—mounting two or more ploughshares on a wheeled frame. An early example is shown in the *Farmers' Advocate* (March 1878: 72), with three small shares on a three-wheel tricycle-type frame, though without a seat for the driver. It was difficult to set the shares to the right depth, or to release a share if it caught on a rock or root. A ploughshare also generates a sideways thrust as it pushes the earth over, so wheeled ploughs tended to drift sideways. The early sulky gang ploughs were therefore ineffective and unpopular.

Efficient mechanisms for setting the depth of each share accurately and for allowing any one ploughshare to move up independently if it hit an obstruction were also developed in the late 1880s. With the evolution of angled wheels to counteract the sideways thrust, it became possible to remove the long "landside" of the walking plough, which had contributed so much to the draft. Since the shares were on a stable frame, they could be tilted slightly "nose down." Only the front edge then touched the bottom of the furrow, further reducing the draft and making it feasible for four horses to pull two fourteen-inch bottoms at about 2 mph.

Disc ploughs began to appear after the turn of the century. With any fixed mouldboard plough, the main draft was created by the friction of the soil against the ploughshare and the landside. A circular saucer-shaped share, free to rotate, reduced this friction considerably. Lateral thrust was resisted by facing the multiple discs of the plough in opposing directions.

However the disc plough was at first heavy to use, difficult in wet ground, and unable to turn a sufficiently wide furrow to be efficient. Advertisements for these ploughs did not appear until after 1900, although journals were discussing the implements in the late 1890s.[6]

Seeding

Before the late nineteenth century seed was always "broadcast" by hand or with a rotary broadcast seeder (Rogin 1931: 206). While the seed was dispersed rapidly, coverage was extremely uneven and much of the seed did not get covered with earth and therefore did not germinate. During the 1870s the endgate seeder—a broadcast seeder fixed to the back of a wagon—was developed, which could cover about twenty to thirty acres per day (Hurt 1979: 28; Rogin 1931: 211). The work went quickly, but coverage was very uneven and much seed was wasted.

An important improvement in seeding technology was the development of the seed drill, which pushed the seeds into the ground to an accurate depth and at even spacing. This improved both coverage and the proportion of the seed germinating. The first drills appeared in the United States in the late 1840s, but Rogin (1931: 207) noted that drilling did not supersede broadcast seeding until about 1885. Early drills were heavy and prone to jamming, but by the mid-1880s became more popular. The *Nor-West Farmer* (April 1886: 441) noted: "Seed drills ... now ... supplanting the broadcast seeder. It is expected the change will result in a marked advance in the early maturity of grain."

In the late 1880s a free-running wheel was added behind each shoe to press the soil back over the small furrows, and during the 1890s there were several further improvements. The Experimental Farm for Manitoba tried out these press drills from 1890 onwards, and found that use of a drill reduced growing time by four days (over the broadcast seeder) and increased yield by seven percent[7] Press drills that used wheels to push the soil down were heavy and cumbersome and therefore slow to catch on. An alternate developed later in the 1890s was to attach a short length of chain behind each shoe. This was almost as effective, and was far lighter to pull. Pulling an early ten-foot drill typically required two horses, which could cover about fifteen acres a day. By 1910 most seed drills used discs to open the furrows, which reduced the draft, enabling the size of the drill to be increased to between twelve and fourteen feet, covering about eighteen acres a day (Davidson and Chase 1920: 119; Gehrs 1919: 85; Kranich 1923: 90).

Harrowing and Cultivating

After the ground had been ploughed, the clods of earth had to be broken up to make seeding easier, to destroy weeds, warm the soil and enhance water retention. For ground that had been ploughed in fall, harrowing was the only operation required before seeding. Before the advent of the seed drill, it was necessary also to harrow after broadcasting the seed, to cover it over. In 1880 most harrows were of the simple "peg tooth" type.[8] During the 1880s and 1890s two new types of harrow became popular. The spring tooth harrow, such as that shown in the *Farmers' Advocate* (20 December 1901: 819), was particularly efficient at pulling up weeds, and coped well with any obstructions in the soil. For 1900 and 1910 the productivity of harrowing increased to about twenty acres per day, and a second harrowing operation after spring ploughing was no longer needed because of the greater efficiency of the new implements. The disc harrow first appeared in the late 1870s, but was not at first practicable.[9] By the early 1890s, though, both the disc harrow and the duck-foot cultivator became popular.[10] These were more efficient rather than faster, and were particularly effective for root crops that needed more cultivation. Table 1 summarizes labour requirement per acre and the cost of implements. Over time less labour was needed, but the cost of implements increased substantially.

Fall Activities

Reaping and Binding Grain

Machines for cutting and collecting grain were first developed in the 1830s, and by the late 1870s the best technology available was the self-raking reaper

TABLE 1: SPRING LABOUR REQUIREMENTS
AND COSTS OF PLOUGHS, HARROWS AND SEEDERS

Productivity*	1880	1890	1900	1910
Plough(days/acre)	0.476	0.476	0.208	0.208
Harrow(days/acre)	0.083	0.083	0.063	0.063
Seeder(days/acre)	0.111	0.111	0.1	0.056
Costs				
Stubble Plough	$25[a]	$22[b]	$18[c]	$13[d]**
Sulky Gang Plough	—	—	$95[e]	$95
Peg Tooth Harrow	$24[f]	$24	—	—
Spring Tooth Harrow[g]	—	—	$25[h]	—
Disc Harrow	—	—	—	$40[i]
Seedbox Seeder and Cultivator[j]	$75[k]	—	—	—
Shoe drill(8 ft.)[l]	—	$65[m]	—	—
Press shoe drill(12 ft.)	—	—	$90[n]	—
Disc drill (12 ft.)	—	—	—	$110[o]
Total cost per farm	$124	$111	$228	$258

*These figures are the inverse of the daily outputs listed in the body of the text.

**A stubble plough is included after 1900 since one would still have been required for some types of work. For the first years of operation a breaking plough would also have been required.

a) Barneby (1884); "Writing Home," Winnipeg Daily Times, 27 June 1879.
b) Manitoba Free Press Weekly, 22 November 1888.
c) Lyle Dick (1981: 193).
d) Farmers' Advocate (Western), 5 July 1905: 992
e) Parson (1981: 152)
f) Manitoba Free Press, 23 November 1878; "Writing Home," Winnipeg Daily Times, 27 June 1887: advertisement of W. H. Disbrowe, Manitoba Free Press Weekly, 23 November 1878.
g) Farmers' Advocate (Western), September 1891: 351.
h) Edmonton Journal, 10 May 1897.
i) Nor-West Farmer, 5 January 1909: 38.
J) Rogin (1931: 209).
k) Manitoba Free Press, 9 February 1877 (advertisement); Lyle Dick (1981: 193).
l) Rogin (1931: 208).
m) Manitoba Free Press, 22 November 1888: 7; Manitoba Free Press Weekly, 22 November 1888: 7; Lyle Dick (1981: 193).
n) Edmonton Journal, 10 May 1897; Lyle Dick (1981: 193).
o) Nor-West Farmer, 5 January 1909: 38 (prize); Lyle Dick (1981: 193).

(David 1966). Drawn by two horses and operated by one man, this type of machine could cut a six-foot swath through the grain. It could cut about ten acres per day, but left the loose stalks on the ground to be tied in bundles later.[11] In 1880 it took a team of five men to cut and bind ten acres of grain in a day.

The first wire binders, which automatically tied the cut stalks into bundles, were sold in the United States in 1873 (Rogin 1931: 110); a few were sold in Canada before 1880. These made subsequent handling of the grain stalks easier and reduced damage to the grain due to rain or frost. Wire binders however had a variety of problems, and by the early 1880s were superseded by twine binders. A typical 1880s twine binder cost about $290 (Hutchinson 1935: 704), needed three or four horses and one driver, and cut and bound about thirteen acres per day.

Early binders were heavy, cumbersome machines. Denison (1948: 85) notes that the binder "didn't immediately replace all other kinds of harvest machinery, since it was costly and complicated, and farmers were unsure of its strength, reliability and durability." During the 1880s and 1890s improvements to metallurgy and machining reduced the draft, which made it possible for the width of cut to be increased. By 1910 a new binder pulled by four horses could cut an eight-foot swath.

TABLE 2: HARVEST LABOUR REQUIREMENT AND
COSTS OF REAPING AND BINDING EQUIPMENT

Productivity	1880	1890	1900	1910
Reap (man days/acre)	0.5	0.100[a]	0.067	0.057[b]
Stook (man days/acre)	0.1	0.1	0.067	0.067
Transport (man days/acre)	0.486	0.486	0.353	0.353
Costs				
Self Rake Reaper (6 ft.)	$200[c]	—	—	—
Twine Binder	—	$260[d]	$155[e]	$155
Wagon	$85[f]	$85	$80[g]	$60[h]
Total	$285	$345	$235	$215

a) Zintheo (1917), 2214.
b) Gray (1967: 177).
c) Hutchinson (1935: 475).
d) Ibid., 704.
e) *Edmonton Journal*, 10 May 1897.
f) *Manitoba Free Press*, 10 March 1877.
g) *Edmonton Journal*, 10 May 1897.
h) *Farmers' Advocate (Western)*, 5 July 1905: 992.

The rapid adoption of the twine binder resulted in shortages of twine and high prices. Twine cost about 18 cents per lb. in the early 1880s, but the price fell by 1910 to 7 cents.[12] Before 1900, sacks for the grain had also to be bought, costing about three cents per 100 lbs. of grain, or about 30 cents per acre of wheat. By 1910, most of the grain was handled in bulk, eliminating the need for sacks.

Collecting and Transporting the Grain

After the bundles were dropped, they were collected in groups of eight to ten and stood on end for about ten to fifteen days for the grain to dry. When first cut, the heads of grain had a very high water content, which made the grain very susceptible to frost damage. Reaping therefore had to be finished ten to fifteen days before the expected date of the first frost. After drying, the grain was either threshed immediately, if the travelling threshing outfit arrived in time, or was collected from the fields to a central stack for threshing later.[13] Fall ploughing could not begin until the fields were clear of the sheaves. The technology for this part of the work did not change between 1880 and 1910, although a greater output was achievable after 1890 when the binders were fitted with bundle carriers, which held up to six sheaves. Releasing the sheaves in one pile greatly reduced the work of collection.

CHANGES TO OTHER ASPECTS OF FARMING TECHNOLOGY
Breaking

Every homesteader faced the heavy task of breaking and backsetting the hard sod before starting any cropping. This occupied a large part of each summer during the first few years on a new homestead, and resulted in smaller crops until the farmer had prepared as much land as he could subsequently crop each year. The only important technical change to this operation was the availability after the 1890s of large steam traction engines. These could pull up to eighteen ploughshares on a frame, breaking up to twenty acres per day. This meant that a farm could produce a full set of crops very quickly, but the high cost of contract breaking meant that few could afford it.[14]

Maintenance of the Summerfallow

With the development of dry-farming techniques in the late 1880s and 1890s, other implements became necessary, mainly to control the density of the topsoil. During the 1890s the most common implement for this was the land roller, such as that advertised in the *Farmers' Advocate (Western)* (July 1891: 270). By 1900 this had been developed into the subsurface packer, a typical example of which was advertised by the Brandon Machine Works in the *Farmers' Advo-*

cate (Western) (5 July 1900: 359). Use of a roller reduced the blowing of topsoil, induced more effective capillary action in the soil, and made reaping and ploughing easier due to the more evenly compacted soil. The pulverizer and compressor performed a similar function to the roller and subsurface packer.[15]

Haying

Every farmer spent a week or two in summer preparing a stock of hay for winter feed. Larger mowers,[16] the side delivery rake[17] and the hay tedder[18] increased productivity in cutting and turning the hay. From about 1900, hay loaders greatly reduced the work of pitching the hay into wagons,[19] and the work of building a stack was also decreased by the development of the over-head horsefork and slings.[20]

Dairy and Beef Cattle

Many farmers kept cows, since these generated food or income throughout the year, and the labour requirement was spread more evenly than that for grain farming. More scientific feeding enabled farmers to increase the lactation period (Spector 1977: 37). The butterfat tester, available by the late 1880s, enabled farmers to select more productive cows, obtaining more milk for the feed used (Schiebecker 1975:184). Improvements in the rearing of beef cattle increased the profitability of that activity, though there is little evidence in census data of an increase in dairy and beef farming in Manitoba and Saskatchewan.

Threshing

All grain had to be reaped and dried before it was damaged by the first frost, but once it had been stacked, threshing could be carried out later in the year. Threshing productivity was therefore not a determinant of farm size, although most settlers had to "meet their notes" on the equipment they had purchased, and threshed their grain as early as possible.

Mechanical separators had existed since the 1850s, but there were several important labour saving devices added to this equipment that greatly reduced the labour requirement. Bigger, more reliable steam engines drove larger separators, and better bearings meant that the cylinder could turn faster. Metal teeth reduced breakages and improved separation.[21] The volume of straw produced by a large separator was a substantial problem, and in the 1880s a team of about six men was needed to avoid overwhelming the thresher. First a belt elevator and later the "wind stacker"—a large fan blowing the straw through a long tube—were added, reducing the labour required to one man by 1900.[22]

Early separators frequently clogged, as the sheaves were thrown in unevenly. The self feeder and band cutter developed in the late 1880s reduced this problem, saving the work of one or two men, and enabling a larger cylinder to be used.[23] Automatic grain baggers eliminated the work of another man, and between 1900 and 1910 usage of these devices became almost universal. Kranich (1923: 221) suggested that by 1920 over 95 percent of threshers were equipped with all three. By 1910 most of the grain was handled in bulk, which eliminated the work of bagging and the subsequent multiple handling of the grain as it was moved to market.

POWER: OXEN, HORSES, AND STEAM

Farm implements had to be pulled back and forth across the fields over large distances. For example, in 1890, ploughing the typical Brandon field crop acreage of fifty-two acres with a fourteen-inch ploughshare involved walking a distance of almost 370 miles. Many of the early implements were heavy and imposed a considerable draft on the motive power used to pull them.

From the earliest days of the Red River Settlement the primary source of power had been oxen. These huge beasts could pull great loads, although only at a slow speed—about 1.5 miles per hour. Oxen were well suited to pulling a plough or a wagon, but for more complicated equipment such as seeders or binders were less appropriate. If a horse felt that an obstruction was stopping the equipment, it stopped pulling, whereas an ox would lean its weight into the harness. This was useful for pulling a plough through roots, but could result in damage to a binder or mower caught on a rock. Oxen were therefore almost indispensable during the initial period of breaking in a new prairie farm, but were less useful after that. An ox could pull as much as a horse, but at two-thirds the speed, therefore generating only two-thirds the horsepower (Davidson and Chase 1920: 287.) On the very bad roads of the early prairie days an ox-drawn wagon could cover only about twelve miles per day (Careless 1973: 88; de Gelder 1973).

Typical farm horses weighed 1,000 to 1,600 lbs. with the average increasing over the period from about 1,100 lbs. in 1880 to 1,400 lbs. by 1910. They could pull about one-tenth to one-seventh of their own body weight continuously, and up to half their weight for very short periods. Horses could work a ten-hour day, similar to oxen, and could keep up a steady speed of about 2.5 mph, compared with about 1.5 mph for oxen, but their feed and care were more involved and expensive. The typical pattern of ownership was for a new settler to buy a yoke of oxen when he arrived and to keep them for three to five years (Morton 1938: 164; Neatby 1979: 24). Another pair of oxen was sometimes added, but four oxen were difficult to harness and handle for field operations, since they are not prepared to back up. Towards the end of the first four or five years a pair of horses was usually bought, and after the 1890s a second pair was added when the oxen were sold (See Table 3.)

TABLE 3: THE AVERAGE NUMBER OF HORSES ON MANITOBA FARMS[a]
AND THE COST OF A PAIR OF HORSES, WITH HARNESS

	1880	1890	1900	1910
Average Number of Horses	1.63	2.74	4.05	4.81
Cost	$350[b]	$400[c]	$275[d]	$525[e]

a) Manitoba data only are used because in other areas horse ranching distorts the number of draft animals. Canada, *Census of Canada*, 1890, 1900, 1910.

b) *Manitoba Free Press Weekly*, 10 March 1877 (report of minister of Agriculture and Immigration 1876) $300/span; and 9 June 1877: native horses $60 to $100, imported $10 to $200; Hamilton (1875), 254-55: $300 to $400/span; Ankli and Litt (1978:51): ($66, $175); Tyman (1972: 44) for 1880: pair of horses $200 to $500, harness $12 to $60.

c) *Farmers' Advocate*, January 1883: 5: horses $450 to $600/pair; *Nor-West Farmer*, January 1886: 353: horses $25 to $190; 28 April 1887: $150 to $400/team; 16 July 1887, good average working team $400.

d) *Nor-West Farmer*, 6 February 1899: 102, 20 October 1899: horses $65 to $175, good pairs $200 to $265, old drivers $40; 20 October 1899: fully broken horses $125 to $175; *Edmonton Journal*, 10 May 1897: tea, of horses $125, harness $32.

e) The price of a horse appears to have increased substantially during the first decade of the twentieth century, probably due to the increased demand derived from the large number of new farms. See "the Horse Market," *Nor-West Farmer*, 5 July 1906: 579: "Not in the memory of many living has there been such a universal scarcity of horses in proportion to the demand as exists at present." See also *Manitoba Free Press Weekly*, 15 May 1907: teams 3,000 lb. to 3,600 lb. weight—$500 to $800, 2,400lb. to 2,800 lb.—$450 to $500; de Gelder (1973: 72)—4 horses $1000; Sheperd (1965:49)—$200; Canada, *Census of Canada*, 1910, average for horses over three years old for Saskatchewan $180, Manitoba $187, Alberta $150.

The power of the horse was generally taken through a harness to pull a moving load directly behind the animal. Some farm implements, though, were stationary, requiring rotary power. For these, the tread and the sweep were developed. The tread was like a short escalator, with one or two horses walking uphill pushing the treads down. A one-horse tread was an efficient means of capturing the power of a horse, since it used the horse's weight rather than its pull, to turn the wheel. An 1,100 lb. horse in a tread could generate almost 1.5 hp, compared with about .75 hp pulling an item of equipment (Davidson and Chase 1920: 293, 296.) It was not feasible, however, to make a tread for more than three horses. Treads were therefore suitable only for small threshing machines.

A sweep consisted of from four to twelve long arms projecting from a central capstan, to each of which was attached a horse that walked around in circles. Power was taken off by a rotating shaft running along the ground. More horses could be harnessed to a sweep but it wasted much of their energy due to the incorrect line of draft (Davidson and Chase 1920: 296). Sweeps were not used as frequently as treads, since there were never enough spare horses. Table 3 shows the average number of horses on Manitoba farms over the period and the cost of a pair of horses with harness.

Steam engines became important, initially for driving stationary equipment such as threshing machines. Though too cumbersome to replace the horse in most field operations, the enormous power of the steam traction engine ensured its dominance for a few applications. The *Manitoba Free Press Weekly* for 17 July 1880 noted that up to 1879 only 252 engines had been built by one of the largest Canadian manufacturers, the Waterous Steam Engine Works Co. of Brantford, Ontario. None of these were traction engines and few would have found their way to the Prairies. Early steam engines needed a team of horses to move them from place to place, since they were "portable" rather than "locomotive."[24]

By 1890 there had been significant progress. Barger and Landsberg (1942: 199) suggest that by then the use of steam engines for threshing was standard in the United States. An advertisement in the *Farmers' Advocate* (June 1886: 189) by the Waterous Steam Engine Works Co., shows that some engines were self-propelled by that date. McKinley (1980) estimates that by 1890 there was a total of 3,000 steam tractors throughout the United States, producing 8-12 hp each.

Steam engines developed very rapidly during the 1890s, and by 1900 large, reliable traction engines of up to 110 hp were available. However, they never became agile enough to carry out all field operations. The major technological improvements to the steam engine had been made by 1900, after which improvements consisted primarily of refinements that made the engines more efficient and reliable.

NEW STRAINS OF CROPS

Before 1870 several varieties of wheat were grown, mostly brought up from the United States, and not necessarily well suited to prairie conditions. From 1870 to 1907 the most important variety was Red Fife, which matured relatively quickly, and was suited to the soil and weather conditions on the Prairies. Both individual farmers and the experimental farms put a great deal of effort into the search for a better strain.

The development of new strains of wheat became important after 1900. Several varieties of imported seeds and of crossbred types were tried. None was really successful until 1908, when Marquis wheat, first tried experimentally in about 1897, was made available on a commercial scale.[25] The reports of the experimental farms show that Marquis ripened six to eight days earlier than Red Fife. This shorter growing period, combined with higher yields, resulted in its rapid adoption. Marquis wheat came too late to be a cause of the wheat boom, but it increased profitability once it was available.

IMPLICATIONS
Farm Size

In this section the labour productivities derived in the earlier sections are used to estimate the effect of technological change on farm size between 1880 and 1910, using Brandon and Indian Head as examples. Spring typically began at both places on 6 April (Hurd and Grindley 1931: 14, fig. 12), and field work could start on about 15 April. The first killing frost of fall came on 17 September in Brandon and at Indian Head on 12 September (Hurd and Grindley 1931: 13, fig. 10). Red Fife wheat took 122 days to mature at Brandon and 129 at Indian Head. By 1900 the use of seed drills accelerated germination of the seed, reducing the growing time to 119 days at Brandon and 127 at Indian Head.[26] Once it was available, Marquis took 109 days at Brandon and 122 at Indian Head.[27]

The annual schedules for the two farms for the four dates examined are prepared on the basis of maximizing the area of wheat grown. The shorter ripening period for oats meant that it was sometimes possible to seed some after the wheat was all in, and reap it before the wheat was ripe. Any land used for oats had of course to be ploughed and harrowed also. The area of oats may in practice have been greater than is modelled here, but that would have been at the expense of wheat acreage, and would therefore have had a minimal effect on total farm area.

For each model the area of spring ploughing has been adjusted so as to result in all the available time being used up, with the areas of ploughing, harrowing, seeding and reaping being equal. As the area of spring ploughing increased, seeding was delayed, thereby delaying the start of harvest and reducing the area of land that could be reaped. Since all grain had to be reaped at least ten days before the first frost in fall, the period from when the grain was ripe to the first frost was normally fully occupied with harvesting. However, in the early days, fall ploughing was critical, so it was essential to seed the wheat as early as possible in spring, so harvesting could be completed earlier. In order to start the fall ploughing the fields had to be clear, so the sheaves had to be collected and either stacked or threshed.[28]

TABLE 4: LABOUR AVAILABILITY

Location	Farmers	Farmers' Sons	Labourers	Spring Total	Excess Farmers	Harvest Excursion	Harvest Total
B R A N D O N							
1880	1.00	0.29	0.19	1.48	0.20	0.00	1.68
1890	1.00	0.38	0.22	1.60	0.29	0.00	1.89
1900	0.96	0.26	0.23	1.45	0.00	0.22	1.67
1910	1.00	0.14	0.24	1.38	0.00	0.12	1.50
I N D I A N H E A D							
1880	0.82	0.18	0.21	1.21	0.00	0.00	1.50
1890	1.00	0.36	0.15	1.51	0.25	0.00	1.76
1900	0.98	0.24	0.18	1.40	0.00	0.27	1.67
1910	1.00	0.11	0.34	1.45	0.04	0.12	1.61

Notes: Figures derived from Canada, *Census of Canada*, 1880, 1890, 1900 and 1910, and Haythorne(1933). Census data do not contain sufficient information to determine how many labourers worked on farms as opposed to, for example, railway construction, so these proportions had to be estimated. For details see Ward (1990). The 1900 census contains very little information on labour which, other than the harvest excursionists, has therefore been interpolated between the 1890 and 1910 figures. In 1890 the number of farmers was greater than the number of farms. The excess farmers are assumed to have been looking for land, and to have worked as harvest help.

When first cut, the grain had a high water content, and had to be dried before it could be handled further. This was achieved by standing the bound sheaves on end in groups (stooks) in the field to dry for ten to fifteen days. After that they could be either collected and threshed immediately, or collected and stacked to await threshing later in the year. The timing of threshing was not technically critical, but since the grain could not be sold until it had been threshed, and most farmers had mortgages and implement payments to make, and store credit to be paid off, there was significant financial pressure to thresh early. Since that took labour away from other critical operations in the fall, the estimates here ignore threshing before the onset of winter, which will result in slight overestimates of the areas of crops.

Farmers lost time to inclement weather, so in spring a deduction of 31 percent is made for that and for Sundays, when many farmers did not work, even at harvest time.[29] At harvest time 27 percent is deducted for inclement weather, and during the fall ploughing period 35 percent. The amount of labour available can be estimated from census data, and is listed in Table 4. Table 5 summarizes the estimates made.

TABLE 5: ESTIMATES OF FARM SIZES

Activity	1880	1890	1900	1910
B R A N D O N				
Spring plough	None	15-23 April 8.3 days, 19 ac	5-22 April 7.25 days, 35 ac	15-26 April 11.3 days, 52 ac
Harrow	15-18 April 3.8 days, 39 ac	23-29 April 5.2 days, 69 ac	22-29 April 6.75 days, 107 ac	26 April-4 May 8 days, 121 ac
Seed and harrow wheat	19-26 April 7.5 days, 39 ac	29 April-22 May 12.2 days, 69 ac	29 April-10 May 11 days, 107 ac	4-11 May 7.1 days, 121 ac
Seed and harrow oats	26-27 April 1.5 days, 8 ac	None	None	None
Reap and stook oats	15-18 August 4 days, 7 ac	None	None	None
Reap and stook wheat	19 Aug.-7 Sept. 18.7 days, 40 ac	29 Aug.-7 Sept. 9 days, 69 ac	27 Aug.-6 Sept. 11.2 days, 107 ac	25 Aug.-7 Sept. 13 days, 121 ac
Collected and stack grain, fall plough	8 Sept.-27 Oct. 51 days, 46 ac	7 Sept.-28 Oct. 51 days, 50 ac	7 Sept-28 Oct. 51 days, 73 ac	8 Sept-28 Oct. 51 days, 69 ac
Total cropped area	46 acres	69 acres	107 acres	121 acres
Census average size	31 acres	69 acres	143 acres	207 acres
I N D I A N H E A D				
Spring plough	None	15-18 April 3 days, 6.6 ac	15 April 0.5 days, 2.5 ac	15-18 April 2.9 days, 14 ac
Harrow	15-17 April 2.8 days, 28 ac	18-23 April 4.6 days, 58 ac	15-21 April 5.1 days, 78 ac	18-24 April 5.8 days, 92 ac
Seed and harrow wheat	18-20 April 2.5 days, 11 ac	23-27 April 4.5 days, 24 ac	21 April-4 May 13.4 days, 78 ac	24 April-5 May 11 days, 92 ac
Seed and harrow oats	20-23 April 4 days, 17 ac	27 April-4 May 7 days, 38 ac	None	None
Reap and stook oats	13-24 August 12 days, 17 ac	20-25 August 5 days, 34 ac	None	None
Reap and stook wheat	25 Aug.-2 Sept. 8.2 days, 12 ac	30 Aug.-2 Sept. 3.6 days, 24 ac	25 Aug.-2 Sept. 8 days, 77 ac	24 Aug.-2 Sept. 9.25 days, 92 ac
Collected and stack grain, fall plough	2 Sept.-27 Oct. 55 days, 41 ac	3 Set.-27 Oct. 54 days, 51 ac	2 Sept.-27 Oct. 55 days, 75 ac	2 Sept.-27 Oct. 55 days, 78 ac
Total cropped area	39 acres	58 acres	77 acres	121 acres
Census average size	16 acres	30 acres	53 acres	158 acres

Note: Total cropped area is the minimum of the areas covered by ploughing, harrowing, seeding and reaping. After 1900, the use of seed drills obviated the need to harrow immediately after seeding.

The estimates depict the potential size of a fully developed farm that used the latest equipment, and represent what settlers might have realistically expected of a new homestead. It is not meant to imply that every farmer would have updated all his equipment every time something new was developed.

Both estimated farm sizes and census averages grew substantially over the period. The estimated potential sizes are initially higher, suggesting that in 1880 and 1890, most farms were not fully developed. For Brandon in 1900, and for both places in 1910, the census average area is greater than the estimates, which may indicate that the increases in labour productivity have been estimated too conservatively, or that other factors were involved. For 1910 the estimated potential size for Indian Head is little greater than that for 1900, because the introduction of Marquis wheat was offset by a fall in the quantity of labour.[30]

The Costs of Operating a Farm

In this section labour and machine costs per acre on a fully developed farm are calculated, to show the effects of the technical changes that had occurred. The only additional data added at this point are farm wages. Table 6 uses the data derived in the earlier sections of the paper to generate these per acre costs. Both labour and equipment costs per acre fall, except that the equipment cost per acre rises in 1910 due to the increased price of horses.

CONCLUSIONS

Towards the end of the nineteenth century there was a significant amount of technical change in almost all aspects of field cropping technology. It would appear likely that this was induced by the agricultural potential of the large areas of land that comprised the next stage of expansion on the North American continent. As the new equipment was brought to the Prairies and adapted to local conditions, productivity increased. Given the short seasons and scarcity of labour on the Prairies, these improvements resulted in increases in the area of crops that the typical farm could grow. The major improvements in the 1880s and 1890s increased farm size and output by 250 percent without commensurate increases in costs, greatly increasing the attractiveness of prairie farming, and contributing to the start of the wheat boom.

TABLE 6: LABOUR AND EQUIPMENT COSTS PER ACRE

	1880	1890	1900	1910
P/t harvest work (days)	0	0	10	8
Harvest wages/day (incl. board)	1.50	2.00[a]	1.60[b]	2.35
Total p/t harvest wages	0.00	0.00	16.00	18.80
Annual farmers' wages[c]	260.00	235.00	225.00	250.00
Annual cost sons and f/t labour	54.60	178.60	132.75	95.00
Total wage cost	314.60	413.60	373.75	363.00
Total equipment cost[d]	409.00	456.00	463.00	473.00
Cost of horses[e]	350.00	400.00	550.00	1050.00
Annual allowance[f]	91.08	102.72	121.56	182.76
Labour cost per acre	6.84	5.99	3.49	3.00
Equipment cost/acre	1.98	1.49	1.14	1.51

Note: These figures are calculated for the estimated acreage at Brandon, and not intended to be comprehensive. The equipment cost includes capital only, not maintenance or operating.

a) *Farmers'Advocate (Eastern)*, January 1885: 12, May 1885: 13; *Manitoba Free Press Weekly*, 2 October 1890; *Brandon Weekly Mail*, 15 July 1885.
b) *Farmers'Advocate (Western)*, January 1892: 5, 5 October 1900: 543; *Manitoba Free Press Weekly*, 15 September 1898; *Edmonton Journal*, 27 September 1897.
c) Based on a farm labourer's wage in Ontario, as a realistic opportunity cost. For more details see Ward (1900: 180-82).
d) Includes all items listed in Tables 1 and 2 for spring and fall equipment for grain farming.
e) For 1880 and 1890 two horses are included, and for 1900 and 1910 four are included.
f) Amortized at 8 percent over ten years.

NOTES

This article first appeared in *Prairie Forum* 20, no. 1 (1995): 19-36. The author was grateful for the assistance of Angela Redish, Ken Norrie, participants at seminars at the University of British Columbia and the University of Toronto Economic History Workshop, and two anonymous referees. Remaining errors are the author's own responsibility. Part of this work was carried out under a ssHRC Doctoral fellowship.

1   Canada, *Census of Canada*, 1880, 1900.
2   Other causes include the time it took to break in a new farm and the problem of land speculation in the early 1880s.
3   See census data in Table 4.

4   For an interesting analysis of changing farm size see Ray D. Bollman and Philip Ehrensaft, "Changing Farm Size Distribution on the Prairies Over the Past One Hundred Years," *Prairie Forum* 13, no. 1 (1988): 43-66.

5   Ploughing was in fact carried out in both spring and fall, but the technology was the same.

6   *Farmers' Advocate, 21* August 1907: 1300; *Farmer's Advocate (Western),* 5 July 1907: 490; 3 May 1905: 659; 22 November 1905: 1695; *Nor-West Farmer,* July 1898: 305; 20 August 1901: 531; 5 September 1906: 809, 812; 20 September 1901, "Farm Implements." See also J. B. Davidson and L. W. Chase, *Farm Machinery and Farm Motors* (New York: Orange Judd, 1920), 67 and W. G. Broehl, *John Deere's Company* (New York: Doubleday, 1984), 208.

7   *Report of the Experimental Farm for Manitoba,* for example, 1890: 249; 1892: 196; 1893: 236.

8   For example, see the illustration in the *Farmers' Advocate (Eastern),* May 1881: 126.

9   See, for example, the *Farmers' Advocate, (Western),* 20 February 1899: 103.

10   See, for example, the Cockshutt diamond point cultivator in the *Farmers' Advocate (Western),* February 1890: 268 and April 1891.

11   Another machine, the McCormick harvester, carried three men, one driving and two tying the grain into bundles before it was dropped. G. Quick and W. Buchele, *The Grain Harvesters* (St. Joseph, MI: American Society of Agricultural Engineers, 1978), 74, suggest that two men on the machine could do the work of four or five men following behind on foot, who would have to first walk around collecting the stalks together before they could bind them into bundles.

12   About six pounds of twine were needed to bind an acre of grain. For details see Ward, "Extensive Development of the Canadian Prairies" (Ph.D. dissertation, University of British Columbia, 1990), 106.

13   For example, the *Farmers' Advocate,* 3 January 1906: 57, notes in an item on "Stook versus Stack Threshing": "threshing from the stack was cheaper, but the risk of crop losses meant that it was better to stack." J. Bracken, *Crop Production in Western Canada* (Winnipeg: Grain Growers Guide, 1920), 121, recommended that farmers stook then stack after ten days.

14   The *Edmonton Journal,* 10 May 1897 quoted a cost of $5.25 per acre for breaking and back-setting.

15   Such as that advertised by the Watson Manufacturing Co. in the *Farmer's Advocate,* 7 September 1904: 1316.

16   *Manitoba Free Press,* 22 November 1887; *Farmer's Advocate (Western),* 5 September 1893; Marvin McKinley, *Wheels of Farm Progress* (St. Joseph, MI: American Society of Agricultural Engineers, 1980), 19.

17   *Farmers' Advocate (Western),* 5 July 1905: 992.

18   Ibid., March 1891: 116; 5 July 1905: 992.

19   See, for example, ibid., 5 July 1905: 992.

20   Ibid., March 1891: 108; 20 June 1899: 339.

21   *Farmers' Advocate (Eastern),* July 1885: 205.

22   In the *Nor-West Farmer* for April 1898 the J. I. Case Co. listed the wind stacker among other recent improvements.

23 A good cutaway view of this attachment is given in the Farmers' Advocate for 1 July 1904: 787.
24 The Rumely portable steam threshing engine needed eight horses to move it. See McKinley, Wheels of Farm Progress, 27.
25 Many other varieties were also tried, but they tended to suffer from plant diseases such as smut and rust. Marquis was even less susceptible to these problems than Red Fife.
26 These durations are derived from the annual reports of the experimental farms, 1890 to 1914.
27 Reports of the experimental farms, Brandon and Indian Head, 1887 to 1914.
28 I am grateful to an anonymous reviewer for help in clarifying the logical sequence in which these constraints worked.
29 Little work was done on Sundays. For example, George Shepherd, *West of Yesterday* (Toronto: McClelland and Stewart, 1965), 62, notes being "shut down by a Mountie" for threshing on a Sunday. See also *Nor West Farmer*, January 1904: 29, and W. P. Rutter, *Wheat Growing in Canada, the United States and Argentina* (London: A. and C. Black, 1911), 117.
30 Census areas for 1890 and 1900 were inconsistent and information is available only at a very high level of aggregation. For 1900 the entire Territories were divided into only four districts. Census averages cover extremely large areas in which many farms were far below their potential size, having been newly established.

REFERENCES

Ankli, R. E. 1974. "Farm-Making Costs in the 1850s." *Agricultural* History 48, no. 1: 51-70.

Ankli, R. E. and R. M. Litt. 1978. "The Growth of Prairie Agriculture." In Akenson, D. H. (ed.), *Canadian Papers in Rural* History 1:35-64. Gananoque: Langdale Press.

Barger H. and H. M. Landsberg. 1942. *American Agriculture 1899-1939:. A Study of Output, Employment and Productivity.* New York: National Bureau of Economic Research.

Barneby, W. H. 1884. *Life and Labour in the Far, Far West.* London: Cassill.

Bollman, Ray D. and Philip Ehrensaft. 1988. "Changing Farm Size Distribution on the Prairies

Over the Past One Hundred Years." *Prairie Forum* 13, no. 1:43-66.

Borins, S. F. 1982. "Western Canadian Homesteading in Time and Space." *Canadian Journal of Economics* 15: 18-27.

Bracken, J. 1920. *Crop Production in Western Canada.* Winnipeg: Grain Grower's Guide.

Briggs, Harold E. 1932. "Early Bonanza Farming in the Red River Valley of the North." *Agricultural History* 6, no. 1: 26-37.

Broehl, Wayne G. 1984. *John Deere's Company.* New York: Doubleday.

Careless, J. M. 1973. *The Pioneers: The Picture Story of Canadian Settlement.* Toronto: McClelland and Stewart.

David, Paul. 1966. "The Mechanization of Reaping in the Ante-Bellum Midwest." In Rosovsky, H. (ed.), *Industrialization in Two Systems.* New York: John Wiley and Sons.

Davidson, J. Brownlee and on Wilson Chase. 1920. *Farm Machinery and Farm Motors.* New York: Orange Judd.

de Gelder, W. 1973. *A Dutch Homesteader on the Prairies*. Translated by H. Ganzevoort. Toronto: University of Toronto Press.

Denison, Merrill. 1948. *Harvest Triumphant: The Story of Massey-Harris*. Toronto: Mc-Clelland and Stewart.

Dick, Lyle. 1981. "Estimates of Farm-Making Costs in Saskatchewan, 1882-1914." *Prairie Forum* 6, no. 2: 183-201.

Dick, Trevor O. 1980. "Productivity Change and Grain Farm Practice on the Canadian Prairie 1900-1930." *Journal of Economic History* 40: 105-10.

———. 1982. "Mechanization and North American Prairie Farm Costs, 1896-1930." *Journal of Economic History* 42, no. 1: 199-206.

Engerman, S. L. and R. E. Gallman (eds.). *Long Term Factors in American Economic Growth*. Chicago: University of Chicago Press. National Bureau of Economic Research, Studies in Income and Wealth No. 51.

Gehrs, John H. 1919. *The Principles of Agriculture*. New York: Macmillan.

Gray, J. H. 1967. *Men Against the Desert*. Saskatoon: Modem Press.

Green, A. G. 1986. "Growth and Productivity Change in the Canadian Railway Sector, 1871-1926." In Engerman and Gallman, *Long Term Factors in American Economic Growth*, 779-812.

———. 1994. "The Nature of Technological Change in the Canadian Railway Sector: 1875-1930." Paper presented at the Nineteenth Conference on the Use of Quantitative Methods in Canadian Economic History, McGill University, April 1994.

Hamilton, James Cleland. 1876. *The Prairie Provinces: Sketches of Travel from Lake Ontario to Lake Winnipeg, and an Account of the Geographical Position, Climate, Civil Institutions, Inhabitants, Productions and Resources of the Red River Valley*. Toronto: Belford Brothers.

Haythorne, G. V. 1933. "Harvest Labour in Western Canada: An Episode in Economic Planning." *Quarterly Journal of Economics* 67: 533-44. Hurd, W. B. and T. W. Grindley. 1931. *Agriculture, Climate and Population of the Prairie* Provinces of Canada. A Statistical Atlas Showing Past Development and Present *Conditions*. Ottawa: Dominion Bureau of Statistics.

Hurt, Leslie J. 1979. *The Victoria Settlement: 1862-1922*. Edmonton: Alberta Culture, Historical Resources Division. Historic Sites Service Occasional Paper No. 7.

Hutchinson, William T. 1935. *Cyrus Hall McCormick*. 2 vols. New York: D. Appleton-Century.

Ingles, Ernest B. 1973. "Some Aspects of Dry-Land Agriculture in the Canadian Prairies to 1925." M.A. thesis, University of Calgary.

Johnson, Paul C. 1976. *Farm Inventions in the Making of America*. Des Moines: Wallace-Homestead Book Co.

Kislev, Yoav and Willis Peterson. 1982. "Prices, Technology, and Farm Size." *Journal of Political Economy* 90: 578-95. Kranich, F. N. G. 1923. *Farm Equipment for Mechanical Power*. New York: Macmillan.

Lewis, F. D. 1981. "Farm Settlement on the Canadian Prairies 1898-1911." *Journal of Economic History* 41: 517-35.

Lewis, F. D. and D. R. Robinson. 1984. "The Timing of Railway Construction on the Canadian Prairies." *Canadian Journal of Economics* 17: 340-52.

Marr, W. and Percy, M. 1978. "The Government and the Rate of Prairie Settlement." *Canadian Journal of Economics* 11, no. 4: 757-67.

McInnis, Marvin. 1986. "Output and Productivity in Canadian Agriculture, 1870-71 to 1926-27." In Engerman and Galan, *Long Term Factors in American Economic Growth*, 737-70.

McKinley, Marvin. 1980. *Wheels of Farm Progress*. St. Joseph, MI: American Society of Agricultural Engineers.

Morton, A. S. 1938. *History of Prairie Settlement*. Vol. 2 in *Canadian Frontiers of Settlement Series*. Toronto: Macmillan.

Neatby, L. H. 1979. *Chronicle of a Pioneer Prairie Family*. Saskatoon: Western Producer Prairie Books.

Norrie, K. H. 1975. "The Rate of Settlement on the Canadian Prairies 1870-1911." *Journal of Economic History* 35: 410-27. Parson, Edna T. 1981. *Land I Can Own*. Ottawa: E. T. Parson.

Quick, Graeme and Wesley Buchele. 1978. *The Grain Harvesters*. St. Joseph, MI: American Society of Agricultural Engineers.

Rogin, Leo. 1931. *The Introduction of Farm Machinery*. Berkley: University of California Press.

Rutter, W. P. 1911. *Wheat Growing in Canada, the United States and Argentina*. London: A. and C. Black.

Schlebecker, J. T. 1975. *Whereby We Thrive: A History of American Farming 1607-1972*. Iowa City: Iowa State University Press.

Shepherd, George. 1965. *West of Yesterday*. Toronto: McClelland and Stewart.

Spector, D. 1977. *Field Agriculture In the Canadian Prairie West, 1870 to 1940, With Emphasis on the Period 1870 to 1920*. Ottawa: Parks Canada, Department of Indian and Northern Affairs, National Historical Sites and Parks Branch. Manuscript Report No. 205.

Tyman, J. L. 1972. *By Section, Township and Range: Studies in Prairie Settlement*. Brandon: Assiniboine Historical Society.

Ward, Tony. 1990. "Extensive Development of the Canadian Prairies." Ph.D dissertation, University of British Columbia.

Zintheo, C. J. 1917. "Machinery in Relation to Farming." In Bailey, L. H. (ed.), *The Cyclopaedia of American Agriculture*. 4 vols. Toronto: Macmillan of Canada.

9. Tractors and Combines in the Second Stage of Agricultural Mechanization on the Canadian Plains

R. Bruce Shepard

The adoption of new technology in agricultural areas is a multi-step process. When a new practice is introduced the initial rate of acceptance is slow, followed by increasingly rapid adoption. Indeed, it is in the last stages of the cycle in which most adoptions occur.[1] Modern Canadian Plains agriculture is an example of this process because it evolved from a lengthy period of mechanization stretching from the turn of the century to the 1950s. There were three incremental steps: up to the end of World War I, the twenties, and during World War II.

The first step had been the early acceptance of mechanical power sources, specifically steam and gasoline engines. Agricultural mechanization was thus well under way by the twenties, and the proliferation of the tractor together with the appearance of the combine indicated the future direction of Plains agriculture. This trend continued until it was cut short by the onslaught of the Great Depression, only to be resurrected and taken to the third and culminating stage during World War II.

The second stage of this process, particularly the late 1920s, was of pivotal importance because that was when the beginnings of modern Canadian Plains agriculture were first evident. While there were other agricultural machines involved, this period witnessed the marriage of the tractor and the combine, which are the two leading elements of mechanized agriculture. Yet it is important to realize that this critical second stage was based upon an earlier experience with agricultural engines. Even if they did not own one of the early engines, farmers had been exposed to their capabilities and this experience had illuminated the potential for further mechanized agriculture.

Previous examinations of the mechanization of Canadian Plains agriculture have underestimated the importance of the early steam and gasoline engines. They have concentrated upon the adoption of the tractor in the twenties, and have missed the multi-step nature of the mechanization process. In addition, they have not included the combine in their assessments of the twenties.[2] The purpose of this paper is to correct this imbalance, and to broaden our view of the mechanization of Canadian Plains agriculture.

Canadian Plains farmers inherited the North American tradition of mechanized agriculture. This tradition began in the 1830s with the introduction of the mechanized reaper. By the 1850s the reaper was popular in southern Ontario, and in the mid-western United States. This lag in acceptance was due to farmers sharing the machines until the price came down. As it did, and as more farmers brought more land under cultivation, more reapers appeared. In the interim, improvements were made, making the basic unit ever more efficient.[3]

As they moved out on to the Great Plains, American farmers took their reapers with them, and began adding new equipment to their operations. Of particular importance was the development of mechanical power sources. By the middle of the 1870s steam power was being increasingly used on American farms and, by the mid-1880s, the self-propelled steam traction engine was being used successfully. Initially steam power made its greatest contributions during the harvest. The success of the reaper had created a demand for more power at harvest time to thresh the crops. Steam engines provided a steady, reliable form of power which replaced the animal power methods employed previously. The advent of the traction engine also allowed the application of mechanical power to field work. Huge gang plows were soon being hitched to these steam behemoths, and millions of acres of the Plains were broken.[4]

While steam engines helped to expand the American frontier, ultimately there were limits to that expansion. The amount of arable land was fixed, and by the 1890s it was clear to many Americans that their era of free homesteading was coming to a close. Irrigation offered some possibilities, yet it was expensive and often limited by terrain. As the turn of the century approached, thousands of American farmers turned northwest to the Canadian Plains—the last large area of arable land on the continent.

The migration of American farmers to the Canadian Plains was paralleled by the expansion of American farm implement companies into the area. Plains farmers in Canada took advantage of this situation, and quickly accepted the new machinery. The early development of the Canadian Plains was therefore unique, because it coincided with the increased mechanization of agriculture. Indeed, the Canadian Plains could be called a mechanized agricultural frontier.[5]

This early agricultural mechanization continued through to the 1920s, and was a logical result of the increasing investment in farm implements by Canadian Plains farmers. Unfortunately the kinds of machines farmers were buying is not known because the federal census did not record such specifics at the time. The census did record farmers' overall investment in machinery, and a general trend is discernible. Between 1901 and 1921 Manitoba farmers tripled their machinery investment. In Saskatchewan and Alberta during the same period farmers increased their agricultural implement investments five times over.[6]

Such large purchases would indicate that Canadian Plains farmers were interested in labour-saving machines. They also bought such large quantities of machinery because it allowed them to quickly bring more land into production. Essentially the Canadian Plains became an agricultural heartland in just a few short decades, an accomplishment which would not have been possible without agricultural machines. Table 1 gives a clear indication of how great the strides were.

TABLE 1. OCCUPIED, IMPROVED, AND FIELD CROP AREAS IN SELECTED YEARS (IN ACRES)[7]

	*OCCUPIED	IMPROVED	FIELD CROPS
MANITOBA			
1901	8,843,347	3,995,305	2,756,106
1911	12,184,304	6,476,169	5,161,858
1916	13,436,670	7,187,737	5,116,661
1921	14,615,844	8,057,823	5,857,635
SASKATCHEWAN			
1901	3,833,434	1,122,602	655,537
1911	28,099,207	11,871,907	9,136,868
1916	36,800,698	19,632,206	13,973,382
1921	44,022,907	25,037,401	17,822,481
ALBERTA			
1901	2,735,630	474,694	188,476
1911	17,359,333	4,351,698	3,378,365
1916	23,062,767	7,510,303	5,505,872
1921	29,293,053	11,768,042	8,523,190

*The federal census from which these figures are taken defined "occupied" as the acreage which had been purchased or homesteaded, "improved" as the acreage which had been cultivated, and "field crops" as the seeded acreage.

Steam traction engine with plow and packer, district of Oxbow, Saskatchewan, circa 1910. (Courtesy of the Saskatchewan Archives Board/R-A108.)

These gains were the result of increasing mechanization, and not merely from more farms coming into production. This is clear when one compares the improved acreage to the number of farms between 1901 and 1921 in Saskatchewan, the leading agricultural province. Four years before it became a province, there were 13,445 farms in Saskatchewan. There were over one million improved acres, and the average improved crop acreage per farm was just eighty-four acres. By 1911, the number of farms had increased to over ninety-five thousand. Improved acreage was almost twelve million acres, and the average improved and field crop acreage per farm had increased to 125 acres. In 1921, there were just under 120,000 farms in the province. Improved acreage was over twenty-five million acres, and the average per farm was now 209 acres.[8] There were more farmers, and they were farming more land, but each farmer was handling more land because of agricultural mechanization.

Agricultural engines played an important role in the evolving mechanization, and the resulting gains. Emil Julius Meilicke, a German immigrant who had farmed in Wisconsin and Minnesota before homesteading south of Saskatoon, used machine power to establish his new farm. As Meilicke recalled,

> When I was still breaking my farm in 1904, I hired two tractors and other machinery to speed up the work. People thought I was crazy to go to all this expense, but I got my farm into cultivation at once, and in 1905 ... we harvested over 46,000 bushels of wheat.... It brought me sixty-four cents a bushel. My extravagance paid.[9]

There were many others who recognized the benefits of machine power. One observer, who travelled through Saskatchewan and Alberta in 1909, estimated that one-half of the breaking in those two provinces that year was done by steam traction engines.[10] These engines, and their successors the gasoline and kerosene tractors, also provided the power to thresh millions of bushels of wheat, oats, and barley. Like their American counterparts, Canadian Plains farmers found that the steady power of an engine was ideal for the demands of threshing.

Canadian Plains agriculturalists recognized that mechanical power sources were the way of the future, and purchased accordingly. Canadian suppliers could not keep up with the demand. In 1907 Canada imported 528 engines for agricultural purposes. By 1911 this figure had climbed to over 2,000 units, with a value of $3.5 million, imported principally from the United States. In 1919 nearly 15,000 tractors were brought in from the United States; and in the next ten years close to 125,000 such units were brought in from south of the border.[11]

This early mechanization of Canadian Plains agriculture had certain interesting side effects. The layout of farm yards was affected by the early mechanization. Saskatchewan farmers, for example, were advised by their provincial Department of Agriculture to plan their farm yards with machinery in mind. Specifically, they were advised that,

> Means of ingress and egress must be provided on either side to give access to the fields. In threshing time it must be remem-
> bered that a large traction engine pulling a separator, a caboose
> and probably a tank, cannot turn a sharp corner and negotiate
> a narrow gateway without considerable risk of running into a
> gatepost.[12]

While mechanization had a considerable impact on farming operations, farmers did not replace their horses right away. Horses continued to be used extensively, even exclusively, on many farms. New mowers and reapers might appear, but it was still horses which pulled them. Yet neither horses, nor animal and human power combined, could account for the great strides in bringing land into production. Whether during breaking or at harvest time, once harnessed the agricultural engine had an important, and increasing role in supplying power on the farm.

Even longtime admirers of the horse recognized the importance of the new power sources. A. F. Mantle, Saskatchewan's Deputy Minister of Agriculture from 1910 to 1916, argued that,

> We do not relish the thought of horseless farms but welcome
> the advent of the tractor that will relieve our horses of the
> slavish part of their work, permit us to reduce their numbers,
> and enable us ... to plow or summerfallow deeper and better ...
> In threshing operations man was displaced by the horse—and
> has never regretted it (at least those of us who have ever swung
> the flail have no regret!); the horse in turn was displaced by the
> steam traction engine—and neither the horse nor man regret it;
> now the steam engine is being largely displaced by the gasoline
> tractor—and no fireman regrets it; perhaps soon the tractor will
> give place to the electric motor! Why then need there be any
> sentimental regrets or doubts about displacing the horse as a
> source of power for breaking sod and plowing summerfallow?[13]

Farmers had several incentives to move from animal to mechanical power,
not the least of which were the problems associated with using temperamen-
tal animals. During the early years of farming in Alberta farmers claimed to
break sod by hand because their horses were too wild. According to one report
about such animals,

> Thus his errant ways seldom allowed him to go straight very
> long and as he meandered the plowing took on much the
> appearance of being done haphazard or several different ways
> at once. Then the plowman turned the team loose or tied them
> up and turned over the wandering sods himself.[14]

The transition from horse to mechanical power was not accomplished
without considerable debate, such as arose at the Dry Farming Congress at
Lethbridge, Alberta, in 1912. At that gathering a representative of the Rumley
Company, the manufacturer of a popular line of engines, made the statement
that tractors would soon replace horses for heavy work on the farm. He was
immediately challenged by several farmers whose experience with horses had
been positive. The Rumley representative countered these arguments with statistics which positively compared the cost of keeping an engine to the upkeep
of horses. He also claimed that, since crop production did not keep pace with
settlement, sooner or later it would be necessary to convert completely from
horses to tractors.[15]

What the Rumley representative meant by this last statement was that horses
were themselves an incentive to get power equipment. Horses consumed the

production of many acres of land. By converting to tractors, the land normally set aside for pasture and feed could be planted with cash crops.[16]

In spite of the obvious benefits of power equipment, and their desire to have it, farmers did not adopt these new machines for general farm use all at once. Steam and gasoline engines made tremendous contributions in plowing and threshing, but horses were still used for most other farm tasks. The problem was that much of this power equipment suffered from serious defects in the eyes of farmers. Steam traction engines, for example, were prone to starting fires, and needed a great deal of water to operate, something not always plentiful in some parts of the Canadian Plains. The early gasoline engines were generally hard to start and, like their predecessors, were often huge machines whose weight bogged them down in wet fields, or pushed them through culverts and wooden bridges. The smaller units which followed in the evolution of the farm tractor were often too small for the heavy work done on the Canadian Plains.[17]

Farmers themselves ran risks when operating the early equipment. Scarcely a threshing season went by without reports in the newspapers about boiler explosions or other such accidents which maimed and even killed the operators. Most farmers recognized the dangers, and sought expert advice. The colleges of agriculture of the Canadian Plains universities responded, establishing short courses which taught the fundamentals of steam and gasoline engineering. In his 1912 Report to the President, the Dean of Saskatchewan's College of Agriculture reported that,

> Short courses of four days' duration in gas traction engineering were held at Tantallon, Lemberg, Davidson, Abernethy, Nokomis, Marcelin and Govan under capable instruction and later a short course was held at the College under the supervision of Prof. Greig. All the short courses held were well patronized and the instructors well received.... The numbers reached directly through these short courses totaled about 1,530.[18]

In spite of the problems and dangers, farmers still purchased the new machines. The simple fact was that the equipment made their work easier, and increased their chances of planting and harvesting crops. The attraction of the new machinery was so great that some farmers bought too much of it. An acute British observer, travelling on the Canadian Plains prior to World War I, noted that the "machinery man" was often a villain in the "Western drama." According to Elizabeth Mitchell,

The great firms of Ontario, or American firms with agencies in Canada, send out skilled salesmen into the country districts, and individual farmers are persuaded to buy a great many things beyond the strictly necessary ploughs and binders, and, in the absence of capital, to mortgage their farms to the machinery company. Then the valuable machine is perhaps left exposed to the weather because there is no shed big enough for it; it goes wrong in some detail, and the farmer is no skilled mechanic to understand the trouble; and there is no machinery shop within hundreds of miles. In some such way as this, the farmer often gains very little and is left burdened with debt.[19]

The implement dealers held a different view of the matter. Responding to a Saskatchewan government initiative to regulate farm equipment selling, F. H. Crane, manager of the Saskatoon branch of Canadian Fairbanks-Morse Company, argued that,

It must be admitted, however, that many times a farmer in his enthusiasm will purchase of his own free will and accord, that is, without being pressed to do so by the wholesaler or the dealer, and to obtain credit he makes a false statement as to his financial standing .... When you consider that less than one half of the farmers' notes which fell due last fall, have been met, you can realize the vast amount of money which is outstanding on our books at the present time... .[20]

Steam engines and tractors were to play an additional role in the early settlement period, that of quickly breaking huge tracts of ranch and grazing land for wheat production. All that was lacking for this to take place was an incentive, and the high wheat prices of World War I provided that essential ingredient. Such was the case of C. S. Noble of Claresholm, Alberta. Noble had migrated from North Dakota to the future province of Alberta in 1902. He was immediately successful in farming and real estate. The high wheat prices during the war led him to purchase the Cameron Ranch, situated near the confluence of the Oldman and Little Bow Rivers. He immediately set to breaking the twenty thousand acres of rangeland. His son later recalled the event:

Headquarter buildings were erected and four additional camps with barns and living quarters were set up at convenient points to work outlying parts of the farm. Lumber and supplies were moved out from Nobleford by wagon, trucks, and steamers with wagon trains.

In 1918, 10 steam tractors ran night and day to break up the land. Dams had been built across draws to hold water and large horse-drawn tanks carried it to the steamers. Coal was hauled for them from the Taber mines. Nearly 400 acres were broken each 24 hours. It was a memorable sight and the smoking monsters broke up the whole tract in record time.[21]

While wartime demands for wheat spurred mechanization, the post-war collapse of the international wheat market postponed further development. Following wartime controls, the Winnipeg Grain Exchange reopened to trading in wheat futures on 18 August 1920. The price of No. 1 Northern wheat averaged just over $2.73 per bushel in September 1920. A year later, however, the price had dropped to under $1.50 per bushel, and in October 1921 it had fallen again to under $1.20 per bushel.[22] In just over one year Canadian Plains farmers had seen the price for their principal crop drop by almost $1.60 per bushel. Such a catastrophic decline had an immediate effect on farmers' purchases, including machinery. Even though they may have wanted one, tractors became a machine farmers learned to do without. The connection between falling wheat prices and decreased tractor sales was direct, as shown in Table 2.

These figures support the argument that the mechanization of Canadian Plains agriculture was an ongoing process, yet also reveal that the pace of change was understandably subject to the farmers' economic condition. Canadian Plains farmers were intent on mechanizing, and did so when they had the means. When farmers were under duress, such as in the early twenties and with the start of the Great Depression, they held back on their purchases and waited until conditions improved.

While subject to the vagary of the wheat market, Canadian Plains farmers could not ignore the internal logic of a process such as mechanization. Once under way, mechanization had a dynamic all of its own. For example, one of the consequences of mechanization was the loss of the skills required to farm with horses. This reinforced the general trend toward more power machinery because young farmers were increasingly oriented toward it. This was apparent to contemporaries such as Evan Hardy of the College of Agriculture at the University of Saskatchewan. In 1925 Hardy observed that,

The horse pulling competitions are reviving a slowly dying interest in teamstership. Many of our men and boys, growing up or working on the farm, have been thinking more of the automobile truck and tractor than of training and driving horses. The lack of expert teamstership was evident throughout the contests....[23]

The increasing prosperity of the late twenties allowed many farmers to look at machinery other than tractors. The combine had appeared, and increased the overall trend toward mechanization. Canadian Plains farmers were interested in the possibilities of the combine because it combined the reaping and threshing tasks. As has been noted, the mechanical reaper was an early implement on the Canadian Plains. It only did part of the harvesting job, however, cutting the grain and tying it into sheaves. The sheaves still had to be "stooked" and left to dry, after which they were collected and threshed. Each step was labour-intensive and time consuming. Canadian Plains farmers also turned to machines to ease the threshing burden. In 1898 there were 368 threshing outfits in the entire North-West Territories, but by 1908 there were 3,219 in Saskatchewan alone.[24]

TABLE 2: TRACTOR SALES ON THE CANADIAN PLAINS, 1919-1931[25]

(Number of units sold)

YEAR	MANITOBA	SASK.	ALBERTA	TOTAL
1919	3,627	3,514	1,703	8,844
1920	3,671	4,229	2,379	10,279
1921	1,057	1,655	716	3,428
1922	1,361	2,475	386	4,222
1923	911	2,524	731	4,166
1924	465	1,213	434	2,112
1925	1,008	2,176	869	4,053
1926	1,498	3,704	1,311	6,513
1927	1,414	5,727	2,885	10,026
1928	2,209	8,703	6,231	17,143
1929	2,423	6,906	5,228	14,557
1930	1,541	4,350	3,100	8,991
1931	186	267	334	787

The combine allowed the grain to be cut and threshed in one operation. Such relatively complicated pieces of machinery had evolved over many decades. A patent for a combined harvester and thresher was issued in the United States as early as 1828. A combined harvester was tested in Michigan in the late 1830s, and spawned other units which operated in the state until at least 1853. A year later one such machine was shipped to California, but was sold in 1856 and apparently abandoned shortly thereafter. Stripper type combines were also in use in Minnesota in 1884.[26]

The early experience with combines in California was successful, and during the 1880s a number of firms began producing the machines commercially. Combines were in regular use in the state during the next decade, and had begun to spread northward. In 1912 a representative of a Spokane, Washington, combine manufacturer attending the Lethbridge, Alberta, Dry Farming Congress boasted that combines were rapidly being introduced into Oregon, Washington, Idaho, Utah, and Montana. He suggested that it was just a matter of time before the machines would be in general use in those states. He also explained that while most of the then current models were designed to be pulled by horses and powered by drive wheels, some had gasoline engines to power them and the new models could be pulled by tractors.[27]

California was not the only grain-growing area to witness the evolution of a combine. In 1843 a stripper type machine was developed in Australia. It was modified in the 1880s by Hugh Victor McKay, and by 1910 McKay's Sunshine factory was shipping units as far away as Argentina. The success of the McKay combine attracted the attention of Canada's Massey-Harris Company, which developed its own version of the machine. Massey-Harris marketed its combines internationally, but did not sell them in Canada.[28]

Apparently Massey-Harris did not feel that there would be a market for combines on the Canadian Plains. Combines had been developed in areas where the climate permitted the uniform ripening of the grain, allowing what has come to be known as straight combining. The climate of the Canadian Plains did not favour this practice and, as has been shown, farmers in the area relied on their mechanical reapers and steam or gasoline driven threshers. This is not to suggest that there were no experiments with the combine on the Canadian Plains. In 1910 two farmers near Welby, Saskatchewan, imported a Holt twenty-foot ground-drive combine from California. They successfully handled six hundred acres of wheat during each of the three harvests between 1910 and 1912. In 1913 they planted their acreage to flax, and the combine was used on it. It is important to note that this rig was pulled by a tractor; a Hart-Parr 30-60 tractor, to be precise. When more combines appeared in the twenties there was a corresponding increase in the need for tractors to pull them.[29]

In 1912 the Holt Company claimed to have three of their combines operating in Alberta; one at Strathmore, one at Tilley, and one at Bassano. According to one contemporary source, "Some doubts are expressed as to the ability of the California method to conditions prevailing here, but the Holt Co. have the faith of conviction and assert that they will win their way to favor."[30] Farmers on the Canadian Plains were doubtless concerned about having to wait until the crop was completely ripe before combining it. Their experiences with early frosts were probably a factor in the farmers' thinking. This problem was overcome by the development of the swather, and by the appearance of combines designed to pick up the windrows.

In 1910 two South Dakota brothers, Ole and August Hovland, built a swather as well as a thresher designed to pick up the windrows. Little came of it until Helmer H. Hanson and his brother Ellert, who had known the Hovlands, moved to Lajord, Saskatchewan. In 1926 they developed two twenty-foot swathers, and rigged a combine to pick up the windrows. Their progress was observed by officials of the International Harvester Corporation, which perhaps explains why that company was the first to market swathers in 1927.[31]

The development of the swather removed many of the farmers' concerns about combining, and from modest beginnings the sale of combines grew rapidly. In 1922 the International Harvester Corporation experimented with one of its units on the federal experimental farm at Cabri, Saskatchewan, while the Massey-Harris Company tested a twelve-foot, motor-driven combine at the Swift Current, Saskatchewan, Experimental Station. This latter test was so successful that the federal government purchased the unit the same year. The J.I. Case and International Harvester companies also sold a few machines in the province.[32] Once under way, combine sales soared, only to decline with the coming of the Depression. Table 3 shows the increase in sales in the late twenties and the impact of the Dirty Thirties:

TABLE 3: COMBINE SALES ON THE CANADIAN PLAINS, 1926-1931[33]

(Number of units sold)

YEAR	MANITOBA	SASK.	ALBERTA	TOTAL
1926	2	148	26	176
1927	21	382	195	598
1928	206	2,356	1,095	3,657
1929	158	2,484	858	3,500
1930	134	939	541	1,614
1931	33	92	54	179

Combines offered farmers several advantages over traditional harvesting equipment. According to E. S. Hopkins, Dominion Field Husbandsman in

1928, a twenty-bushel crop, grown on six hundred acres, could be harvested by a binder and threshed by a separator for 17 ½ cents per bushel. The same crop could be harvested by a combine for 9 ⅓ cents per bushel. Farmers could also reduce their crop loss by harvesting with a combine. In 1928 the Swift Current Experimental Station ran tests which showed a 1.16 percent loss with straight combining, and a 3.58 percent loss with a traditional binder and separator.[34]

What this meant to a farmer can best be appreciated by examining the case of an individual farmer. Anthony Tyson farmed near Neidpath, Saskatchewan, southwest of Swift Current. A partial account of his farm statistics appears as Table 4:

TABLE 4: ANTHONY TYSON'S FARM STATISTICS, 1917-1929[35]

YEAR	ACRES	BUSHELS THRESHED	WAGES	THRESHING
1917	120	1055	$343.50	$158.25
1918	155	950	57.75	176.00
1919	250	210	*.00	38.00
1920	250	00 (Hail)	283.00	8.00
1921	215	2,631	390.20	430.45
1922	215	5,400	607.00	923.00
1923	230	5,280	410.00	800.00
1924	250	4,150	362.50	566.80
1925	255	4,716	320.00	773.40
1926	272	4,678	284.00	701.70
1927	280	6,162	435.00	1149.30
1928	270	8,100	568.00	1409.56
1929	255	2,400	366.00	338.76

* It is not clear why Mr. Tyson did not pay any wages for farm help in 1919. Perhaps he did not need any help that year because of the light crop, although the following year he was hailed out and still paid out nearly $300.00 for help. It is possible that in 1920 he had a hired man for other duties.

Using Mr. Hopkins' estimate of 9 ⅓ cents per bushel to harvest a crop with a combine, had Anthony Tyson purchased a combine to harvest his 1927 crop, he would have spent $537.07 on the harvesting. His cost in 1928 would have been $753.30, and in 1929 it would have been $223.20. Over this three-year period he would have spent $1,549.57 using a combine, compared to the $2,897.62 he actually spent. The saving would have been $1,348.05. In addition, Mr. Tyson spent $1,369.00 on labour in this three-year period. He may have been able to halve this figure had he used a combine during the harvest, thereby saving an additional $684.50. His total savings, therefore, would have been $2,032.55. With the cost of a combine at the time running between $1,200.00 and $3,000.00,[36] Mr. Tyson would have been close to paying off an initial investment in just three years.

These calculations are general and the situation with the combine hypothetical. Still, they do emphatically illustrate the gains to be made by obtaining a combine. Farmers such as Mr. Tyson could do their own rough calculations, and the result was the increasing use of combines.

As Mr. Tyson's records indicate, the cost of farm labour was an important consideration in farming mathematics. Labour was not cheap, when it could be found. There was never enough labour for farmers, particularly at harvest time. Excursion trains were organized to bring badly needed help from Eastern Canada to the Plains. Still, there never seemed to be enough help.[37]

Furthermore, the wages paid were only part of the cost of labour. At harvest time the farm family was expected to provide board and lodging to the temporary workers. In addition to being costly, harvest labour put a serious burden on domestic arrangements in the farm home. As one contemporary observer noted, "The Combine is a farm machine which reduces the labour for women on the farm. The drudgery of cooking and serving the harvest hands is eliminated with the reduced need for labour to do the harvesting." The farm wife may have played an unheralded role in increasing mechanization, since the number of harvesters required for the two similar sized crops of 1924 and 1929, went from 21,000 men to virtually nil at least in part because of the use of combines.[38]

While the combine helped the farmer with some of his problems, particularly during the harvest, it also created new ones for him. The combine was most effective when the crop had reached uniform ripeness. This meant careful attention to spring seeding to insure that the entire crop reached maturity at the same time. A farmer might leave a green field to ripen, but this increased the risk of frost damage. As a contemporary observer concluded,

> ...the use of the Combine is not a cure-all for farm ills. The use of it may assist, however, in solving some of the problems of the harvest. The successful Combine users are those who farm throughout the year with the use of the Combine as the goal.[39]

The growth in combine use in the late twenties thus heralded a major change in farming methods on the Plains. Traditionally, the peak work period of the farmer's year was the fall—harvest time. The combine shifted this period to the spring. By eliminating the need for large crews to operate the binders and separators, and increasing the need to plant early so as to insure a ripe crop in time for combining, the combine altered the farmer's work year drastically.

The combine also reinforced the trend toward power equipment. Not only did farmers now need tractors to pull the combines, they also saw the need for power to pull larger tillage and seeding implements in the spring. Unfortunately there are no statistics available on how much of this machinery Canadian Plains farmers were purchasing, but it must have been substantial because implements designed to be pulled by horses could rarely be used with tractors. Fortunately there are statistics on another farm machine, the truck, and it is possible to glimpse the impact of the combine.

The combine spurred the use of trucks because combines could hold a wagonload of grain in their tanks. Rather than haul the combine to a storage area for unloading, many farmers saw the logic of unloading the combine in the field and hauling the load out by truck. This method also saved the farmer from hiring additional and expensive horses, wagons, and men to drive them, to haul the grain to an elevator. This effect is documented by the increase in the number of truck registrations in the late twenties. In 1926, for example, Saskatchewan had 8,688 trucks. The following year saw an increase to 11,346, and by 1929 there were 18,671 trucks registered in the province.[40]

Combines also had an impact on the sale of other types of machinery, notably threshers. Unfortunately records on such sales are incomplete. In 1928, the first year in which thresher sales figures are available, threshers outsold combines 6,247 to 3,657 on the Canadian Plains. The following year, however, there were 3,500 combines sold compared to 2,095 threshers. In 1930 threshers regained the advantage with 2,046 sales to 1,614 combine sales. In just three years combine sales had gone from half that of threshers to essentially equal footing.[41]

Thresher sales were also affected by the coming of the Great Depression, and in 1931 only 445 units were sold on the Canadian Plains compared to 179 combines. Combines continued to take second place to threshers in sales during the early thirties, but by the latter half of the decade combines had gained the lead. By 1940 combines were consistently out-selling threshers by a wide margin in the three Plains provinces. In 1945 the trend was firmly established, and combines outsold threshers 5,940 to 88. The transition begun in the late twenties had been resurrected by profits and good crops during World War II. By 1951 there were nearly 43,000 combines in Saskatchewan alone.[42]

The mechanization of Canadian Plains agriculture prior to 1930 was pronounced. This trend was so strong that this mechanization could be described as a characteristic of the type of farming in the area. Looking back upon the period, the 1955 Saskatchewan Royal Commission on Agriculture and Rural Life observed that,

Threshing using an Avery engine, circa 1920. (Courtesy of the Saskatchewan Archives Board/R-A2177.)

A strong trend towards tractors and combines was evident in the 1920s. One can assume that, had serious economic depression and drought not occurred during the 1930s, the change from horse to tractor power would have been complete by the early 1940s.[43]

Canadian Plains farmers who did not keep pace with mechanization paid a harsh price, however, and that was being forced to give up farming. Mechanization was a mixed blessing, for while it allowed farmers to do more, it eventually meant fewer farmers were required to work the land. The first indication of this process began to appear in the mid-twenties.

Saskatchewan, the leading agricultural province, first noted this effect of mechanization as early as 1926. For the first time since it was settled there was a decline in the number of farms in the province that year. While other factors, such as the fall in wheat prices in the early twenties, were no doubt operating, it appears that some farmers in Saskatchewan were taking over their brethren. They were able to do this because increased mechanization allowed them to work more land.

In 1921 there were 119,451 farms in Saskatchewan, but five years later there were almost two thousand fewer. The entire decline appears to have been in the number of farmers working between 101 and 200 acres of land. In 1921 there

Combining south of Saskatoon, October 1951. (Courtesy of the Saskatchewan Archives Board/R-A11,309-2.)

were 37,059 farmers in Saskatchewan in this category, yet five years later there were only 33,276, a drop of nearly four thousand. Half of this number appear to have expanded their holdings because there was an increase during this period of almost two thousand in the number of farmers who were working 201 or more acres; from 80,570 to 82,520. The remaining two thousand were likely those who left farming. In terms of the percentage of holdings, those in the 101 to 200-acre category fell from just over 32 percent of provincial farmers to just over 28 percent. Farmers holding more than two hundred acres increased from just under 67.5 percent to over 70 percent of all provincial farmers. As might be expected, there was also an increase in the average size of farms, from just over 368 acres to almost 390 acres.[44]

As North America's last major agricultural frontier, the Canadian Plains area benefitted from the mechanization of agriculture under way when it was first settled. Canadian Plains farmers knew the advantages of mechanization, and purchased accordingly when they could afford it. This resulted in phenomenal strides in the amount of land devoted to agriculture during the first decades of the twentieth century. Steam traction engines, in particular, aided in this development, before giving way to the more efficient gasoline-powered tractor. Tractors were supplemented by combines during the twenties as the trend to mechanization continued.

The 1920s, particularly the latter half of the decade, were thus of critical importance to agricultural mechanization in the area because they were the second stage of a process which transformed farming on the Canadian Plains and laid the base for our modern form of agriculture. There were profound changes in the implements farmers used and in their traditional work year, with a premium being placed on early planting, and a lessened harvest burden. Not all farmers benefitted to the same degree, or mechanized at the same rate. Hundreds of them paid a harsh penalty for delaying mechanization, and were forced off the farm. They were soon to be joined by thousands more when full mechanization resumed during World War II, the effect of which was seen in the late 1940s and during the 1950s.

NOTES

This article first appeared in *Prairie Forum* 11, no. 2 (1986): 253-71.

1   Herbert F. Lionberger, *Adoption of New Ideas and Practices* (Ames, Iowa: Iowa State University Press, 1960), 12, and 13.
2   Robert Ankli, H. Dan Helsberg, John Herd Thompson, "The Adoption of the Gasoline Tractor in Western Canada," in Donald H. Akenson, ed., *Canadian Papers in Rural History*, Vol. II (Gananoque, Ontario: Langdale Press, 1980), 9-39; John Herd Thompson, *The Harvests of War: The Prairie West, 1914-1918* (Toronto: McClelland and Stewart, 1978), 63.
3   Alan L. Olmstead, "The Mechanization of Reaping and Mowing in American Agriculture, 1833-1870," *Journal of Economic History* 35 (2) (June 1975): 327-52; Richard Pomfret, "The Mechanization of Reaping in Nineteenth-Century Ontario: A Case Study of the Pace and Causes of the Diffusion of Embodied Technical Change," *Journal of Economic History* 36 (2) (June 1976): 399-414.
4   R. Douglas Hurt, *The Dust Bowl: An Agricultural and Social History* (Chicago: Nelson-Hall, 1981), 22-23; E. M. Dieffenbach and R. B. Gray, "The Development of the Tractor," in Alfred Steffered, ed., *Power to Produce: U.S. Department of Agriculture 1960 Yearbook of Agriculture* (Washington: Government Printing Office, 1960), 26. See also, Reynold *Steam Power on the American Farm* (Philadelphia: University of Pennsylvania Press, 1953), throughout.
5   L. H. Thomas, "A History of Agriculture on the Prairies to 1914," *Prairie Forum* 1 (1) (April 1975): 31-44; R. Bruce Shepard, "The Mechanized Agricultural Frontier of the Canadian Plains," *Material History Bulletin*, National Museum of Man, Ottawa (Spring 1979): 1-22.
6   R. Bruce Shepard, "The Mechanized Agriculture Frontier," 1-22.
7   Canada, *Census of the Prairie Provinces, 1916* (Ottawa 1918), 291; Canada, *Census of Canada, 1921*, vol. 5 *Agriculture* (Ottawa 1925), 5.
8   Canada, *Census of the Prairie Provinces, 1916* (Ottawa, 1918), 291; Canada, *Census of Canada, 1921*, vol. 5, *Agriculture* (Ottawa 1925), 5; Canada, *Census of Saskatchewan, 1926* (Ottawa 1927), 198. The averages are my own calculations.

9   Emil Julius Meilicke, *Leaves From the Life of a Pioneer; Being the Autobiography of Sometime Senator Emil Julius Meilicke* (Vancouver: Wrigley Print Company, 1948), 135. A copy of this book is in the Shortt Collection, University of Saskatchewan Archives, Saskatoon, Saskatchewan (hereafter cited as U. of SA). Interestingly, Meilicke did not use mechanical power sources after this initial positive experience because he felt that they were too expensive.

10   A. Fred Mantle, "The Traction Plow in Western Canada," *Thresherman's Review* (October 1909), 7, cited in Wik, *Steam Power*, 150.

11   Canada Year Book, 1911 (Ottawa 1912), 158-59; W. G. Phillips, *The Agricultural Implement Industry in Canada: A Study in Competition* (Toronto: University of Toronto Press, 1956), 65.

12   Saskatchewan Archives Board (hereafter cited as SAB), Saskatoon, Government Pamphlet Collection, Ag. 8, No. 67, "Planning the Farmstead and Buildings," Province of Saskatchewan, Department of Agriculture, 1912(?).

13   SAB, Regina, F. H. Auld Papers, Box 2, File 9, Part H, "Progress in Western Agriculture" (1911 or 1912), 2.

14   Lethbridge, Alberta, *Herald*, 21 October 1912.

15   Ibid., 23 October 1912.

16   John T. Schlebecker, *Whereby We Thrive: A History of American Farming, 1607 to 1972* (Ames, Iowa: Iowa State University Press, 1975), 203. See also, James H. Gray, *The Roar of the Twenties* (Toronto: Macmillan, 1975), 50; and James H. Gray, *Men Against the Desert* (Saskatoon: Western Producer Prairie Books, 1967), 11.

17   R. Bruce Shepard, "The Mechanized Agricultural Frontier," 1-22; Ankli, Helsberg, and Thompson, "The Adoption of the Gasoline Tractor," 9-39.

18   U. of SA, Saskatoon, College of Agriculture Files, I, Dean's Correspondence, A. "Reports to the President, 1912-28," 30 June 1912.

19   Elizabeth B. Mitchell, *In Western Canada Before the War: Impressions of Early Twentieth Century Prairie Communities* (Saskatoon: Western Producer Prairie Books, 1981), 161-62.

20   SAB, Regina, Department of Agriculture Records, Statistics Branch (Selected, 1910-1916), #1670, "Cost of Farm Machinery," F. H. Crane to the Department, 13 May 1913.

21   S. F. Noble, "Dr. Charles S. Noble," Lethbridge, Alberta, *Herald*, 28 December 1967.

22   Vernon Fowke, *The National Policy and the Wheat Economy* (Toronto: University of Toronto Press, 1978), 177.

23   U. of SA, Saskatoon, Hardy Papers, "Horse Pulling Contests, 1925-1955," clipping from the *Western Producer*, 9 July 1925.

24   John Archer, *Saskatchewan: A History* (Saskatoon: Western Producer Prairie Books, 1980), 147.

25   *Canadian Farm Implements* (January 1947), 8.

26   L. H. Thomas, "Early Combines in Saskatchewan," *Saskatchewan History* 8 (1) (Winter 1955): 1-5; Herbert F. Miller, Jr., "Swift Untiring Harvest Help," in Alfred Steffered, ed., *Power to Produce,* 165-66; Thomas Isern, *Custom Combining on the Great Plains: A History* (Saskatoon: Western Producer Prairie Books, 1982), 12.

27   T. Isern, *Custom Combining,* 12; Lethbridge, Alberta *Herald,* 24 October 1912.

28   Richard J. Friesen, *The Combine Harvester in Alberta: Its Development and Use, 1900-1950,* Background Paper No. 9 (Edmonton: Reynolds-Alberta Museum, 1983), 16-17.

29 Thomas, "Early Combines in Saskatchewan," 1-5.
30 Lethbridge, Alberta, *Herald*, 25 October 1912.
31 Helmer H. Hanson, *History of Swathing and Swath Harvesting* (Saskatoon: Western Producer Prairie Books, 1967), throughout; Thomas Isern, "Adoption of the Combine on the Northern Plains," *South Dakota History* 10 (2) (Spring 1980): 101-18.
32 J. S. Taggert and J. M. Mackenzie, "Seven Years' Experience With the Combined Reaper-Thresher," *Experimental Farm Bulletin 114-136*, New Series (Ottawa, 1929), 1, 5 and 23; Thomas, "History of Agriculture," 31-44; Isern, *Custom Combining*, 18.
33 *Canadian Farm Implements* (January 1947), 8.
34 Taggert and Mackenzie, "Seven Years' Experience," 17-18.
35 Edna Tyson Parson, *Land I Can Own: A Biography of Anthony Tyson and the Pioneers Who Homesteaded with Him at Neidpath, Saskatchewan* (Ottawa: Westboro Printers Limited, 1981), 71.
36 W. W. Swanson and P. C. Armstrong, *Wheat* (Toronto: Macmillan, 1930), 58.
37 John Herd Thompson, "Bringing in the Sheaves: The Harvest Excursionists, 1890-1929," *Canadian Historical Review* 59 (4) (1978): 467-89.
38 U. of SA, Publications Collections, Ag. Bulletin No. 38, 1927, Evan A. Hardy, "The 'Combine' in Saskatchewan," 5; Swanson and Armstrong, *Wheat*, 58-59.
39 U. of SA, Saskatoon, Hardy, "The 'Combine' in Saskatchewan," 10.
40 G. T. Bloomfield, "'I Can See A Car in That Crop': Motorization in Saskatchewan, 1906-1934," *Saskatchewan History* 37 (1) (Winter 1984): 3-23.
41 Friesen, *The Combine Harvester in Alberta*, 72
42 Ibid.
43 Saskatchewan, *Royal Commission on Agriculture and Rural Life*, Report No. 2, *Mechanization and Farm Costs* (Regina, 1955), 30-31.
44 Canada, *Census of Saskatchewan, 1926* (Ottawa, 1927), 198, 208, 211-12.

Ranching

10. **Eastern Capital, Government Purchases, and the Development of Canadian Ranching**

A. B. McCullough

ollowing the American Civil War the ranching industry on the American Great Plains enjoyed a period of extraordinary growth. Stories of huge profits fuelled an investment boom by eastern American and British investors. The boom began in Texas and moved rapidly north; by 1882 the ranching frontier had reached the Foothills country of Alberta. Many ranges in the United States were dangerously overstocked. The disastrous winter of 1886 broke the boom, but by that time ranching was well established and survived on a sounder basis.

The "beef bonanza" in the United States was paralleled, allowing for a time lag of a decade, by a similar boom in the Canadian West. Cattle ranching began on a small scale in the Foothills of Alberta in the 1870s with small, independent ranchers grazing cattle on crown land without the benefit of a lease. In 1880 Senator Matthew Cochrane, a Canadian manufacturer and stockbreeder, successfully lobbied the Canadian government to allow large leases of crown land for ranching in western Canada on very favourable terms. Cochrane and other eastern Canadian and British investors subsequently organized a number of large corporate ranches. Through their capital, managerial talent and political connections the corporate ranches dominated the Canadian ranching country economically, socially and politically until the first decade of the twentieth century.

Canadian historiography of ranching has evolved through two stages and may now be moving toward a synthesis. The first historians who dealt with ranching, L. V. Kelly, C. M. MacInnes, and A. S. Morton and Chester Martin, viewed Canadian ranching as an extension of American ranching practices and attitudes into Canada. Moreover, they saw the ranching era as a brief interlude between the time of the aboriginal buffalo hunters and agricultural settlement. This view of the Canadian ranching frontier as a simple and transitory extension of American ranching was substantially modified by the work of Sheilagh Jameson and L. G. Thomas. Both Jameson and Thomas emphasized the cultured, anglophile society which flourished in the ranching country in the three decades before the Great War. They characterized it as significantly more hierarchical, ordered and respectful of authority than American ranching society.[1] This theme was developed and expanded by David Breen. Breen also argued that Canadian ranching developed within a policy and legal framework set by the central government and enforced by the North-West Mounted Police (NWMP), whereas American ranching developed in an environment which was less regulated and more responsive to local pressures.[2]

Breen's well-documented and carefully argued thesis has become the new orthodoxy, but it is now undergoing some modifications. In particular, the work of Simon Evans has emphasized that there were regional variations in Canadian ranching, and that on the short-grass plains of southeastern Alberta and southern Saskatchewan American influence was stronger than that of Britain.[3] W. M. Elofson has argued that similar environments and similar technical problems on the American and Canadian ranching frontiers produced similar responses. American technical expertise in ranching was valued and eventually earned American immigrants such as George Lane an important place in Canadian ranching society.[4]

This article focuses on the role of eastern Canadian, British and American investors as sources of capital for the Canadian ranching industry in the years 1880-1900. Although existing historiography acknowledges the role of eastern capital in general terms, the scale of eastern investment in ranching has not been examined in detail. This article begins with a consideration of the financial contributions of three individuals and three ranches—Mattthew Cochrane, president of the Cochrane Ranch; Frederick Stimson, manager of the North-West Cattle Company; and, Duncan McEachran, general manager of the Walrond Ranch—and moves to a more general estimate of the financial significance of eastern-based corporate ranching. Finally, the role of the federal government as a market for beef, and hence a source of operating capital, in the early years of ranching will be considered.

Ranching in the Canadian Foothills began when small herds of breeding cattle were brought into the region from British Columbia and the United States in the early 1870s; David Breen suggests that by 1880 there were 200 small ranchers between the boundary and the Bow River.[5] These early ranchers were a mixed lot: serving members of the North-West Mounted Police, ex-policemen, ex-whiskey traders, missionaries, Americans and Canadians. What united them all was that they were local residents and, so far as is known, owner-operators. They operated on a small scale: it is probable that there were no more than 3,000 cattle in the Foothills area in 1880-81.[6]

The 1880s witnessed a mushroom growth of ranching in the Foothills and a significant change in its social and economic structure. The 1891 census found 128,068 head in the subdistricts, Calgary/Red Deer and Macleod, which included the Foothills ranching country. Of these at least 95,000, or 74 percent, were in forty-two herds of 500 head or more.[7] Such a herd represented an investment of at least $15,000, more than most individuals could dispose of at a time when the working cowboy was fortunate to make $40 a month plus board. At the top, the industry was dominated by ten ranches which, collectively, owned 67,000 cattle and 4,000 horses. All of the ten ranches were incorporated: four in Britain, five in Canada and one in the United States. Together they had a combined authorized capital of more than $2.5 million dollars (see Table 1).

TABLE I: CAPITAL OF INCORPORATED RANCHES

Canadian Ranches	Cattle	Authorized Capital ($)	Paid Up Capital ($)
BARC		200,000	
Bow River Horse Ranch		40,000	
Brown Ranch*	2,658	100,000	
Cochrane Ranche*	12,782	500,000	250,000
Cypress Cattle Co.	1,447	50,000	
Glengarry Ranch*	2,801	120,000	
MCC		100,000	
Mount Royal Ranch		50,000	
NWCC*	10,382	300,000	150,000
Mount Head Ranch		145,500	76,629
Stewart Ranch*	2,353	150,000	
British Ranches			
CACCC*	5,000		
New Oxley*	8,388	583,920	89,495
Quorn*	5,000	340,620	145,980
Walrond*	5,000	486,600	339,932
American Ranches			
Circle*	7,396	500,000	

* One of the ten largest ranches, based on herd size.

BARC - British American Ranch Company; CACCC - Canadian Agriculture, Coal, and Colonization Company; MCC - Benton and St. Louis Cattle Company Military Colonization Company; NWCC - North-West Cattle Company; Circle - Benton and St Louis Cattle Company.

Source: Canada, Secretary of State, *Annual Report*, 1881 to 1890, in Canada, *Sessional Papers*; Great Britain, PRO, BD 31; NA, RG15,Vol. 1220, File 192192,"Stock returns for the year 1891"; Henry C. Klassen,"The Conrads in the Alberta Cattle Business, 1875-1911," *Agriculture History* 64, no. 3 (1990): 34.

Ranching is not inherently a capital-intensive, large-scale industry, although large ranches enjoy significant economies of scale. Ranching in the Foothills probably would have developed without the large infusion of capital which came in the 1880s, but the imported capital forced the growth of the industry and shaped it. The early settlers who had pioneered the ranching industry in the 1870s were overshadowed by, and in some cases placed in bitter opposition to, the large ranches. An evolving system of small, locally owned, ranches was replaced by a system dominated by a few large corporate ranches owned outside of the region. These large ranches continued to dominate the Foothills ranching industry until the early twentieth century.

Four ingredients—grasslands, markets, cattle, and expertise in handling cattle—came together in the 1880s and attracted a fifth, capital, which was essential to the boom in ranching in the Foothills.

The Foothills became available as ranchland in the late 1870s as the result of two interrelated events. The first was the virtual extermination of the buffalo; the second was the signing of Treaty 7 in 1877. The buffalo had provided a demonstration of the suitability of the Foothills as grazing lands; their extermination provided a vacuum into which domestic cattle could be introduced. The destruction of the buffalo also fatally weakened the Plains Indian tribes and forced them to sign treaties ceding their lands to the Canadian government. With the buffalo gone and the Indians largely confined to reserves, the land was available for peaceful and orderly occupation by ranchers.

Indirectly, the destruction of the buffalo helped to provide a market for beef. The Plains Indians had been dependent on the buffalo as a source of food, and with its disappearance they were destitute. Under the terms of the treaties the Canadian government was responsible for assisting them in times of "general famine": to meet this obligation the government began to purchase large quantities of beef about 1880. Initially most of this beef was imported from Montana, but by 1885 Canadian ranches were able to fill many of the contracts. For a decade government beef contracts provided the underpinning for much of the Canadian ranching industry.

In 1880 there were only a few thousand cattle in the Foothills area, but there were hundreds of thousands in Montana, Idaho, Wyoming and Oregon within driving distance of the Foothills. These cattle were reasonably priced and of good quality. The Earl of Lathom, a well-known British breeder and shareholder in the Oxley ranch, commented on the herd the Oxley purchased in Montana: "They compare favourably with the cattle of England. You can't improve them. If you keep them up to their present quality you will do well.[8]

Ranches in the United States provided the fourth ingredient—expertise in handling cattle. Most of the cattle which were driven north to stock Foothills ranches were driven by American cowboys supervised by American foremen. Many of these cowboys, Doc Frields of the Walrond, George Lane and John Ware of the North-West Cattle Company, and W. D. Kerfoot of the Cochrane, for example, remained in Canada at least long enough to train a generation of Canadian cowboys; some of them became Canadian cowboys and ranchers.

Finally, the early 1880s were a period when capital was available. The opening of the Canadian range coincided with what James Brisbin labelled the "beef bonanza." Fanned by reports such as that of a British Royal Commission on Agriculture which stated that annual profits of 33.3 percent were ordinary in ranching, and by a few actual cases of extraordinary dividends (the Prairie Land and Cattle Company paid dividends of 26, 30 and 20.5 percent in 1881, 1882 and 1883).[9] British and eastern American investors poured millions into western land and cattle. While most of this investment went into the western United States, some found its way to Canada. Canada was also enjoying its own investment boom during the early years of the National Policy, and many capitalists in eastern Canada looked to western Canada for investment opportunities.

By 1881 all of the factors which had led to massive investment in ranching in the United States were in place in Canada; all that remained was for a legal framework to be established. In the United States ranching had been developed outside of any formal legislated framework; ranchers simply occupied vacant rangeland and used it as their own. The first occupiers (other than the Indians) were generally accorded customary rights to their range, but as the county filled up disputes developed and overgrazing became a serious problem. Free grass had its attractions, but it did not offer the security of tenure which major investors desired. By the 1880s major corporate ranches in the United States were beginning to purchase and lease their range.[10]

The first small ranches in the Foothills were developed on free-grass principles. However, in 1881 the Canadian government introduced regulations which brought the development of ranching under its control and clearly distinguished the development of ranching in Canada from its development in the United States. Several factors contributed to this decision: a different tradition towards public lands than was present in the United States; the existence of a strong central presence (most notably the NWMP) in western Canada; the example of the American approach which had led to overstocking and disputes over land; and the need for revenue. In spite of these incentives, the Canadian government did not develop appropriate grazing lease regulations until it was forced to by a dynamic capitalist, Matthew Henry Cochrane.

Senator Matthew Henry Cochrane (1823-1903), circa 1885-88. Cochrane was appointed to the Senate of Canada by John A. Macdonald in 1872, a position he served in until his death. The Town of Cochrane, Alberta, is named in his honour. (Courtesy of the Glenbow Archives/NA-239-25.)

MATTHEW HENRY COCHRANE

If the Canadian ranching industry had a godfather, it was Cochrane. Born in 1823 near Compton, Quebec, Cochrane made a fortune manufacturing footwear in Montreal. In 1864 he bought a large farm east of Montreal and entered the business of purebred stock breeding with a specialization in Shorthorn cattle. This was not an unusual move for a successful businessman, but Cochrane pursued it with unusual vigour and deep pockets. He bought the best stock available in Britain, often at record prices, and sold its progeny in Canada, the United States and Britain.[11] In the late 1870s Cochrane became involved in the growing trade of feeding beef cattle for live export to Britain.[12]

Cochrane's contacts in international stock breeding and in the export trade would have made him well aware of the boom in western ranching,[13] and his contacts as a Conservative senator put him in a position to influence the government's policy towards ranching. In December 1880 Cochrane wrote Sir John A. Macdonald outlining his plans for raising stock in the Foothills to fill government contracts in the area and for export to foreign markets. In a later letter he expanded on this proposal, saying that he would put 8,000 head of breeding cows and 300 to 400 purebred bulls on his lease in 1881, and that he expected to invest $500,000 in the first two years of operation.[14]

Following negotiations with Cochrane, the government issued regulations which allowed ranchers to lease up to 100,000 acres for a period of twenty-one years at a rate of 10 per acre. The lessee was required to stock the lease at the rate of one animal for every ten acres within three years. An amendment to these regulations in 1882 allowed leaseholders to import animals to stock their lease duty free; since the import duty on livestock was 20 percent, this was an important concession.[15]

The lease regulations were crucial in shaping the development of ranching in the next decade. They provided a security of tenure which was helpful in attracting large, non-resident investors; the fact that the security proved illusory was not, in the initial phase, relevant. Because leases were centrally administered they gave an advantage to those with good contacts in Ottawa and put the smaller ranchers who were already established at a disadvantage. By the spring of 1883 Cochrane had obtained control of three leases totalling 189,000 acres on the Bow River west of Calgary and had imported 13,266 head of cattle from the United States.[16] Clearly he had made a large investment in ranching; the company books do not survive, but enough is known to estimate the extent of the investment. The Cochrane Ranche Company Limited was incorporated with an authorized capital of $500,000 and by the end of 1882 $250,000 was listed as paid up. In addition the company had borrowed $110,000.[17] Most of this money had been spent on cattle. The original herd of 6,799 head cost about $142,000 delivered on the Bow River. In 1882 the company paid $107,250.00 for 4,290 cattle delivered to its lease; it also imported 500 slaughter steers at $40 a head to fulfill a contract for beef in late 1882.[18] Assuming that the remaining 1,677 cattle, of the 13,266 which were entered at the border, were purchased for about $25 a head, the Cochrane Ranche invested a total of $311,175 in cattle in its first two years of operations; this amounts to 86 percent of its paid up capital and debt.

The Cochrane suffered very heavy losses during the winters of 1881-82 and 1882-83. James Walker, the company's manager, estimated the losses during the first winter at 1,000 head; Senator Cochrane estimated losses during the second winter at 3,000 head or $100,000.[19] As a result of the losses, Cochrane arranged for new leases to the south of Fort Macleod and reorganized the company, reducing the capital from $500,000 to $250,000.[20] The move to a new range, the reorganization, beef sales, better management and better luck apparently turned the company around. In 1889 it increased its capital from $250,000 to $400,000 on the grounds that it had fully recovered from its earlier losses and restricted in its operations and unable to pay dividends which had been legitimately earned.[21] The company continued to operate, apparently successfully, until 1906 when Senator Cochrane's heirs decided to sell out.

When Cochrane shifted his herd south to Fort Macleod he did not abandon his Bow River range; he organized a new ranch, the British American Ranch Company (BARC), to take over the northern leases. The BARC was capitalized at $200,000 and focussed on raising horses and sheep.[22] The BARC's leases were close to Calgary and came under severe pressure from settlers as early as 1885; as a result the company gave up some of its leases and in 1888 sold its remaining stock and interest in leases to the Bow River Horse Ranch for $36,479.20 plus $67,068 in shares in the new company.[23]

Cochrane is remembered primarily as the individual who pushed the federal government into developing the system of grazing lease regulations which shaped the development of ranching during the 1880s and who, through his personal and financial commitment, demonstrated the viability of large corporate ranching in western Canada.

FREDERICK SMITH STIMSON AND THE BAR U

Frederick Smith Stimson is not so well known as Senator Cochrane, although he was the key figure in a ranch, the North-West Cattle Company (NWCC) or Bar U, which was almost as large as the Cochrane and was, arguably, more successful.

Frederick Stimson was born in Compton, Quebec, in 1845.[24] His father was a prosperous merchant, farmer and money lender. At his death in 1863 his estate had a net value of $101,495.50; Fred Stimson inherited a quarter of this and took over the management of the family farms.[25] By 1871 he was one of the largest farmers in the district with 1,000 acres, 124 cattle and twenty-three horses; his farm was comparable in size with that of his neighbour, Senator Cochrane, although he was not nearly so well known or wealthy.[26]

Stimson probably discussed the prospects of ranching with Senator Cochrane. He certainly would have discussed ranching with his wife's brother-in-law,

William Winder, who was a NWMP officer in the Foothills country from 1874 until 1881, when he retired from the police force and organized a ranching company. Fred Stimson's brother, Charles, was one of the investors in the Winder Ranch.[27]

When Senator Cochrane went west in June 1881 to inspect the ranching country, Frederick Stimson and William Winder were part of his party.[28] Although Stimson was well-to-do he was not so wealthy that he could finance a large ranch on his own, and before leaving for the West he had apparently talked to the Allan family of Montreal about the possibility of investing in a western ranch. Sir Hugh and Andrew Allan were among the wealthiest men in

Fred Stimson, circa 1900. (Courtesy of the Glenbow Archives/NA-117-1.)

Canada. They had wide-ranging investments in textiles, insurance, banking, iron and steel, and shipping.[29] As owners of the Montreal Ocean Steamship Company they carried large numbers of livestock to Britain and had a vested interest in the development of Canadian ranching.

After two false starts, Stimson and the Allans organized the North-West Cattle Company. It had a nominal capital of $150,000 divided into 1,500 shares. By 1884 the capital was fully paid up; the Allan family were the dominant shareholders with 858 shares, while Frederick Stimson and his family held 289 shares.[30] Andrew Allan became the president of the company; Frederick Stimson became the resident manager.

The company acquired two leases totalling 114,000 acres in 1882. Stimson established headquarters on Pekisko Creek, where he was to live for most of the next twenty years. He also purchased a herd of 3,000 head in the Lost River district of Idaho for "less than $20" a head and had them driven to the NWCC lease.[31] This purchase, totalling about $60,000, appears to have been the company's major investment in American cattle although it did import a number of purebred bulls.

The company increased its nominal capital to $300,000 in 1884, and two years later used some of its additional shares to purchase a neighbouring ranch, the Mount Head. The Mount Head had been incorporated in Britain in 1883 with a nominal capital of £30,000 sterling, about $145,500. By 1886, 158 of its shares (worth $76,630) were fully paid up; it had a 44,000 acre lease to the south of the NWCC and about 1,600 head of cattle[32] The NWCC paid $50,000 in the form of 500 of its own shares for the Mount Head; it also agreed that a portion of the NWCC's clear profit of $133,204.25 would be distributed as a stock dividend subsequent to the purchase.[33]

The combined capital of NWCC and Mount Head represented an investment of between $200,000 and $250,000; over the years a few more shares were taken up, and by the time the company was wound up it had a total capital of $262,600.[34] Throughout the late 1880s and early 1890s the NWCC typically ran a herd of 10,000 cattle and perhaps 800 horses.[35] The cattle herd was probably worth about $200,000. Towards the end of the century encroaching settlement forced the company to reduce its herd and to buy some of its grazing land.

The reference to a clear profit of $133,204.25 in the first five years of operation suggests that the NWCC was a profitable operation, and in fact the ranch had the reputation of being the best managed of all the large ranches. Much of the success can be attributed to Stimson, who was the resident manager from 1882 until the ranch was sold in 1902. The sale was prompted by the death of Andrew Allan in 1901 and perhaps by a feeling that growing agricultural settlement was pushing out ranchers. The purchasers paid $220,000 for 3,090 cattle, nearly 500 horses, and between 18,000 and 19,000 acres of deeded land, as well as buildings, equipment and hay.[36] Stimson was not consulted in the deal and felt the ranch had been undervalued by at least $100,000. He tried, unsuccessfully, to block the sale through the courts; when he failed he left the country and moved to Mexico.[37]

DUNCAN MCEACHRAN AND THE WALROND

Duncan McEachran was born in Campellton, Scotland in 1841. He graduated from the Royal Veterinary College in Edinburgh in 1862 and moved to Canada, where he helped to organize veterinary colleges in Toronto and Montreal. At a time when the transatlantic trade in live cattle was just beginning, he recognized the need to prevent the importation of livestock diseases into Canada from Europe. He persuaded the Canadian government to establish a livestock quarantine system at Quebec City, and in 1876 he was appointed the inspector of stock at the quarantine station.

It is probably as the government inspector of imported livestock that McEachran met Senator Cochrane and became involved in his plans to establish a ranch in western Canada. He was a major shareholder and general manager of the Cochrane Ranche for several years but he had personal and business disagreements with the senator, and about 1884 he left the company.[38]

In 1882, well before he left the Cochrane, McEachran began negotiations with Sir John Walrond to establish another ranching company.[39] Walrond, a member of the British gentry from Devon, assembled a group of investors and organized the Walrond Cattle Ranche Limited. The company had a nominal capital of about £100,000 ($486,600) divided into about forty shares; the largest shareholders were Sir John with five shares, T. H. G. Newton with six, and Duncan McEachran with 3.2.[40] McEachran was hired as the Canadian manager, and in 1883 he travelled west to establish the ranch. The company secured leases totalling about 260,000 acres on the upper Oldman River and purchased two herds of American cattle in 1883 and 1884; by 1891 it reported that it had 10,359 cattle and 549 horses.[41]

In most cases we have only the sketchiest knowledge of the financial history of the large ranches; few if any of their detailed financial records have survived. The Walrond is the exception in that a cashbook (held by the Glenbow Archives) covering its history from 1883 to 1897 has survived. The book provides a fairly complete record of infusions of capital into the company, and of receipts from sales and major expenses. Although not complete, it provides a good basis for constructing a financial model of a large, corporate ranch with about 10,000 head of cattle.

The Walrond had a nominal capital of $486,600; in 1887, $339,932.01 of the capital was described as paid up.[42] Whether this much capital was actually invested is not clear; the cash book only shows a total of $262,014.88 in capital between 1883 and 1888. Almost all of this was entered in 1883 and 1884; 84 percent of it was paid out in the same years for the purchase of 6,471 cattle in the United States.[43] The only other major expenditure for cattle was a payment of $21,427 for stockers purchased in Ontario in 1895. Total cattle purchases over the years 1883-97 were $257,315.33.

The ranch began to sell cattle in 1885. The first sales were made to individuals and firms who had contracts to provide beef to the Indian Department and the NWMP; beginning in 1887 the Walrond began to contract directly with the Indian Department. In 1888 the company began to make large sales to non-government purchasers: in some cases the Walrond sold its own cattle in Britain; in others it sold to Gordon, Ironside and Fares who specialized in the export trade; and in still others it sold to Pat Burns, who supplied the domestic market in Alberta and British Columbia. Over the period covered by the account book cattle sales totalled $676,789.51. Of these, 47 percent were made directly to the Indian Department; 15 percent of sales were made to individuals and firms known to have been government contractors. Government purchases declined in relative importance during the 1890s, but it seems clear that during the first decade of the Walrond's existence they were essential to the ranch's survival. To what extent this was true of the ranching industry as a whole will be considered below.

The Walrond was moderately successful; between 1883 and 1895 it paid eight dividends of 5 percent and one of 7 percent. A decision was made to wind up the company in 1895 and in 1897 its assets were sold to the New Walrond.[44] Some of the British shareholders withdrew with their capital intact, and others elected to take shares in the new company. Eastern Canadian shareholders took a leading position in the company and McEachran became the president. Following the hard winter of 1906 the company sold off its remaining cattle, although it continued as a landholding company for forty years.[45]

ESTIMATES OF INVESTMENT IN RANCHING

Individuals such as Cochrane, Stimson, McEachran and the ranches with which they were associated played major roles in the development of Canadian ranching in the 1880s and 1890s. The large corporate ranches dominated the ranching industry: in 1891 the ten largest ranches owned 67,000 out of the 128,068 beef cattle in the ranching country. What is more difficult to estimate is the size of the investment by eastern capitalists in western ranching.

The ten largest companies operating in 1891, owners of 52 percent of the beef cattle in the ranching country, had a combined authorized capital in excess of $2.5 million. Their paid up capital was probably half that; for the six companies for which figures are available, paid up capital was only 48 percent of authorized capital. Even paid up capital is not an accurate measure of actual investment in ranching: shares can be issued as "paid up" for any reason and do not necessarily represent money paid in to the company.

A more realistic way to estimate investment in ranching is to evaluate the ranching assets and the value of cattle imported to stock ranches in the decade 1881-91. In 1881 there were 5,690 beef cattle in the North-West Territories; of these perhaps 3,000 were in the ranching country. By 1885 the Alberta district (which included Fort Macleod, Calgary and Edmonton) reported 57,464 beef cattle. The 1891 census found 134,483 beef cattle in Alberta—95 percent of them in the ranching districts of Fort Macleod and Calgary/Red Deer (see Table 2). This growth was beyond what might be expected on the basis of natural increase, but can be explained as the result of large imports from outside of the region. These imports represent investment in ranching.

TABLE 2: CATTLE IN THE NORTH-WEST TERRITORIES, 1881-91

1881	Oxen	Milk Cows	Horned Cattle	Total
North-West Territories	3,334	3,848	5,690	12,872
Foothills	—	—	—	3,000
1885				
North-West Territories	?	?	69,557	86,958
District of Alberta	486	3,334	57,464	61,284
1891				
North-West Territories	7,583	37,063	187,211	231,857
District of Alberta	811	10,969	134,483	146,263
Sub-district of Calgary and Red Deer	251	5,104	47,311	52,666
Fort Macleod	249	2,940	80,757	83,946
Edmonton	311	2,925	6,415	9,651

Source: *Census of Canada*, 1891. Bulletin No. 7. For the 1881 estimate see Macoun, *Manitoba and the Great North-West*, 271.

Canadian customs records provide details of imports of cattle into the North-West Territories during the early ranching period. Between 1881, when large-scale ranching got under way, and 1887, when imports were virtually shut off, 39,006 head valued at $981,237, were imported under the provision which allowed leaseholders to import stock duty-free for their leases (see Table 3).

During the same period an additional 21,876 head, valued at $550,175, were imported; most of them paid duty, but a few were imported duty-free as settler's effects and some were imported under a provision for the improvement of stock. Many of the 21,876 cattle were imported to fulfill beef contracts, but an equal number may have been imported to stock ranches operated by non-leaseholders.[46] Assuming that half of the 21,876 cattle were imported to stock ranches, they then represent an investment of over $250,000 in ranching.

TABLE 3: CATTLE IMPORTED FROM THE UNITED STATES
TO THE NORTH-WEST TERRITORIES, 1878-1892

Year	Number	Value($)	Duty	Average Price
1878-79	1,905	28,050	2,805	14.72
1879-80	653	15,882	3,176	24.32
1880-81	2,588	52,205	10,441	20.17
1881-82	5,299	97,015	19,403	18.31
1881-82 I	1,480	21,296		14.39
1882-83	5,911	145,264	29,052	24.58
1882-83 R	10,847	252,927		23.32
1882-83 S	519	12,323		23.74
1883-84	4,718	148,519	29,702	31.48
1883-84 R	2,495	74,425		29.83
1883-84 S	609	16,624		27.30
1884-85	1,097	34,771	6,958	31.70
1884-85 R	10,510	252,455		24.02
1884-85 S	216	7,230		33.47
1885-86	1,444	49,586	9,917	34.34
1885-86 R	1,809	55,197		30.51
1885-86 S	7	335		47.86
1886-87	539	16,170	3,234	30.00
1886-87 R	13,345	346,233		25.94
1886-87 S	37	1,042		28.16
1887-88	17	648	123	38.12
1887-88 S	15	300		20.00
1888-89	25	520	104	20.80
1889-90	2,467	33,191	6,647	13.45
1890-91	365	5,320	1,596	14.58
1891-92	956	12,823	3,846	13.41

I = Imported for improvement of stock; R = Articles imported for ranches into NWT (eliminated late 1886); S = Imported as settler's effects into NWT(not in 1889 and subsequent reports).

Source: Canada, *Sessional Papers*, 1880, No. 1, "Tables of Trade and Navigation"; CSP 1881, No. 2; CSP 1882, No. 1; CSP 1883, No.2; CSP 1883, No.2; CSP 1884, No.1; CSP 1884, No. 1; CSP 1885, No. 1; CSP 1886, No. 1; CSP 1887, No. 1; CSP 1888, No. 3; CSP 1889, No. 1; CSP 1890, No. 2; CSP 1891, No. 4; CSP 1892, No. 5; CSP 1893, No. 5; CSP 1894, No. 6; CSP 1895, No.6; CSP 1896, No. 6, CSP 1897, No. 6; CSP 1898, No. 6; CSP 1899, No. 6; CSP 1900, No. 6; CSP 1901, No. 11.

The customs records provide evidence of an investment of about $1.23 million in cattle in the years 1881-87. This should be taken as a minimum figure because, especially for non-leaseholders, smuggling would have been an attractive alternative to paying a 20 percent duty. There is some evidence that imports exceeded what was reported in the returns. In 1886 William Pearce reported that 34,000 cattle had been brought into the grazing district in 1886; 26,000 of them came from the United States. The customs returns reported imports of 3,260 in 1885-86 and 13,921 in 1886-87.[47] If Pearce, who is generally regarded as the best-informed federal official on the situation in the ranching country, was right, then in 1886 the customs returns under-reported the importation of cattle by 100 percent. Pearce's report also reminds us that not all cattle bought to stock the ranching country came from the United States. British Columbia had well-stocked ranges which supplied horses and cattle to the Foothills country in the 1870s and 1880s.[48] On the basis of Pearce's report one might surmise that 25 percent of the cattle brought into the Foothills country in 1886 came from other parts of Canada, most probably British Columbia; unfortunately, detailed evidence as to the internal trade in Canadian cattle is almost totally lacking.

A detailed listing of livestock imports into the District of Alberta from 1 June 1880 to 30 April 1885 provides evidence that the large eastern based ranches were the leading importers of cattle. The list provides the names of importers of 50,302 cattle; of these, 40,343 were imported by nine large ranches (see Table 4). Seven of these were among the ten incorporated ranches which dominated the ranching industry. Two, the Halifax ranch and the Military Colonization Company, were mid-sized, eastern based ranches. To this list of large imports by identifiable corporate ranches one might add the 7,000 head which the Powder River Cattle Company (the 76 Ranch) moved from Wyoming in 1886 and subsequently sold to the Canadian Agricultural Coal and Colonization Company.[49]

Not all of these cattle were imported to stock ranches; several importers were also government contractors and would have imported some cattle for immediate slaughter, not for stocking ranches. Even with this qualification it seems likely that the 40,343 cattle imported by the nine large ranches plus the 7,000 imported by the Powder River Company represented an investment in excess of $1 million in ranching.[50]

Again, $1 million is a minimum figure; we have no information on purchases by two of the ten largest ranches (the Glengarry and the Brown) or by a number of mid-sized, eastern based ranches such as the Winder, the Alberta and the Cypress.

In the open-range type of ranching which prevailed in the Foothills country in the 1880s, cattle were the principal asset of any ranch. Professor W. Brown

TABLE 4: IMPORTS OF CATTLE BY MAJOR COMPANIES, 1881-84

	Number of Cattle	Years	Cost ($)
Cochrane	13,266	1881-82	311,175
Circle (Baker and D.W. Davis)	6,235	1881-83	
Halifax Ranch	875	1883	
Military Colonization Co.	620	1884	
NWCC	3,014	1882	60,000
Oxley	6,139	1883-84	
Quorn (Moore & Martin, Baxter)	1,039	1883-84	
Stewart Ranch	2,584	1882-83	
Walrond	6,571	1883-84	219,308
Total	40,343		

*These are costs which can be verified from sources other than the customs records. See sections on the Cochrane, the NWCC, and the Walrond.

Source: NA, RG15, Volume 1182, File 11007, Statement showing.... The figures for the Walrond are from A. A. Den Otter, "Transportation, Trade and Regional Identity in the Southwestern Prairies," 9 and GAA, Walrond Account Book, p. 29.

of the Ontario Agricultural College estimated the cost of establishing and carrying a 250-cow ranch for two years at $15,765; of this $10,050 was spent on cattle, $2,550 on buildings and equipment, $3,000 on wages, board, and personal expenses, and $165 on land and rent. Professor Brown's hypothetical ranch was a small one; larger ranches were known to have a lower level of operating expenses to total investment. In 1887 C. W. Martin estimated the value of his Sheep Creek (Quorn) Ranch at $145,500 divided as follows: 2,425 head of cattle, $97,000 (66 percent); 215 horses, $24,250 (16 percent); land improvements and equipment, $74,250 (16 percent).[51] In 1883 and 1884 the Walrond invested $219,308 (i.e., 83.5 percent) of its available capital, $262,000, in cattle. The Cochrane Ranche purchased about $311,175 worth of cattle, equal to 86 percent of its combined capital and debt of $360,000.

If one assumes that the large ranches invested 80 percent of their capital in cattle and horses and 20 percent in land, equipment and improvements, then it is fair to estimate the minimum investment of the larger eastern ranches at $1.25 million exclusive of operating expenses. Investments by mid-sized, eastern based ranches probably brought the total to $1.5 million.

Before leaving the question of the capital which eastern ranches put into the Foothills in the 1880s, it maybe useful to make a comparison with British investment in American ranching. In 1883 the nine largest British ranches operating in the United States had a total ordinary paid up capital of $8,734,567 plus debentures and preferred capital totalling $5,886,886. They estimated that they owned 584,927 horses and cattle. In addition there were numerous smaller British cattle companies in the United States. W. Turrentine Jackson estimated the total British investment in American ranching during the 1880s at as high as $45,000,000. In absolute terms, British investment in the American ranching industry was many times higher than eastern investment in the Foothills industry, but in relative terms, it was less significant. In 1883 there were 23,000,000 cattle west of the Mississippi in the United States; the nine largest British companies owned 2.5 percent of the total; by 1891, ten corporate ranches owned over 50 percent of the cattle in the Foothills ranching country.[52]

GOVERNMENT BEEF PURCHASES

The Canadian government played an important role in establishing ranching. By negotiating treaties with the Plains Indians and by supporting an effective police force, the government provided a more secure environment for ranching than had generally been available in the United States. The government's grazing lease policy provided at least an illusion of secure access to rangeland which helped to attract investors. Support for the construction of the Canadian Pacific Railway provided access to eastern Canadian and British markets for beef. Finally, for more than a decade after ranching was established on a large scale, the government, through its purchases of beef, provided a large local market for ranchers.

All of these contributions by the government to the success of ranching have been recognized in existing historiography. Some, for example the link between the NWMP and ranching, have been examined quite closely.[53] This portion of the present article will consider the role of government as a market for beef in more detail than has been done before; it will argue that government contracts were essential to the success of Foothills ranching during its first decade.

The federal government emerged as a purchaser of beef in the Foothills in 1874 and 1875 when the NWMP established posts at Forts Macleod, Calgary, and Walsh. For the next ten years there were typically 150 to 200 policemen stationed at these three posts. As part of their rations the men were entitled to 1.5 pounds of beef, or a substitute, daily. So long as game was available the police did not purchase much beef, but in 1876-77 they bought 272,124 pounds for $9,200.70.[54] In 1880-81 the force contracted for the delivery of 157,080 pounds, with 71,400 pounds to be delivered at Fort Walsh and 30,600 pounds at Fort Macleod.[55] During the ten years after 1885 the police contracted for an average of about 160,000 pounds of beef annually in the Foothills country at an average cost of $14,150 (see Table 5).

TABLE 5: ESTIMATES OF PURCHASES OF BEEF
BY THE GOVERNMENT IN THE TREATY 7 AREAS, 1881-1900

Year	Department	Pounds	Value ($)
1881	ID	983,110	58,986
	NWMP	102,000	4,691
1882	ID		270,282
	NWMP		
1883	ID		303,415
	NWMP	96,000	7,680
1884	ID		325,072
	NWMP	136,000	19,040
1885	ID		261,705
	NWMP	180,550	25,770
1886	ID	3,030,610	349,702.19
	NWMP	192,223	21,150.02
1887	ID	2,403,622	237,932.67
	NWMP	154,858	15,703.37
1888	ID	2,136,231	168,832.04
	NWMP	94,972	9,200.77
1889	ID	2,119,394	147,246.99
	NWMP	260,269	22,199.89
1890	ID	1,872,300	151,980.44
	NWMP	174,296	16,868.91
1891	ID	1,711,587	144,441.63
	NWMP	170,139	15,836.33
1892	ID	1,423,427	109,797.11
	NWMP	166,394	14,681.43
1893	ID	1,508,882	98,519.58
	NWMP	157,971	12,176.00
1894	ID	1,466,067	84,699.17
	NWMP	154,623	11,924.22
1895	ID	1,281,206	69,166.27
	NWMP	146,412	9,203.45
1896	ID	1,290,187	67,595.44
	NWMP	107,375	7,181.18
1897	ID	1,174,878	59,293.18
	NWMP	140,565	9,899.88
1898	ID	1,335,998	74,486.54
	NWMP	106,516	7,897.47
1899	ID	1,022,345	62,856.77
	NWMP	84,500	6,840.19
1900	ID	1,329,798	85,135.26
	NWMP	65,637	5,345.11
Total, 1886-1900		27,283,282	2,097,793.50
Average, 1886-1900		1,818,885	139,852.90

ID - Indian Department; NWMP - North West Mounted Police.

Source: ID, 1880-85: Department of Indian Affairs, Annual Reports, Accounts. The figures for 1881 is for all deliveries of beef in the Manitoba and North-West Territories Superintendency. The figures for 1882 to 1885 are 91 percent of the total amount paid in Treaty 7 under the heading "Supplies for Destitute Indians". It is based on the ratio of beef to total supplies in 1885-87.

NWMP, 1880-85: Canada, *Sessional Papers*, 1880, No. 80; Report of the Auditor on *Appropriation Accounts for the Year Ended 30th June 1883*, p. 353; ibid. 1884, pp. 242-43, 250; *Annual Report of the Auditor General*, 1885. The figures are contract figures for deliveries at Fort Walsh, Macleod, Calgary and Medicine Hat.

1886-1900: Canada, *Sessional Papers*, 1887-1901, Public Accounts and Annual Reports of the Department of Indian Affairs. Figures are the principal contracts for the police at Calgary, For Macleod and Lethbridge, and the contracts for supplying beef to "destitute" Indians at the Blackfoot, Blood, Stoney, Sarcee and Piegan reservations. Smaller contracts for police outposts, for Indian residential schools, and for Indian farms have not been recorded in the table. In addition the table does not include the purchases of livestock to distribute to aboriginal people. Thus the figures are minimum government purchases for the area. Nevertheless they represent much the largest part of NWMP and Indian Department purchases in the Treaty 7 area.

In the late 1870s the Indian Department emerged as an important market for beef. Under the terms of treaties signed with the Plains Indians in 1874, 1876 and 1877 the government agreed to provide aboriginal people with livestock to assist them in becoming farmers and with food in the event of a general famine. With the extermination of the buffalo, all of the Plains tribes were faced with famine and called upon the government for assistance. In 1880-81 the government bought 983,110 pounds of beef on behalf of aboriginal people in the Manitoba and North-West Territories Superintendency; over the years 1882-85 the government bought about $290,000 worth of beef annually for distribution in the Treaty 7 area. Purchases for Treaty 7 (which approximates the ranching district in the 1880s) peaked at $350,000 in 1886, but they remained at over $100,000 per year until 1893. From 1886 until 1900 the federal government purchased an average of $139,852.90 worth of beef annually for use by aboriginal people and by police in the Treaty 7 area. This figure does not include minor purchases of cattle for distribution to Natives, and of beef for residential schools or for small police outposts.

How important were government purchases in relation to the entire market? The local non-Native population was small in the 1880s and cannot have provided a major market. The construction crews on different railway projects provided important but temporary markets. New markets were opened with the completion of the Canadian Pacific Railway from Winnipeg to Calgary in 1883 and from Montreal to Vancouver in 1885. Some cattle were shipped to Winnipeg by rail in 1884 and more were sent east in 1885; nevertheless, the Fort Macleod *Gazette* reported that in 1885 the Indian contracts still made up the most important part of the market.[56]

As government purchases peaked at $370,852.21 in 1886, a new and lucrative market opened in Britain. About 700 head from western ranches were shipped to Britain in 1887; 4,500 head went overseas in 1888, and the market continued to grow.[57] At approximately $45 per head 4,500 export steers brought in $202,500, compared to $178,032.81 for police and Indian Department contracts in 1888. The government's role as a buyer continued to decline. The 1891 census reported that during the previous year 16,210 cattle had been sold or an in the Calgary/Red Deer and Macleod sub-districts[58]; in 1890 and 1891 the government purchased an average of 1,964,161 pounds of beef for Treaty 7; this was probably equivalent to 2,500 head of cattle or 15 percent of the cattle sold in the two sub-districts. Although the government was not the major market for beef in 1891 it was still important; for some ranches, such as the Walrond, the government remained the major market until at least 1893.

Not all of the government's expenditures on beef went into the pockets of Canadian ranchers. During the 1870s and the first half of the 1880s most of the beef which the government purchased was imported from the United States. The principal beneficiaries of the early contracts were I. G. Baker and Company of Fort Benton, and American ranchers. Canadian firms began to benefit from government contracts as early as 1882 when the Cochrane Ranche sub-contracted with Baker and Company to supply 64,000 pounds of beef to the police at Calgary. In 1884 the minister of Agriculture reported that the North-West Cattle Company, the Oxley and the Stewart ranches were supplying beef to the police, the CPR and the public.[59] Even though Baker remained the largest single government beef contractor in the Treaty 7 area until 1890, Canadian ranchers benefited indirectly from Baker's contracts as soon as they had cattle available for sale: in 1885 the Walrond sold $54,100 worth of cattle to Baker, presumably to fill a beef contract.[60] After 1886 very few cattle were imported and, whether Baker or Canadian firms held the contracts, the cattle which filled them would have come from Canadian ranches.

CONCLUSION

Eastern capital and government purchases were not essential to the development of the ranching industry in the Foothills; ranching on a small scale began to develop in the 1870s, and it was the initial success of small owner/operator ranches in the area which tempted eastern capitalists to enter the field. There is no reason to believe that ranching would not have continued to develop without large-scale eastern investment and perhaps without government support; however, the economic and social structure which might have developed in the absence of eastern investment would have been quite different.

Eastern investment, supported by government purchases, forced the pace of development by five to ten years. From the time when the first cattle were introduced into the Foothills, about 1874, to the arrival of the first Cochrane herd, the cattle population of the ranching country grew from nil to 3,000. By 1885 there were 57,464 beef cattle in the District of Alberta. The explosive growth can be credited almost exclusively to the importation of at least 40,000 head of beef cattle by nine large eastern ranches. Doubtless the growth would have occurred without eastern investment, but it would have been substantially slower.

By forcing the pace of development, eastern capital helped to forestall a move by American ranchers into the Foothills country. By 1885 American ranges as far north as Montana were overcrowded; in 1886 a number of American ranches, including the Powder River Cattle Company, took out leases in Canada and moved some cattle north. The established ranchers, M. H. Cochrane, D. McEachran, H. Montague Allan (of the NWCC) and others, successfully lobbied the government to have the privilege of duty-free importation of stock for a lease withdrawn so as to prevent any further American importation.[61] The disastrous winter of 1886 relieved the pressure on American ranges and caused most of the American companies to withdraw from Canada; the cancellation of the duty-free privilege made it difficult for them to return.

The presence of the large corporate ranches helped to establish the ordered, cultured, Anglophile and class-conscious society which, historians such as Breen have argued, distinguished Canadian from American ranching society. The shareholders, executives, and managers of the corporate ranches were drawn from comfortable, sometimes wealthy, backgrounds; some of them represented the new monied aristocracy of eastern Canada; some represented the landed gentry and or aristocracy of Britain. By upbringing, education, and inclination they were leaders; and their economic strength put them in a position to shape ranching society. Their background and attitudes were different from those of the settlers who started ranching on a small scale in the Foothills in the late 1870s. The early settlers, one suspects, were closer to the frontier ethos which influenced American ranching.

Finally, although the big ranches had a role in developing a ranching society in the Foothills which can be distinguished from ranching society in the United States, it should be remembered that there were many corporate ranches in the United States with eastern American, British and even Canadian roots. The "beef bonanza" of the early 1880s was a continental and intercontinental phenomenon. The Cochrane, the Walrond, the North-West Cattle Company and the other Canadian corporate ranches were a part of this continental development even as they contributed to a distinctive Canadian variation of ranching society.

NOTES

This article first appeared in *Prairie Forum* 22, no. 2 (1997): 213-35. The research for this article was done as part of the program to develop the Bar U Ranch National Historic Site. An early version of the article was delivered at the Canadian Historical Association Annual Conference, June 1994, Calgary, Alberta.

1 L. V. Kelly, *The Rangemen* (Toronto: W. Briggs, 1913); C.M. MacInnes, *In the Shadow of the Rockies* (London: Rivingtons, 1930); A. S. Morton and C. Martin, *History of Prairie Settlement and Dominion Lands Policy* (Toronto: Macmillan, 1938); Sheilagh Jameson, "The Era of the Big Ranches," *Alberta Historical Review* 18 (Winter 1970): 1-9; Lewis G. Thomas, "The Rancher and the City: Calgary and the Cattlemen, 1883-1914," *Transactions of the Royal Society of Canada*, VI, Ser. IV (June 1968): 203-15.

2   For a discussion of the early historiography of Canadian ranching see David H. Breen, "The Ranching Frontier in the Prairie West: An Historiographical Comment," in L. G. Thomas (ed.), *Development of Agriculture on the Prairies* (Regina: University of Regina, 1975); and Breen, *The Canadian Prairie West and the Ranching Frontier, 1874-1924* (Toronto: University of Toronto Press, 1983).

3   Simon Evans, "The Passing of a Frontier Ranching in the Canadian West, 1882-1912," (Ph.D. dissertation, University of Calgary, 1976); Simon Evans, "The Origin of Ranching in Western Canada: American Diffusion or Victorian Transplant?" *Great Plains Quarterly* 3, no. 2 (1983): 79-91; Simon Evans, "American Cattlemen on the Canadian Range, 1874-1914," *Prairie Forum* 4, no.1 (1979): 121-35.

4   W. M. Elofson, "Adapting to the Frontier Environment: The Ranching Industry in Western Canada, 1881-1914," *Canadian Papers in Rural History*, Vol. 8 (Gananoque, ON: Langdale Press, 1992), 307-27.

5   David H. Breen, *The Canadian Prairie West and the Ranching* Frontier, *1874-1924*, 11.

6   *Census of Canada. 1880-81,* Vol. 3,132-33; John Macoun, *Manitoba and the Great North-West* (Guelph: The World Publishing Company, 1882), 271.

7   *Census of Canada, 1891.* Bulletin No. 7; Library and Archives Canada (hereafter LAC), RG15, Vol. 1220, File 192192, Stock returns for the years 1890 and 1891.

8   John Roderick Craig, *Ranching with Lords and Commons: or, Twenty Years on the Range* (New York: AMS Press, 1971), 105.

9   Edward Everett Dale, *The Range Cattle Industry: Ranching on the Great Plains From 1865 to 1925* (Norman: University of Oklahoma Press, 1960), 79; *Burdett's Official Intelligence for 1890* (London: Spottiswoode & Co., 1890), 872-73.

10   Ernest S. Osgood, *The Day of the Cattleman* (Chicago: University of Chicago Press, 1954), 87-88. By 1891 the Scottish-owned Matador ranch in Texas owned 445,000 acres and leased 220,000. William M. Pearce, *The Matador Land and Cattle Company* (Norman: University of Oklahoma Press, 1964), 51.

11   *Dictionary of Canadian Biography, Volume 13, 1901 to 1910* (hereafter *DCB)* (Toronto: University of Toronto Press, 1994), 208-09; "A Gold Mine in a Cow," *The Illustrated Journal of Agriculture* (August 1879): 62.

12   Canada, Department of Agriculture, *Report of the Minister of Agriculture for the Dominion of Canada for the Calendar Year 1879,* "Reports of Tenant Farmer's Delegates," p. 141.

13   He was one of the organizers of the Dominion Cattle Company which purchased a large ranch in the Texas Panhandle. *Canada Gazette* 16, no. 3 (22 July 1882): 122; Edward Brado, *Cattle Kingdom: Early Ranching in Alberta* (Vancouver: Douglas and McIntyre, 1984), 20.

14   LAC, RG15, Vol. 1209, File 142709, Part 1, Cochrane to Macdonald, 17 December 1880; ibid., Cochrane to Minister of the Interior, 10 February 1881.

15   Canada, Department of the Interior, *Annual Report of the Department of the Interior for the year Ended 30th June, 1881,* Dominion Land Regulations, p. 6; Privy Council No. 183, 3 February 1885.

16   Cochrane, and other rancher; circumvented the 100,000-acre limit by taking leases out in different names. Glenbow Archives (hereafter GA), M1303, Cochrane Ranche Notebook; LAC, RG15, Vol. 1182, File 11007, Statement showing the number of Horses, Cattle and Sheep ... entered... .

17   Montana State Archives, Power Papers, MC55448, 22, McEachran to Power, 18 December 1882.

18   GA, M1303, Cochrane Ranche Notebook.

19   William Naftel, "The Cochrane Ranche," *Canadian Historic* Sites, No. 6. (Ottawa: Parks Canada, 1977), 15; LAC, RG15, Vol. 12, File 142709, Part 1, Cochrane to McPherson, 7 June 1883.

20   LAC, RG95, Vol. 2292, Cochrane Ranche.

21   Ibid., Petition, 8 February 1889.

22   LAC, RG95, Vol. 2508, British American Ranch Co., Ltd.; LAC, RG15, Vol. 1202, File 137261, Part 1, Stock return, September 1885.

23   LAC, RG15, Vol. 1202, File 137261, Part 1, Indenture of Sale, 11 July 1888.

24   Archives Nationales de Quebec (hereafter ANQ), Hull, Registres d'État Civil, Protonotaires de Sherbrooke, film No. 125.1, Anglican, Compton, 1842, p. 3; LAC, Reel T-6551, Fourth Census of Canada, 1901, Alberta, Central Alberta, Pekisko, 202-29, Schedule 1, p. 2, line 16.

25   ANQ Sherbrooke, CN501-0028, Greffe de D. M. Thomas, No.404, Inventory of the Estate of Arba Stimson and Mary Smith, 29 February 1864; ANQ Sherbrooke, CN501-0023, Greffe de C. A. Richardson, No. 8149, Last Will and Testament of Arba Stimson.

26   LAC, Reel C-1090, 1871 Census of Canada, Quebec, Compton County, Compton Township, Div. 2. Schedule 4, p. 13, line 1; ibid., Schedule 5, p.13, line 1. The returns for Cochrane's farm are in Div. 1, Schedule 4, page 2, line 18 and Schedule 5, p. 2, line 18.

27   *Macleod Gazette, 10* November 1885, p. 3, "Capt. Winder's Death"; A. B. McCullough. "The North-West Mounted Police at Fort Walsh: a Statistical Study," *Manuscript Report Series,* No. 213, (Ottawa: Parks Canada, 1977), 170; ANQ Hull, Registres d' État Civil. Dénomination Non-catholiques, Film No: 125.2, Church of England, Compton, 13 December 1869, p. 8; LAC, Reel C-1276, Canada Census 1861, Canada East, Compton County, Compton Township, No. 168, p. 37, Lines 24-30; RG15, Vol. 1218, File 175296, Barry to Russell, 25 August 1882.

28   *Montreal Herald and Daily Commercial Gazette,* 17 June 1881, "Cattle Ranching in the North-West."

29   *DCB, XI, 1881-1890,* pp. 5-15; *DCB, X11, 1901-1910,* pp. 13-14. When Sir Hugh Allan died in 1882 his estate totalled between 6 and 10 million dollars.

30   LAC, RG95, Vol. 2616, The North-West Cattle Company (Limited), Riley to Secretary of State, 3 September 1884.

31   Montana Historical Society, T. C. Power Papers, MC55, Vol.166, File 6, Stimson to Power, 26 April 1882; ibid., Vol. 167, File 4, 28 July 1882. Brado, *Cattle Kingdom,* 110, specifies $19 per head.

32   Great Britain, Public Record Office (hereafter PRO), BD/31/3171/18402, Summary of Capital and Shares of the Mount Head Ranch, 5 March 1886; Department of Indian and Northern Affairs library, Orders in Council re: Department of the Interior, Grazing lease to T. D. Milburne, 11 April 1882; Nanton and District Historical Society, *Mosquito Creek Roundup: Nanton and Parkland* (Nanton, AB: Nanton and District Historical Society, c. 1975), 34-35.

33   LAC, RG95, Vol. 2616, North West Cattle Company, Agreement between Milburne, acting on behalf of The Mount Head Ranche Company, and the NWCC, 26 October 1886.

34 Cour Superieure, Montreal, 23 May 1901, No. 46, NWCC in liquidation, & H. M. Allan, liquidator, and F. S. Stimson, plaintiff. Statement of Hugh Robertson, Manager of Royal Trust, liquidator, 7 May 1907.
35 Canada, *Sessional Papers*, 1886, No. 20b. *Mosquito Creek Round Up*, 262, states that the Mount Head had 2,500 cattle when it was bought by the NWCC.
36 Calgary *Herald*, 5 February 1902; Cour Superieure, Montreal, 23 May 1901, ibid.
37 Brado, *Cattle Kingdom*, 120-23.
38 Brado, *Cattle Kingdom*, 75; A. McTavish and C. Reilly, "Frank White's Diary," *Canadian Cattlemen* (March 1946): 245.
39 Montana State Archives, Power Papers, MC55448, 22, McEachran to Power, 18 December 1882.
40 PRO, BT31/3925/24835, Memorandum of agreement between the Walrond Ranche Limited and several persons, 9 December 1887; *Burke's Peerage, Baronetage and Knightage, 1903*, pp. 334, 1544.
41 LAC, RG15, Vol. 1220, File 192192, Stock Return for the Year 1891.
42 PRO, BT31 /3925/24835, Summary of capital and shares, 28 June 1888.
43 A. A. Den Otter, "Transportation, Trade and Regional Identity in the Southwestern Prairies," *Prairie Forum* 15, no.1 (Spring 1990): 9. GA, Walrond Account Book, 1883 and 1884.
44 The Walrond account book records dividend payments to Canadian shareholders.
45 PRO, BT/31/3925/24835, Walrond Ranche, Limited; LAC, RG95, Vol. 818, New Walrond Ranche.
46 In 1890 a return of stock in the ranching area found that non-leaseholders held 14,981 cattle and leaseholders held 91,822. LAC, RG15, Vol. 1220, File 192192, Stock Returns ...1890.
47 Canada, Department of the Interior, *Annual Report of the Department of the Interior for the year 1886*, 18-20.
48 Hull and Trounce brought 500 head of horses and 3,000 cattle from British Columbia in 1886. See Kelly, *The Rangemen*, 194.
49 Beinecke Library, Yale University, Zc47, Moreton Frewen to Shareholders, 25 July 1888; University of Alberta Archives, MG9/2/5/1, Pearce Papers, Settlement, General, 12 Inspection of Kaye farms, 1889-90, Pearce to Dewdney, 12 February 1889.
50 Calculated at the average price declared for cattle imported to stock ranches in the appropriate years.
51 Macoun, *Manitoba and the Great North-West*, 275; PRO, BT31/3863/24365, Quorn Ranch.
52 Maurice Frink, W. Turrentine Jackson, and Agnes Wright Spring, *When Grass Was King: Contributions to the Western Range Cattle Industry* (Boulder, CO: University of Colorado Press, 1956), 186, 223, 233.
53 D. H. Breen, "The Mounted Police and the Ranching Frontier," in Hugh A. Dempsey, *Men in Scarlet* (Calgary: McClelland and Stewart West, 1974), 115-37.
54 A. B. McCullough, "Prices, Transportation Costs and Supply Patterns in Western Canada, 1873-85," *Microfiche Report Series*, No. 78, (Ottawa: Parks Canada, 1982), 268.
55 Canada, *Sessional Papers*, 1882, No. 80, Contract with I. G. Baker & Company for North-West Mounted Police supplies, 1880-51, p. 3.
56 *Macleod Gazette*, 24 January 1885, "The Price of Beef"; Canada Department of Agriculture, *Annual Report of the Minister of Agriculture for 1884*, p.136.

57   Canada Department of Agriculture, *Annual Report of the Minister of Agriculture for 1887*, p. 198-99; Canada Department of the Interior, *Annual Report of the Department of the Interior for the year 1888*, pp. x and 10.

58   *Census of Canada, 1891*, Vol. 4, Table 3, p. 221.

59   GA, Cochrane Ranche Notebook, M1303, "Mounted Police Contract, Calgary"; Canada Department of Agriculture, *Annual Report of the Minister of Agriculture for 1884*, p.136.

60   GA, Walrond Account Book, p. 38. The account does not identify this as a cattle sale, but the size of the transaction is such that it could be nothing else.

61   LAC, RG15, Vol. 1204, File 141376, Part 1, Cochrane et al. to White, c. 9 October 1886; ibid., W.E. Cochrane to minister of the Interior, 21 October 1886.

11. American Cattlemen on the Canadian Range, 1874-1914[1]

Simon M. Evans

This paper is concerned with the range cattle industry which dominated the land-use pattern of the western Canadian prairies for a thirty-year period between the disappearance of the buffalo and the arrival of large numbers of homestead settlers during the first decade of the twentieth century. It was the cattlemen who occupied the foothills of the Rockies and penetrated Palliser's triangle. During the period of their dominance a new veneer of population was spread unevenly over the land, new patterns of circulation and interaction were established, new elements were added to the local society, and attitudes towards natural ecosystems were profoundly altered.

Settlement studies of the prairies have paid but scant attention to this period of pastoral occupancy. This neglect is surprising since, for a brief period after confederation, the grasslands of Alberta and Assiniboia constituted a region of considerable geo-political significance. It was the rancher as well as the policeman and the engineer who extended the Canadian *ecumene* during the 1870s and 1880s, and in doing so made the dream of a nation extending from sea to sea a reality.

The origin of the range cattle industry which was established in southern Alberta during the last two decades of the nineteenth century has been variously interpreted. A. S. Morton and J. F. Booth regarded the development as an extension of the "cattle kingdom" of the United States.[2] This view has been endorsed by J. H. Warkentin[3] and by W. L. Morton who remarked:

> The advance into the plains, led by the spearhead of the Canadian Pacific Railway, had begun on a broad front of settlement to Indian Head, to Regina, until the dusty core of Palliser's triangle was reached, and the farming front from the east was stopped by the ranching front advancing from the south.[4]

L. G. Thomas and D. H. Breen, on the other hand, have stressed the contrasts which existed north and south of the border.[5] Indeed they view the Alberta ranches as estates transplanted from the settled farmlands of eastern Canada or the shires of Britain.

To a considerable extent this apparent dichotomy is a product of a scale problem. In overview the similarities between developments on the Canadian prairies and the economy which had evolved in Texas and diffused northwards appear overwhelming. On the other hand, as the scale of inquiry is increased, so the differences in events and practice north and south of the border became more apparent.

Thus the general context of this study is an inquiry into the nature of the Canadian ranching frontier. To what degree was it unique and differentiated from the much larger range cattle industry to the south? The investigation of this broad theme requires detailed consideration of a number of issues,[6] but the immediate aim of this paper is to assess the direct contribution of

A cattle roundup on the Greeley and Parsons ranches in the Cypress Hills area in 1897. In the bottom left of the photograph a steer's brand is being examined. The photograph was taken by Geraldine Moodie, the wife of John Douglas Moodie, an officer with the North-West Mounted Police. (Courtesy of the collection of David McLennan, Regina, Saskatchewan. Sourced at the Southwestern Saskatchewan Oldtimers Association Museum in Maple Creek Saskatchewan.)

American cattlemen and cattle companies to the evolution of ranching on the grasslands of the Canadian west. How many American ranchers moved into western Canada? Where did they come from? Where and when did they settle? What were the sizes of their herds? Answers to these questions provide a useful foundation from which investigation of more complex issues may proceed.[7]

Analysis of lease agreements between the Canadian government and prospective ranchers has allowed the spatial patterns of ranching in the western Canadian prairies to be charted with some accuracy. Each lease authorized by the Canadian Department of the Interior had to be granted by the Governor General in Council, and these documents represent a contractual agreement between the lessee and the government. They specify the names and addresses of the lessees and the locations of their leases in terms of township and range.[8] From this material maps were constructed of leased land from 1882 to 1892. Thereafter lists of lessees were published in the Sessional Papers until 1906, although the details of lease locations were kept at local land offices. In addition, large-scale movements of American stock on to Canadian grassland called for comment by officers of the North-West Mounted Police, the Department of Agriculture, and the Department of the Interior. Finally, stock movements were featured in the columns of local newspapers on both sides of the border, while the views and aspirations of prominent ranchers were reported in some detail.

DIFFUSION OF RANCHING FROM MONTANA:
THE ROLE OF FORT BENTON TRADING COMPANIES, 1874–1882

The great trading companies of Fort Benton, Montana, dominated the economic life of southern Alberta from their base on the Missouri River until the Canadian Pacific Railway reached Calgary in 1883.[9] The small-scale activities of early settlers and men who had served a term in the North-West Mounted Police were entirely overshadowed by the diversity and scale of the operations of the I. G. Baker Company and its rival, the T. C. Power Company.[10] The strength of these companies lay in their control of the shortest water route to eastern metropolitan centres. From Fort Benton, at the head of navigation, a network of trails reached westward to the gold camps of the Bitterroot, and northward to the Canadian border (Figure 1).[11] Their hold over transportation links was translated into control of wholesaling and retailing functions, and this in turn meant that the companies played an important role in banking. They were the only organizations capable of supplying the North-West Mounted Police with the goods and services which they desperately needed on their arrival at the Oldman River in 1874.[12] Care was taken to cultivate these early contacts, and almost all the supply contracts approved by the Canadian Department of

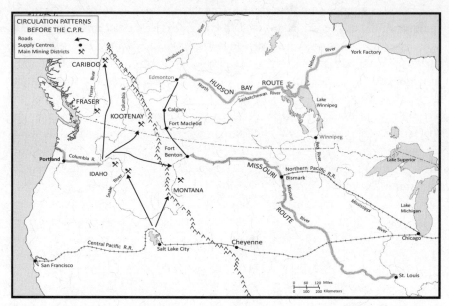

Figure 1. Transportation patterns prior to the coming of the railway. (Prepared by Diane Perrick, Canadian Plains Research Center.)

the Interior between 1874 and 1883 accrued to the Fort Benton companies.[13] Indeed, it has been estimated that well over half the money appropriated in Ottawa to police the North-West Territories ended up in the bank accounts of the Fort Benton merchants.[14]

The presence of commercial giants such as these could not fail to exert a powerful influence on the development of the range cattle industry. Even before the arrival of the police, the I. G. Baker Company managed large numbers of work oxen. When it secured ongoing contracts to supply beef to the police and the Indians, the company added to its herds.[15] By 1878 it was handling some $500,000 worth of cattle annually, and supplying government agents as far away as Saskatchewan.[16] In 1882 range interests were formally separated from trading concerns, and the Benton and St. Louis Cattle Company was incorporated.[17] This company was known in Canada as the "Circle Outfit" because of the brand which was used on its Canadian cattle. It maintained a continuous presence on the Canadian range until 1912. Many early ranchers purchased their first store cattle and "she-stock" from this company's herds, while small ranchers in Canada received short-term loans and sub-contracts from the parent company.

Like any fledgling industry in a remote underdeveloped region, the cattle industry of the Canadian plains was desperately short of skilled labour. The Benton companies employed large numbers of "key personnel," men who understood weather and grass, who could ford swollen rivers and control prairie fires, and who knew much of the ways of cattle and horses. They were employed as bull-whackers, mule-skinners, coach-drivers, and cowboys. Many of these employees settled north of the border and played their part in the region's subsequent history.[18]

Thus the range cattle industry of the western prairies during these formative years depended to a considerable extent on Montana trading companies to provide stock, capital, expertise, and vital transportation linkages.[19]

The process of gradual diffusion of men, herds, and techniques from Montana into the foothills and plains of western Canada was disrupted during the early 1880s. The government of Canada promoted the development of a Canadian cattle industry as one of several tactical thrusts which were to contribute to the grand strategy of the National Policy. At the same time the completion of the Canadian Pacific Railway revolutionized circulation patterns. Exceptionally favourable terms for leasing large acreages were used as an inducement to draw risk capital into the underdeveloped West. The investment community of eastern Canada snapped up the lure, and ranch companies, hastily formed in Montreal, Ottawa, Toronto, and London, England, poured capital into stocking the Alberta range. The character of ranching in the foothills was transformed, and the industry was dominated by major Canadian and British companies for the next two decades.[20] Leases covering some four million acres were granted in 1882, and in the next two years ten cattle companies established a hold over the choice grazing lands of the chinook belt. Not one of these companies was American.

The promulgation of the lease legislation had little or no direct impact on well-established Montanan ranching interests. I. G. Baker continued to hold vital supply contracts and to import large herds. No attempt was made to take out leases because the border was virtually open, and enormous tracts of land were available free. Nevertheless, the relative importance of the American contribution was eclipsed by the spectacular inflow of cattle to stock the newly acquired Canadian and British ranches.

THE "BEEF BONANZA" AND AMERICAN EXPANSION, 1885-1886

Conditions on the Great Plains of the United States prompted American cattlemen to evaluate carefully the potential of the Canadian range during 1885. The zenith of the "Beef Bonanza" had been reached.[21] Competition for young stock and unoccupied range raised production costs, while overproduction resulted in declining prices. Buoyant optimism was replaced by a widespread feeling of unease.[22] A reform administration had demonstrated its intentions of bringing imperious western cattle kings to heel. Thousands of miles of illegal fences were being torn down, and a start had been made in clearing illegal stock from Indian lands.

One obvious response to these pressures was to look north to the Canadian range. It was manifestly understocked, and the lease legislation offered a cheap means of obtaining legal range rights which did not exist in the United States. Moreton Frewen, part owner and manager of the Powder River Cattle Company, quotes a graphic summary of the situation by his foreman, E. W. Murphy:

> Boss, can we get clear out of Wyoming before the fall, and save ourselves in Alberta? You will have these southern cattle here in five months as thick as grasshoppers, and this being so, if you lose those five months you had better advertise for skinning outfits; your money will be in green hides the next two winters here, unless I mistake.[23]

On this occasion Frewen listened to good advice and his herds were settled on a lease along the Bow River by the end of the year.

American interests besieged the Department of the Interior with requests for leases during the fall of 1885 and the spring of 1886. By the end of that year they had secured leases covering 721,000 acres, about 19 percent of the total.[24] Conrad Kohrs visited Ottawa and obtained 187,000 acres for his Pioneer Cattle Company. He was also instrumental in obtaining a large lease for Dan Floweree.[25] The Benton and St. Louis Cattle Company took out a 100,000-acre lease to the west of Coaldale. The location of leased land held by Americans is shown in Figure 2. It was concentrated around the Cypress Hills and in the short-grass prairie lands well to the east of the main Canadian and British holdings.

Projected expansion far exceeded that which actually took place. A further sixteen leases amounting to 1.2 million acres were approved by Order in Council, but never taken up. If this wholesale expansion of the American range frontier had reached fruition, the character of the industry in Canada might have altered once again.

The catastrophic impact of the winter of 1886-87 curbed the thrust of American cattlemen on to the Canadian range. Robert S. Fletcher has examined the impact of that winter in Montana, and he concluded that "In 1886 and 1887 nature and economics seemed to conspire together for the entire overthrow of the industry."[26] The summer of 1886 was exceptionally dry, and grasshoppers and fires further reduced the available forage. Cattlemen prayed for a mild winter, for their cattle were in no condition to face an unusually cold season. In the event, the weakened herds were subjected to the worst

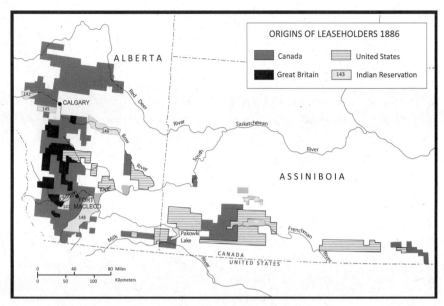

Figure 2. (Prepared by Diane Perrick, Canadian Plains Research Center.)

winter on record. Farmbred "pilgrim" cattle shipped from the mid-west to stock the new leases had no ability to fend for themselves. Southern cattle, thrown on the Canadian range in poor condition in the late fall, suffered terribly. Dan Floweree's herds, wintering to the west of the Cypress Hills, were ravaged. Under the headline "A Woeful Story," the *River Press* estimated that 50 per cent would not begin to cover the losses.[27] Further east, the herds of the Home Land and Cattle Company on their Wood Mountain lease were decimated by the prolonged cold snap experienced in February and early March. The June round-up showed that 4,000 of the 6,000 head of cattle ranged in Canada had perished.[28] The great Niobrara Cattle Company held some 39,000 head of fat cattle in 1886, with an estimated value of one million dollars. After their round-up they could muster only 9,000 head, and their liabilities totalled $250,000. Among their losses was a herd of yearlings ranged to the east of the Cypress Hills.[29]

The bubble had burst and the boom was over. The depleted American herds retreated from Canada to their familiar home ranges, and planned expansion never took place. The deaths of thousands of cattle and record shipments of surviving stock, to meet immediate financial commitments, meant that there was once more room for the remaining ranchers to expand within Montana Territory. In 1888, the massive Indian reservation which had reached northward from the Missouri to the Canadian border was severely reduced in size, and some twenty million acres of grazing land were made available to the stock-man.[30] The combination of boom conditions and crowded ranges, which had encouraged American cattlemen to look north of the line for grass, no longer existed. The invasion of the Canadian range was put off for almost a decade.

DRIFTING AND SMUGGLING: PRESSURE ON THE BORDER: 1895-1902

The cattle industry in Montana recovered quickly from the severe losses of 1887. The more accessible regions of central Montana were being enclosed, and open-range cattle companies were increasingly confined to the area north of the Missouri River.[31] Competition with sheep-farming interests became increasingly bitter because good prices for wool and beef encouraged expansion of both types of enterprise.[32] Ranchers, already acquainted with the geography of the Canadian range, began to exert clandestine pressure on the grasslands across the border after 1895. To the west, the Floweree family of Great Falls ranged its herds north from the Sun River and considered the Milk River Ridge and the Pothole country as part of their habitual range.[33] In 1896, the North-West Mounted Police were forced to hire Indians to help them drive 15,000 head of American cattle back from the ridge across the line.[34] The "D Bar S" firm, which was heir to the famous partnership between

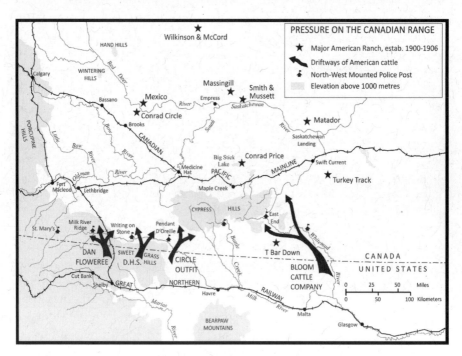

Figure 3. Movement into Canada from the United States. (Prepared by Diane Perrick, Canadian Plains Research Center.)

Sam T. Hauser, Granville Stuart, and A. J. Davis, ran cattle to the Milk River and beyond in the vicinity of the Writing-on-Stone police post. The Circle Outfit ranged northward through Pendant D'Oreille to the shores of Bad Water Lake. Further east, cattle belonging to the Bloom Cattle Company grazed northward up the Whitemud River from their home range near Malta, Montana (Figure 3).[35]

Careful observations by police detachments during 1896 established without a doubt that large numbers of American cattle were grazing on the Canadian range throughout the year. Fall round-ups gathered only such stock as were marketable and left the young steers and "she-stuff" within Canada. The Commissioner of the North-West Mounted Police summarized the situation in 1899:

> The difficulty with American Cattle drifting across the line still continues and I regret to report that it is assisted by American owners and their employees. The Round Up (U.S.) party gather up all their cattle on our side, and take them past our outposts, but as soon as the vicinity of the line is reached, those fit for shipment are carefully picked out and the balance are let go when they promptly return to their accustomed haunts on our side, and continue to annoy our settlers, who have prudently put up hay for their own cattle. As fast as they drive these intruders from their stacks they return.[36]

In 1899 American round-up parties, operating on Canadian soil to the south of the Cypress Hills, handled some twelve thousand head of cattle.[37] Thus the cowboys of the "Turkey Track" and the "T Bar Down" became familiar with the area along the Whitemud River six years before these American companies obtained legal rights to land in Canada.

Attempts to control this persistent drifting of American cattle became one of the most onerous and frustrating duties of the police. They were forced to take up the job of cowboy in addition to their other duties, as the number of civilian line-riders employed was never adequate. No sooner had a herd of strays been gathered and dispatched south of the border than reports would arrive of incursions at some other point. Some Montana companies owned stock on both sides of the border, making the detection of smuggling virtually impossible. Friction between the police and the customs department added to the frustrations of law enforcement; Superintendent Burton Deane, who was stationed at Lethbridge during this period, and whose reports are a particularly useful source, later wrote a book about his experiences and included a chapter entitled "Wholesale Cattle Smuggling."[38] Nor could the police rely on wholehearted support from Canadian cattlemen, many of whom were members of the Montana Stockgrowers' Association. These men were concerned lest zealous enforcement of the letter of the law might provoke retaliatory action against Canadian cattle driven south in winter storms.

In 1885 American cattlemen had been prepared to go to Ottawa to gain official recognition of their presence. As pressure on the open range in Montana mounted, so they once again reviewed the situation north of the line. They found the lease system in disrepute. The leases of the major cattle companies had been cancelled in 1896, and no alternative policy with regard to extensive leases had been implemented. By occupying the understocked range along the border, American ranchers were merely imitating their Canadian counterparts who grazed extensive herds on Crown lands without any restrictions.

THE LAST OF THE OPEN RANGE, 1902-1906

The opening years of the twentieth century witnessed a surge of optimism in the cattle industry. Large leases were once more made available by the Canadian Department of the Interior. Great cattle companies from the United States once more moved on to the Canadian range with the support of the government. There was an influx of cattle from south of the border which bore comparison with the great in-migration of the 1880s. The vast area north of the Canadian Pacific mainline, and on both sides of the South Saskatchewan River, was stocked for the first time.

This areal expansion was pioneered by a number of major American cattle companies. Several of these had originated in Texas and had followed the grass further and further northward, pushed by the ubiquitous "nester" with his barbed-wire fences, and by the sheepman with his contemptible "woolies." These outfits held fast to the methods of the open range. Almost all of their investment was tied up in cattle, and very little in deeded land and improvements. They were footloose and mobile, but at the same time very vulnerable to crowding since they had legal rights to so little. Tony Day and H. W. Cresswell were typical of this breed.[39] Both men started their ranching careers in Texas, and the turn of the century saw them working as partners in Dakota. They held some 20,000 cattle in their grade herd, and maintained an interest in 10,000 first-rate Herefords in the panhandle of Texas. John Clay described their motive for moving to Canada in a later reminiscence: "As settlers drifted in and made ranching more or less a misery, they moved up to Canada."[40] Duty was paid on 30,000 head, and additional "dogies" were acquired from Manitoba. Wilkinson and McCord shipped 4,000 head of cattle and horses from Canyon City, Texas, to Billings, Montana, in eight trains during April

1903.[41] They then trailed them north to Sounding Lake where they established their home ranch. Smith and Mussett moved up from Kansas to cross the South Saskatchewan and stock the range to the north and east of the present site of Empress. The Conrad Price Cattle Company moved north of Maple Creek to the shores of Big Stick Lake and ranged its cattle north into the Great Sandhills country. The Turkey Track and the T Bar Down legalized their *de facto* occupation of Canadian rangelands.[42] The former company grazed some 20,000 head between the Whitemud River and Swift Current, while the latter held 10,000 head south of the Whitemud and the Cypress Hills. After almost three years of careful investigation and negotiation the famous Matador Land and Cattle Company established its Canadian Division on a 150,000-acre lease north of Saskatchewan Landing. Some 3,000 head of steers and spayed heifers were moved on to the range in June, 1905.[43] By 1903 there were twenty-two leaseholders from the United States. Together they held some 60,000 acres, or about 30 per cent of the total leased acreage.[44]

The long-distance movement of large cattle companies from the United States to Canada was paralleled by short-distance migration of smaller ranchers from Montana. More than half the American leaseholders in 1903 came from Great Falls, Butte, Whitlash, and other settlements in northern Montana. Most of these men leased from 2,000 to 10,000 acres. Others filed on homesteads and did not take out leases for some years. T. B. Long was one such person. He came to the Cypress Hills from the overcrowded Madison Valley of Montana in 1904. He recalled his first impressions in the following terms:

> Several years' growth of grass rippled in the wind, knee deep to a horse as far as the eye could see. It is just impossible to describe the amount of grass we saw, and there was free range everywhere.[45]

Long was accompanied by Tom Whitney when they moved their 1,500 head of cattle north from the station at Havre, Montana. They soon met up with Fred Garrison, who had moved from the Madison a year or so previously. It is clear that word-of-mouth communication encouraged small cattlemen to move across the border, while others were drawn into Canada behind the herds of the great companies and stayed to establish their own herds.

The movement of American cattlemen, both large and small, to the Canadian range meant a considerable growth in the flow of cattle northward across the border. Cattle inspectors at western ports of entry recorded imports of 63,000 head in 1903. The total number of "other horned cattle" in the Northwest Provinces more than doubled, from 698,409 to 1,560,592, between 1901 and 1906.[46]

The social significance of this "American invasion" was great. Incoming farm settlers, whether from the eastern provinces, the mid-western states, or from Europe, had their stereotypical images of "cowboys and cattle kings" confirmed by their contacts with the great cattle companies which dominated the dwindling short-grass range during the first decade of the twentieth century.[47] This fact has done much to obscure the fact that a robust and uniquely Canadian variation of the cattle kingdom had taken root in the foothills area twenty years previously.

Scarcely had the newcomers established themselves when they were subjected to another killing winter, that of 1906-07.[48] The most savage impact of this exceptional winter was felt in the country between the Little Bow and the Red Deer River, and northwards to Sounding Lake, just the areas recently occupied by American ranchers. Harry Otterson, the foreman of the T Bar Down, made a trip from the ranch headquarters toward the end of March. He described it thus:

> It surely was a gruesome ride. The cattle were in all stages of dying. The bush was simply lined with dead cattle. The live ones at night would lie down on the dead and many would not be able to get up again, consequently they were literally piled up.[49]

The exact extent of the losses caused by the winter of 1906-07 will never be known, but even the most conservative estimates support the conclusion that about half the working capital invested in the range cattle industry was liquidated.[50] The most obvious effect of the disaster was the failure of a large number of major cattle companies, and the retreat of others to the United States. Smith and Mussett sold out to the Massingill brothers. The pioneers of the Sounding Lake range, Wilkinson and McCord, were forced out of business. The list is long; by 1909 the Turkey Track, the T Bar Down, and the Conrad Price Cattle Company had all closed down their Canadian range interests.[51] Expansion projected for 1907 did not take place. Surviving stock were sold for what they would bring. From 1907 onwards, it was the American farmer, rather than the rancher, who contributed most to the growth and development of the Canadian west.

CONCLUSION

The American cattleman's advance into Canadian grassland was both time-specific and orientated toward particular regions. A period of gradual diffusion of men, cattle, and techniques, under the umbrella of large Montana trading companies, was disrupted by the introduction of lease legislation by the Canadian government in 1882. The advance across the border of 1885 had little lasting effect. A second influx occurred from 1901 to 1905, and followed a number of years of *de facto* occupation of range along the international border. Both movements were spearheaded by a limited number of large companies and were dramatically halted by environmental cataclysms. The later "invasion" was accompanied by a considerable migration of farmer-ranchers from Montana. These men merged with American farmers who were flocking in to take advantage of the opportunities offered by the "Last Best West."

It is difficult, if not impossible, to generalize about the Canadian range cattle industry as a whole. Two very different traditions evolved side by side. One, located in the foothills and valleys flanking the Rocky Mountains, was the creation of the eastern Canadian establishment, and was closely linked to Imperial markets.[52] The short-grass prairies to the east, on the other hand, were occupied briefly by the last survivors of a colourful company which had ridden the trails and followed the grass up from Texas.

NOTES

This article first appeared in *Prairie Forum* 4, no. 1 (1979): 121-35.

1   This paper is based on material gathered during three years of research on the range cattle industry in Canada. See Simon M. Evans, "Ranching in the Canadian West, 1882-1912," Ph.D. thesis, University of Calgary, 1976. The author acknowledged the generous support of the Isaac Walton Killam Memorial Scholarship Programme, which made this research possible. An earlier version of this paper was presented at the Annual Meeting of the Canadian Association of Geographers, University of Regina, 1977.

2   A. S. Morton, *History of Prairie Settlement*, Volume 2 of W. A. Mackintosh and W. L. G. Joerg (eds.), *Canadian Frontiers of Settlement* (in nine volumes; Toronto: Macmillan and Co., 193440); and J. F. Booth, "Ranching in the Prairie Provinces," in R. W. Murchie, *Agricultural Progress on the Prairie Frontier*, Volume 5 of *Canadian Frontiers of Settlement*, p. 53.

3   J. H. Warkentin, "Western Canada in 1886," *Historical and Scientific Society of Manitoba*, III, 20 (1963-64), p. 105.

4   W. L. Morton, "A Century of Plain and Parkland," in Richard Allen (ed.), *A Region of the Mind* (Regina: Canadian Plains Research Center, 1973), p. 170.

5   L. G. Thomas, "The Ranching Period in Southern Alberta," M.A. thesis, University of Alberta, 1935; and D. H. Breen, "The Canadian West and the Ranching Frontier, 1875-1922," Ph.D. thesis, University of Alberta, 1972.

6   Some of the most important include: (a) the role played by governments north and south of the line, (b) the contribution of stockmen's associations, (c) the complex interplay of the Chicago and trans-Atlantic markets, and (d) the influence of the "wild west" image on the attitudes and perceptions of ranchers from eastern Canada and Great Britain. The complexities of the search for origins of ranching in any particular area are exemplified in a recent article by Terry G. Jordan, "Early Northeast Texas and the Evolution of Western Ranching," *Annals of the Association of American Geographers*, 67, No. 1 (March 1977), pp. 66-87.

7   See Robert F. Berkhofer, *A Behavioural Approach to Historical Analysis* (New York: The Free Press, 1969), p. 231.

8   For detailed consideration of these sources and the maps produced from them see Evans, "Ranching in the Canadian West." Copies of the Orders in Council are housed in the Library of the Alberta Department of Lands and Forests, Edmonton, Alberta. Some exploratory use of them was made by A. A. Lupton, "Cattle Ranching in Alberta, 1874-1910," *The Albertan Geographer*, 3 (1966-67), pp. 48-59.

9   The best overview of this period is by Paul F. Sharp, *Whoop-Up Country: The Canadian American West, 1865-1885* (Minneapolis: University of Minnesota Press, 1955).

10   Breen provides a detailed account of early ranching and the contribution made by ex-policemen to its growth. However, he does not consider the role of the Montana-based companies. Breen, "The Canadian West and the Ranching Frontier," pp. 55-143.

11   Gerald L. Berry, *The Whoop-Up Trail* (Edmonton: Applied Art Productions Ltd., 1953).

12   S. W. Horrall, "The March West," in Hugh A. Dempsey (ed.), *Men in Scarlet* (Calgary: McClelland and Stewart, 1974), pp. 23-24.

13   The following figures are indicative of the importance of Montana companies to the police, but they do not include the private purchases of members of the force at I.G. Baker stores.

FINANCIAL YEAR	TOTAL EXPENDITURE, N.W.M.P.	PAYMENT TO I.G.B.
1874-75	$333,583	$23,395*
1875-76	$369,518	$122,771
1876-77	$352,749	$126,243

*The force did not reach the West until September.
Canada, *Sessional Papers*, 1879, xii, Vol. 10, no. 188, "Expenditure for North-West Mounted Police, 1876, 77, 78, and of amounts paid to J.[sic]G. Baker and Co.," pp. 1-168. See also T. Morris Longstreth, *The Silent Force* (London: The Century Co., 1927); and for a delightfully biased account of the early contacts between the company and the police, see Ora J. Halvorson, "Charles E. Conrad of Kalispell: Merchant Prince with a Gentle Touch," *Montana, the Magazine of Western History*, 21 (1971), p. 61.

14   Sharp, *Whoop-Up Country*, p. 222.

15   Government contracts to supply Indians provided a multi-million dollar market for American cattlemen which had much to do with the spectacular growth of the range cattle industry during the late 1860s. See Edward Everett Dale, *The Range Cattle Industry* (Norman: University of Oklahoma Press, 1930), pp. 62-65.

16   Canada, *Sessional Papers*, 1880, xiii, Vol. 3, no. 4, "Report of the Commissioner for Indian Affairs, North-West Territories," p. 87.

17 Montana Stock Growers' Association, *Brand Book, 1885-86*, Historical Society of Montana. Archives, Helena, p. 162.

18 For example, D. W. Davis was an itinerant whiskey trader for the I. G. Baker Company until the arrival of the police. He then became manager of the company's store in Fort Macleod. He served as Member of Parliament in Ottawa from 1887 to 1896. Howell Harris was foreman of the Circle Outfit for many years, and became a respected figure in the Western Stockgrowers' Association. Frank Strong, who rescued Senator Cochrane's herd on the Waterton Range in 1883, was also an employee of the company. See L. V. Kelly, *The Rangemen* (Toronto: William Briggs, 1913), p. 191.

19 Hill suggested that the witnesses of the signing of Blackfoot Treaty Number Seven were symbolic of the three "giant spearheads of the invasion of settlers… ," the Hudson's Bay Company, The North-West Mounted Police, and the Christian Church. He fails to point out that Charles E. Conrad of the I. G. Baker Company was also a signatory. Douglas Hill, *The Opening of the Canadian West* (Don Mills: Longman Canada Ltd., 1976), 137; and Canada, *Sessional Papers*, 1878, xi, Vol. *8*, No. 10, "Annual Report of the Department of the Interior," p. xlvii.

20 For a discussion of the response to the Canadian government's lease legislation, see Simon M. Evans, "Spatial Aspects of the Cattle Kingdom: The First Decade, 1882-1892," in A. W. Rasporich and Henry Klassen (eds.), *Frontier Calgary* (Calgary: McClelland and Stewart West, 1975), pp. 41-56.

21 This phrase became popular in the United States during the 1870s, and was used as the title of a widely read book, General James B. Brisbin, *The Beef Bonanza: or, How to Get Rich on the Plains* (Philadelphia: J. P. Lippincott Co., 1881; republished, Norman: University of Oklahoma Press, 1959).

22 Ernest Staples Osgood, *The Day of the Cattlemen* (University of Chicago, 1929), 216-20; and Maurice Frink, W. Turrentine Jackson, and Agnes Wright Spring, *When Grass Was King* (Boulder: University of Colorado Press, 1956), pp. 50-57 and 224-30.

23 Moreton Frewen, *Melton Mowbray and Other Memories* (London: Herbert Jenkins Ltd., 1924), p. 222.

24 Orders in Council, 1886, *Department of Interior*, Vol. viii.

25 The *River Press* of Fort Benton, the *Stockgrowers' Journal*, the *Rocky Mountain Husbandman*, and the *Weekly Yellowstone Journal* and *Livestock Reporter* all included reports of herds moving towards the Canadian border during July and August, 1886. For instance, *River Press*, August 18, 1886; and for a front-page interview with Conrad Kohrs, *River Press*, September 22, 1886.

26 Robert S. Fletcher, "That Hard Winter in Montana, 1886-1887," *Agricultural History*, 4 (1930), p. 123. See also Ray H. Mattison, "The Hard Winter and the Range Cattle Business," *Montana, the Magazine of Western History*, 1 (1951), pp. 5-21; and Barbara Fifer Rackley, "The Hard Winter, 1886-1887," *Montana, the Magazine of Western History*, 21 (1971), pp. 50-59.

27 *River Press*, February 23, 1887.

28 *Weekly Yellowstone Journal*, August 20, 1887.

29 *Fergus County Argus*, October 27, 1887; the *Macleod Gazette* carried the story of the failure of the famous Swan Land and Cattle Company of Wyoming on its front page, *Macleod Gazette*, June 14, 1887; see also *River Press*, March 2, 1887.

30 Osgood, *The Day of the Cattleman*, p. 223.

31 Robert H. Fletcher, *Free Grass to Fences* (New York: University Publishers Inc., 1960), p. 113; *Fergus County Argus*, August 7, 1890; and *River Press*, April 27, 1892.

32 Robert S. Fletcher, "The End of the Open Range in Eastern Montana," *Missouri Valley Historical Review*, 16 (September 1929), p. 204.

33 William H. McIntyre, "A Brief History of the McIntyre Ranch," *Canadian Cattlemen*, 10 (September 1945), p. 94.

34 D. H. Breen, "The Mounted Police and the Ranching Frontier," in Hugh A. Dempsey (ed.), *Men in Scarlet* (Calgary: McClelland and Stewart, 1974), p. 123.

35 Canada, *Sessional Papers*, 1897, xxxi, Vol. 5, No. 8, "Report on Cattle Quarantine," p. 105.

36 Canada, *Sessional Papers*, 1899, xxxiii, Vol. 12, No. 15, "Report of the Commissioner North-West Mounted Police," p. 3.

37 Canada, *Sessional Papers*, 1900, xxiv, Vol. 12, No. 15, "Report of Superintendent G. B. Moffat, Maple Creek," p. 41.

38 R. Burton Deane, *Mounted Police Life in Canada* (London: Cassell and Co. Ltd., 1916), pp. 154-81.

39 Mary Terrill, "Uncle Tony Day and the Turkey Track," *Canadian Cattlemen*, 6 (June 1943), p. 8.

40 Letter from John Clay to Mary Terrill. Clay was a Scot who lived through the great period of the open range in the American West, and his book is one of the best sources on British-owned ranches. John Clay, *My Life on the Range* (Norman: University of Oklahoma Press, new edition 1962).

41 Margaret V. Watt, "McCord's Ranch, A Chronicle of Sounding Lake," *Canadian Cattlemen*, 15 (November 1952), p. 20.

42 For general coverage see C. J. Christianson, *Early Rangemen* (Lethbridge: Southern Printing Co., 1973); and W. J. Redmond, "The Texas Longhorn on Canadian Range," *Canadian Cattlemen*, 3 (December 1938), p. 112.

43 William M. Pearce, *The Matador Land and Cattle Company* (Norman: University of Oklahoma Press, 1964), pp. 87-93. Blasingame gives an account of the journey from Motley, Texas, via Sioux City and Portal, to a point on the Canadian Pacific Railway to the south of the lease. Ike Blasingame, *Dakota Cowboy: My Life in the Old Days* (Lincoln: University of Nebraska Press, 1958), p. 282.

44 Canada, *Sessional Papers*, 1904, xxxviii, Vol. 10, No. 25, "Report of the Secretary, Timber, Mines, and Grazing," pp. 60-76.

45 T. B. Long, *70 Years a Cowboy* (Regina: Western Printers Association Ltd., 1959), p. 9.

46 Canada, *Sessional Papers*, 1907, No. 17a, "Census of Population and Agriculture of the Northwest Provinces," p. xxiii. "The Northwest Provinces" was the term used in the 1906 Census to describe the Prairie Provinces.

47 See for instance, Wallace Stegner, *Wolf Willow* (New York: Viking Press Edition, 1966), pp. 127-38: and G. Shepherd, *West of Yesterday* (Toronto: McClelland and Stewart, 1965). The impact of this influx of cattle from the United States on the quality of Canadian range herds is discussed in Simon M. Evans, "Stocking the Canadian Range," *Alberta History*, Vol. 26, No. 3, (Summer 1978), pp. 1-8.

48 General accounts of this winter may be found in Stegner, *Wolf Willow;* Blasingame, *Dakota Cowboy;* and R. D. Symons, *Where the Wagon Led* (Toronto: Doubleday Canada Ltd., 1973).

49 This quotation comes from one of the best contemporary accounts of the 1906-07 winter in the short-grass area, Harry Otterson, "Thirty Years Ago on the Whitemud River," unpublished manuscript, Glenbow-Alberta Institute, Calgary, Alberta.

50 See the remarks of the Livestock Commissioner, J.G. Rutherford, *The Cattle Trade of Western Canada* (Ottawa: King's Printer, 1909), p. 8.

51 Otterson remarked: "Practically all the large owners began to make preparations to close out or still further reduce their herds." Otterson, "Thirty Years Ago on the Whitemud River," p. 27.

52 Simon M. Evans, "Canadian Beef for Victorian Britain," *Agricultural History*, 53, no. 4 (1979), pp 748-62.

12. The End of the Open Range Era in Western Canada

Simon M. Evans

INTRODUCTION

The opening years of the twentieth century saw a surge of optimism pass through the cattle industry in western Canada. Vast areas of southern Alberta and Assiniboia were still covered with native grasses. Canadian and British cattle companies continued to flourish using, for the most part, the extensive methods of the open range. Cattle companies from the United States moved onto the Canadian range with the support of the Government, and to the newcomers "the virgin prairie of southwestern Saskatchewan looked like the promised land."[1] There was an influx of young cattle from south of the border which bore comparison with the great immigration of the 1880s.[2] At the same time exports from the Canadian West of mature ranch cattle rose to new heights.

By 1912, wheat was king. An ever growing network of railways had reduced the open range to a series of disconnected islands. The grid pattern of barbed wire fences and road allowances, which had existed for so long only on the maps of the surveyors, became a finite reality. Homestead settlement burgeoned to reach its maximum density in Palliser's triangle. Few cattlemen were complacent enough to expect that their hold over the open range would last forever, but none could have anticipated the speed with which their industry was to be brought low and the grassland violated by the ploughs of ten thousand farmers.

This transformation from open range to a mosaic of farms and ranches was a product of an infinite number of decisions on the part of a heterogeneous collection of people with differing aims, aspirations and goals. They monitored the changing economic milieu according to their varied perceptions and abilities and constantly adjusted their strategies. In most circumstances, land use change over extensive areas is likely to be a slow process taking generations to achieve. In this case the transformation seems to have been precipitate. The aim of this paper is to examine the factors which prompted such rapid reappraisal.

Contemporary observers blamed the end of the open range on the killing winter of 1906-1907. The advantages of hindsight partially confirm their assessment. No industry could sustain the awful losses of that winter without undergoing changes. Nevertheless, this incident, for all its importance, must be reviewed in context. The dominance of ranchers over the grazing lands of the North West rested on three foundations: the support of the government of Canada; the widely held belief that the shortgrass prairies of Palliser's Triangle were too arid for arable farming; and the insatiable demand of the British market for beef. All three of these foundations were being eroded even before the killing winter.

RANCHERS AND THE GOVERNMENT

The government of Canada had encouraged the establishment of large-scale cattle raising in the North West by lease legislation enacted in 1883.[3] During the next twenty years the rancher was assured of the tacit support of the Department of the Interior.[4] The original "closed leases," which precluded homestead settlement, had been cancelled in 1896, but generous settlement terms had enabled most major ranch companies in the foothills to transfer from leased to deeded land with minimum disruption. Moreover, the establishment of stock watering reserves, shelter reserves, driftways, and cattle trails, did much to ensure that ranchers maintained their hold over summer grazing areas far beyond their deeded acreage.[5]

The appointment of Clifford Sifton as Minister of Interior to the incoming Laurier administration in 1896 heralded changes for the ranchers as it did for the Department itself.[6] Sifton seems to have recognized the fact that the process of farm settlement, however vigorously encouraged, would take some time to run its course. In the interim large amounts of land would remain unused. He felt that nothing should be done to disrupt the profitability of the range cattle industry which was making use of this territory. Settlers were therefore encouraged to take out small leases in the vicinity of their homesteads, while large leases were issued to "deserving cases" in increasing numbers during 1902 and 1903. The leased acreage increased from half a million acres in 1901 to about two million acres in 1903. Nonetheless, complaints concerning the complexity and ubiquity of stock watering reserves along the line of the railway between Fort Macleod and Calgary were received during his first year in office and became increasingly strident.[7] There was a danger that this bad publicity might reduce the flow of homesteaders into the region. In 1900 Sifton moved to dismantle the system of water reserves which had been painstakingly put together over the past twelve years. W. M. Pearce, the chief architect of the reserves, was reassigned to become Inspector of Surveys in 1901.[8]

Sifton's attitude towards the cattlemen was pragmatic. On the one hand, they should be encouraged to make the best possible use of unoccupied land. On the other hand, their interests should not be allowed to obstruct the progress of more intensive farm settlement. There was one thing the ranchers could be sure of; Sifton would not tolerate unorganized occupation of crown lands. They could rely on the Department of the Interior to support their rights over those of casual squatters.

The fact that their leases were open to homestead settlement was severely criticized by the cattlemen. Petitions and requests for changes in the homestead and pre-emption clause of the leases were received from the Western Stock Grower's Association and from residents and ranchers of Assiniboia.[9] The rancher's position is graphically explained in a letter from Alexander Mackay, Secretary of the American "Matador Land and Cattle Company," to Sifton, he explained:

> A lease granted with the right of settlement to squatters and others would be of little or no use for the purposes of this company, as the business would be prosecuted on a considerable scale, and if the company were exposed to vexations, annoyances and blackmailing at the instance of squatters and others, their business would not be worth pursuing under these conditions.[10]

Government officials acknowledged the problem, admitting that, "there is no doubt that it is not very satisfactory to the leaseholder to have no assurance that his ranch may not be broken up by homesteads or sales at any time."[11] The whole situation was subject to thorough review during the fall of 1903 and new regulations were introduced which went some way to meet the demands of the cattlemen.[12]

These regulations were in force for only five months before Sifton resigned his cabinet post. He was replaced at the Department of the Interior by Frank Oliver—who for years had championed the squatter and the farm settler against the big cattle companies. Oliver explained his attitude to the Inspector of Ranches unequivocally, "The policy of the Department is to protect the rights of owners of cattle actually in occupation, whether those owners apply for a lease or not."[13] The rights of squatters were to be upheld against those of the cattle companies—a complete reversal of earlier government policy. Oliver's new regulations abolished absolutely the possibility of "closed" leases. All leases were subject to withdrawal and cancellation, while no lease was to be issued until the land concerned had been inspected and proven unfit for agriculture. Moreover, climatic factors alone were not sufficient to justify classifying land as unsuitable for agriculture. Only land which was too gravelly, stony, sandy, or of too rough a surface for agriculture, was to be classed as not fit for agriculture and therefore suitable to be covered by grazing leases. Although the new policy was not retroactive and a number of large leases remained, the implicit recognition of a difference between "land suitable for agriculture" and "the grazing lands" had disappeared, and with it the claim of ranchers to special treatment under the Dominion Land's Act.

"PASTORAL" VS. "AGRICULTURAL"

During the opening years of the twentieth century the range cattle industry was challenged for the first time with competition for space from commercial agriculture which was sustained by eastern capital and spurred by rapidly evolving technology. The rise in the price of wheat, the extension of railway mileage, the wet cycle of rainfall, advances in dry-farming technology and agricultural mechanization, as well as the energetic policy of the Department of the Interior, all contributed to a positive surge of settlement.[14]

This advance can be mapped in a number of ways, for instance by using population data for each township or the cropped area in each township. Another good indicator of the progress of settlement is the expanding railway network. Cash grain farming depended on the railway for access to markets and there was a theoretical distance beyond which it was not economic for the farmer to move his product to an elevator. This critical distance varied with time, the condition of the rural roads, the capital available to the farmer, and the price of wheat. In the early years of the twentieth century it was estimated at about ten miles.[15] However, when the pulse of settlement was running strongly farmers established themselves in anticipation of railway building and for short periods grain was hauled for much greater distances. For this reason the areas marked on maps 1 and 2 as being "beyond the economic reach of the railways" and therefore available for open range grazing, are those areas which were more than 20 miles from the nearest railway line.

To the ranching industry, the expansion of settlement along the railway between Calgary and Fort Macleod was most critical (Figure 1). Superintendent Primrose reported from Fort Macleod in 1903: "The increase of settlement in this district has been enormous, to the north from Macleod to Nanton, to the east to Kipp, to the south to the boundary, and west to the Crow's Nest Pass, nearly every available section of land has either been taken up or purchased."[16] Stopping places on the railway, marked by corrals and a derailed cattle car in 1902, grew rapidly into thriving small towns like Claresholm, Nanton, and

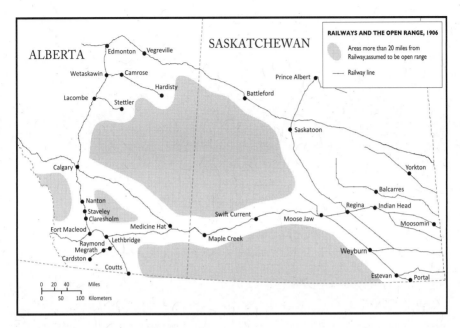

Figure 1. Railways and the open range, 1906. (Prepared by Diane Perrick, Canadian Plains Research Center.)

Stavely. The Mormon settlements of Raymond Stirling and Magrath, blossomed almost overnight. "Last year, at the time of the director's visit, a new town called Raymond was being laid out, and the only object which broke the monotony of the plains was the surveyor's tent. Not an acre of crop was in sight. Within a year a town had sprung up with a population of 600 all comfortably housed. Their crops cover an area of nearly 5,000 acres and grain of all sorts has given very satisfactory yields."[17] A continuous broad band of settlement reached from Calgary to Fort Macleod by 1906.

Not only did the number of settlers increase rapidly, but individually and collectively they posed a more formidable threat to the open range than had their predecessors. During the 1890s, the ranchers had had to face the incursions of homesteaders and squatters most of whom aspired to become small ranchers. They were resented by established cattlemen but at least they shared an interest in stock and the range. After the turn of the century, the great influx of settlers was made up of people of a very different ilk. Many of them were experienced cash grain farmers. Some of them had sold farms in the United States for $15.00 to $20.00 an acre and were purchasing land in Canada for $4.00 to $5.00 an acre. They had capital to invest in machinery and hastened to break up their holdings and to extend the scale of their agricultural operations.[18]

Contemporary observers were unanimous in expressing the view that this increase in settlement meant the end of the ranching industry as they had known it in the foothills. "The farmer and his fences are gradually driving the big ranches further and further back, and it is only a question of years when the real ranch will have ceased to exist and the farm with its small bunch of cattle will have taken its place."[19] Foothills ranchers were cut off from their summer grazing by a belt of farmland some 50-100 miles wide. At the same time they had to compete with settlers for continued use of winter range in the Porcupine Hills and the valleys of the eastern slopes of the Rocky Mountains. Ranchers were faced with two alternatives, either they could intensify production by fencing, winter feeding and cropping fodder, or they could relocate on the shortgrass prairie to the east. In 1906 an era ended when the lands of the Cochrane Ranch, one of the first and perhaps the most prestigious of the major Canadian cattle companies, were sold to the Mormon Church.[20]

MARKETS, PRICES, AND THE MEAT TRADE

The influx of settlers into "cattle country" was not the only problem which faced ranchers in the early years of the twentieth century. Marketing problems and falling cattle prices affected both the "Cattle King" and the small ranchers and eroded their ability to compete with the cash grain farmer. The last year of high cattle prices was 1902, and by 1903 many individuals were prompted by dwindling profit margins to bypass the major cattle buyers and ship direct to Britain.[21] This was not successful. An alternative strategy was to withhold cattle from market in the hope that prices would pick up. The Provincial Legislature of Alberta displayed interest in "the depressed state of the beef market," soon after its establishment in 1905.[22] In a report to the committee on agriculture, the Secretary of the Alberta Stockbreeder's Association summed up the problems facing the cattlemen. He pointed out that because of the rapid settlement of the country, stockmen were being confined to their own landholdings. This involved a much larger capital expenditure for lands, fences, buildings, as well as measures to provide regular winter feeding for cattle. This situation was compounded by a "gradual decline in beef prices during recent years." He argued that wholesale spaying of heifers and disposal for slaughter of breeding cows demonstrated the stockman's loss of confidence.[23] The following year a more formal Commission of Inquiry reached the conclusion that "the rancher and the producer are not receiving sufficient remuneration for their labour and investment."[24]

There was a growing feeling among western Canadian cattlemen that the companies which controlled the export trade were responsible for their problems. In particular, they were convinced that they were being provided with extremely poor service by the Canadian Pacific Railway. The Territorial Purebred Cattle Breeder's

Association petitioned the Dominion Department of Agriculture to investigate the reasons why United States' cattle were landed at British ports cheaper and in better condition than "the bruised and ill-handled Canadian ranch cattle."[25]

The truth of the matter was that the Canadian Pacific railway was totally incapable of handling the increased freight traffic which was being offered for shipment. Not only had eastbound shipments of cattle jumped threefold from 20,000 head in 1898 to 70,000 head in 1903, but also increased quantities of wheat competed for rolling stock, overloading the railway during the critical fall shipping season.[26] Cattle were moved slowly in poorly equipped boxcars, 48 to 72 hours elapsing between stops for food and water.

These ongoing complaints from the Canadian grasslands reflected far reaching changes which were taking place in the international meat trade. Canada's cattle trade was threatened, in the long term, both by keen competition and by substitution based upon improved technology.[27] In 1900, live cattle made up about half Great Britain's beef imports; by 1910 this figure had fallen to 17 per cent. The shipment of live cattle was an inefficient and costly way to deliver meat to a distant market. It meant the transportation of large quantities of offal, and additional costs for labour, insurance and feed. Losses at sea had been cut to a minimum in properly equipped vessels, but considerable shrinkage was inevitable. The fact that sustained the trade in livestock was the British housewife's penchant for meat freshly killed by local butchers. Meat slaughtered in abattoirs adjacent to the ports fetched 1.5d to 2d a pound more than chilled beef. However, as the chilling process was perfected, and the problem of distributing this highly perishable commodity was solved, chilled meat became more and more difficult to distinguish from freshly slaughtered beef, and the comparative advantage of the livestock trade disappeared. Technological innovations in chilling techniques paved the way for a spectacular expansion of exports of beef from Argentina, and of lamb from Australia and New Zealand. Imports of frozen and chilled beef from these sources to Britain doubled between 1900 and 1910.

Thus, the winter of 1906-1907 struck an industry which was already concerned as to its future and extremely vulnerable. The tide of homestead settlement was at last flooding in and the government had neither the will nor an obvious justification for special support of range interests. Dry farming technology had evolved to a point where it was becoming difficult to assume that there would be extensive areas of the Canadian West which were unsuitable for arable farming. Finally, increased costs of production and problems with transportation decreased the Canadian cattleman's ability to compete with both the cash grain farmer for land, and with beef producers elsewhere in the world for markets.

THE WINTER OF 1906-1907

In March Harry Otterson, the manager of the T Bar Down outfit, rode from the ranch headquarters to the winter range along the Whitemud River. Later he described his expedition in these words:

> It surely was a gruesome ride. The cattle were in all stages of dying. The brush was simply lined with dead cattle. The live ones at night would lie down on the dead and many would not be able to get up again, consequently they were literally piled up.[28]

At the same time cowboys were using axe handles to dispatch cattle trapped in deep drifts along the valley floor of the Red Deer River near Dorothy.[29] A. E. Cross recommended that weak cattle in his range herd should be put out of their misery "so that available hay could be used to feed the stronger beasts."[30]

The inhabitants of the small towns of southern Alberta were shocked by the condition of the range stock. The back lanes and alleys of Claresholm harboured several hundred head.[31] Literally hundreds more were killed by trains because they were too weak to struggle out of the drifted cuttings. One night Fort Macleod was invaded by a herd of half-starved cattle, their legs raw and bleeding from breaking through crusted snow. In the morning the bodies of 48 steers were removed from the streets.[32]

Such, in brief but dramatic terms, was the impact of the notorious winter of 1906-07. Wallace Stegner spent that winter on a farm near the Whitemud River. He felt that it signalled the end of the open range in the country to the east of the Cypress Hills. He remarks: "The net effect of the winter of 1906-07 was to make stock farmers out of ranchers. Almost as suddenly as the disappearance of the buffalo, it changed the way of life of the region.[33]

In the spring the losses were evaluated. The main impact of the winter fell on the shortgrass prairie from Sounding Lake southward to the Cypress Hills and the United States border. Here, losses were estimated at between 60 and 65 per cent. Even the leading cattle companies, noted not only for their size, but also for their skillful and experienced management, lost more than half their herds. Newcomers were hit far harder. Smith and Mussett, who had moved from Kansas to range just north of Prelate in 1902, counted 3,500 head in the fall, most of them "pilgrim" stock from Manitoba. There were only 236 survivors in the spring.[34]

In contrast, ranchers in the Bow Valley, west of Calgary, emerged from the winter almost unscathed. Near Red Deer final tallies showed that losses were only marginally greater than normal. In this area, frost-damaged crops provided excellent feed. West of High River, reports from A. E. Cross's A7 Ranch were also reassuring, while at Pincher Creek estimates put losses at 25 per cent.

The exact extent of the losses caused by the severe winter of 1906-07 will never be known, but even the most conservative estimates support the conclusion that about half the working capital invested in the range cattle industry was liquidated. No industry could survive such an ordeal without undergoing profound changes.

The most obvious effect of the disaster was the failure of a large number of major cattle companies, and the retreat of others to the United States. By 1909 the Turkey Track, the T Bar Down, and the Conrad Price Cattle Company had all closed down their Canadian range interests. Veteran Canadian cattlemen like the Maunsell brothers and George Emerson went bankrupt, as did the Prince and Kerr Ranch and the High River Trading Company.[35] Hardened cattlemen suffered personal revulsion during the "carrion spring" and retired as soon as they were able to tie up their affairs.[36]

Some broad based companies could absorb their losses and capitalize on the misfortunes of others. The Winnipeg based firm of Gordon, Ironside and Fares bought out several American interests and consolidated their herds along the Whitemud River. Pat Burns, George Lane and A. E. Cross were all able to expand their operations in 1907.[37] In general, the severe winter encouraged the expansion of corporate control of the cattle industry and replaced eastern and foreign capital with Western Canadian capital.

Among the survivors, the experience of the severe winter speeded up a movement towards more intensive methods of animal husbandry. The capacity for adaptive response to changing conditions on the part of the cattleman should not be forgotten. Exhaustive experiments at government experimental farms and careful monitoring of private operations had proved conclusively that a policy of feeding grain to steers during the winter could pay handsome dividends.[38] Pat Burns had always specialized in winter feeding. By 1908, he was using oats and barley meal in increasing quantities in place of wild hay.[39]

The small stockman was also profoundly affected by the severe winter. For years the homestead farmer, who was characteristically short of both capital and labour, had looked upon a small herd of beef cattle as a sheet anchor for his farming operations. The heavy losses of the winter, taken in conjunction with low cattle prices and rising wheat prices, forced farmers to reassess their position. A great many of the smaller ranchers sold their stock and devoted their entire attention to growing grain. The recorder of brands noted that there had been a decrease of 900 cattle brands in Alberta and Saskatchewan in 1907.[40] The Department of Agriculture urged farmers to stay with cattle, and these sentiments were echoed by Pat Burns who remarked: "Dazzled by dollar wheat, many farmers are now selling their stock to make room for more wheat. Keep your cows for breeding. We are now shipping cattle which should be kept at home."[41] Stock shipments eastward trebled as cattle, breeding cows and calves were sold off for whatever they would fetch.

Thus the winter of 1906-07 disrupted the growth of "mixed farming" and the gradual transfer from extensive methods to more intensive stock farming. Authorities who had noted the decline in the area of the open range, had expressed the hope and the belief that the consequent shortfall in cattle production would be made good by increased output of farmbred stock. Mixed farming was encouraged as it was good for the soil and provided some insurance against natural and fiscal calamity. Official perception of sound agricultural practice proved to be at variance with the aspirations of immigrant farmers.[42] Small-scale stock rearing on farms was only appealing as long as substantial returns rewarded limited inputs of capital and labour. The number of cattle raised on farms remained high while each farm was surrounded by tracts of natural grassland. From 1903 onwards the smaller stock-farmer found it less and less easy to dispose of his product. Export buyers were not impressed with small lots of inferior cattle and local markets were oversupplied. Moreover, the incoming flood of settlers quickly reduced the common grazing land adjacent to farms, and the price of wheat continued to rise. The winter of 1906-07 demonstrated that cattle were a risk unless hay was put up, pastures were fenced, and some shelter was provided. Such measures meant demands on time and capital. The farmer who had maintained a small beef herd weighed his strategy carefully during the summer and fall of 1907. Was it worth his while to re-establish his herd? Or should the money be spent on a contract with a steam-ploughing firm to break additional land for wheat? Evidence suggests that the latter alternative was adopted by many.

Significant changes in the Dominion Land's Act were passed in 1908 and hastened the transaction from pastoral activity to crop farming. Odd numbered sections were offered for sale in the form of pre-emptions or purchased homesteads at $3.00 per acre.[43] An immense reserve of land which had been available for grazing was thrown open to settlement. Forty years later one aspiring rancher recollected the impact of this legislative change:

> In our district [Ghost Pine Creek north-east of Drumheller] the little fellows just getting started in ranching like myself had but two choices. They could sell out entirely and leave the country, or they could turn farmer. The later course meant pre-empting another quarter section and selling of all but a few cows and possibly a few three year old steers as they happened to have. The proceeds from the sale of cattle would have to be invested in another team of horses and in farm machinery.[44]

This was not merely a local phenomenon, it was a widespread trend worthy of mention in the annual review of agriculture in Alberta: "This year men who have lived in the west for years as ranchers raised their first crop, and will raise more next year."[45]

The pace of farm settlement increased in the years immediately following the bad winter. The activity at the Lethbridge and Calgary land offices reached a fevered pitch as the number of entries jumped from about 2,000 in 1906 to more than 5,000 in 1909 (Figure 2). The acreage sown to major crops made an equally obvious upturn (Figure 3).

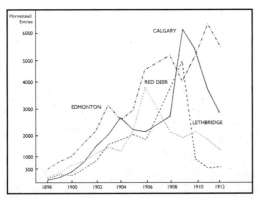

Figure 2. Homestead entries at selected agencies, 1898-1912. Source: Alberta, *Department of Agriculture*, Annual Reports.

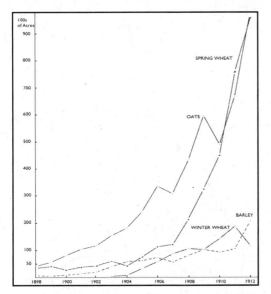

Figure 3. Acreages sown to major crops, Alberta, 1898-1912. Source: Canada, *Department of the Interior*, Cereal Map, 1914.

A massive program of railway building accompanied the expansion of settlement. The area classified as being beyond the economic reach of the tracks was reduced to a few scattered pockets and even these were threatened by planned extensions (Figure 4). Typical of railway building which was critical to range interests were the lines laid north from Lethbridge to Noble, Barons, and Carmangay, and the spur from Suffield to Vauxhall, with its planned extension to Welbeck. The entire area between the mainline and the Crow's Nest branch was lost to the open range. Further north the line from Bassano to Empress was graded, while the link from Drumheller to Hanna and the Saskatchewan border penetrated the heart of the last major range area.

Advancing settlement was spurred by the tide of enthusiasm for dry farming which was generated by the work

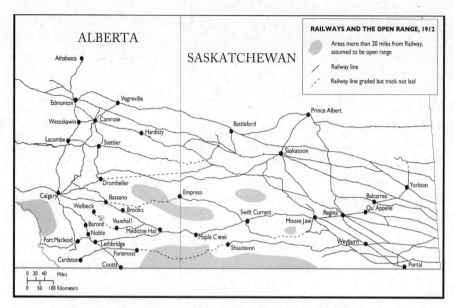

Figure 4. Railways and the open range, 1912. (Prepared by Diane Perrick, Canadian Plains Research Center.)

of Hardy Webster Campbell in the United States. Incoming farmers were encouraged by his optimistic forecast for the shortgrass prairies of southern Alberta and Saskatchewan: "I believe of a truth that this region, which is just coming into its own, is destined to be the last and best grain garden of the world."[46] The fact that the seventh annual dry farming congress was held in Lethbridge in 1912 was fitting culmination of a promotional campaign which had started five or six years before.

Contemporary reports reflect the excitement which the rapid pace of change engendered. There were some eighteen steam ploughs operating in the Lethbridge area in 1908, while along the Milk River newly broken land was yielding 30-40 bushels to the acre.[47] Around Fort Macleod cattle buyers were finding it hard to buy enough cattle to meet their demands. Superintendent Begin reported from Maple Creek:

> Ranching in this part of the province will soon be a thing of the past. Ranchers are going out of business. Most of the land has been opened for homesteaders. Old ranch grounds are gradually being cut up by farmers. Stock cannot anymore roam over the country as hitherto. ...Maple Creek is supposed to be ranching country, with a great number of cattle, but the price of beef in the town of Maple Creek is higher than any place in the province.[48]

A process of enclosure which had started in the valleys of the Bow and the Oldman rivers in the 1890s had reached its logical conclusion. Even in the heart of Palliser's Triangle the settled ways of the farmer replaced those of the pastoralist.

SUMMARY AND CONCLUSIONS

In May 1912 four men met in Calgary and planned the "greatest outdoor show on earth," The Calgary Stampede. Their aim was to memorialize a way of life which had disappeared forever. Each of the four had been intimately involved in the open range phase of the cattle industry and their collective perception that an epoch had ended must carry much weight.[49]

The ending of the open range era was a process which took more than a decade to work itself out. It was not an event which can be positively identified and dated. During the last years of the nineteenth century interested parties in the valleys of the foothills had witnessed the first fences, the last general roundup, and the advent of farm settlement. The quickening pulse of settlement had pushed those in search of unenclosed range from the Little Bow to the Red Deer River during the first years of the twentieth century. Finally, with bewildering speed, farmers had invaded the last bastions of the range cattle industry in the period from 1906 to 1912.

The traumatic impact of the severe winter of 1906-07 must be evaluated in this context. It accelerated trends which were well established before its onset. The attitude of the government to the cattleman changed as the pace of settlement picked up. Competition for space increased production costs and rendered the export trade in live cattle uneconomic, while advances in dry farming technology reduced the comparative advantage of pastoral activity. Some cattlemen held fast to traditional methods and were swept aside by the pace of change. Others, both great and small, displayed remarkable resilience and adaptability. They constantly modified their strategies and intensified their operations. Ironically, 1912 was the start of a period of increased prosperity for the stockmen of western Canada. Demand for cattle continued to exceed supply, and the great United States market was exploited to the full. During the next two decades, much of the short-grass prairie won from the pastoralists by the dry farmers, was abandoned, and neglected fences gave way once more to open grassland. However, the stockmen who took advantage of these territorial and market opportunities had abandoned for the most part both the methods and the philosophy of the open range era.

NOTES

This article first appeared in *Prairie Forum* 8, no. 1 (1983): 71-88.

1   Harry Otterson, "The Southwestern Saskatchewan Range Industry," *Canadian Cattlemen*, 1 (September, 1938), p. 91.
2   Simon M. Evans, "Canadian Cattlemen on the Canadian Range, 1874-1914," *Prairie Forum*, 4, no. 1 (1979) pp. 121-135.
3   Canada, *Statutes*, 1881, ch. 16, sec. 8; Order in Council, May 20, 1881, *Department of Interior*, Vol III, p. 617; and December 23, 1881, Vol. III, p. 805.
4   The role of the Canadian Government is assessed in Simon M. Evans, "Ranching in the Canadian West, 1882-1912," (unpublished Ph.D. thesis, University of Calgary, 1976); and more briefly in "American Diffusion or Victorian Transplant: the Origin of Ranching in Western Canada," *Great Plains Quarterly* 3, no. 2 (1983): 79-91; also D. H. Breen, "The Canadian West and the Ranching Frontier, 1875-1922," (unpublished Ph.D. thesis, University of Alberta, 1972); and "The Turner Thesis and the Canadian West: A Closer Look at the Ranching Frontier," in *Essays on Western History* ed. by Lewis H. Thomas (Edmonton: University of Alberta Press, 1976).
5   Canada, *Sessional Papers*, 1886, XIX, Vol. 6, No. 8, "Report of the Superintendent of Mines," p. 20; and Order in Council, December 13, 1886, *Department of Interior*, Vol. VIII, p. 895.
6   D. J. Hall, "Clifford Sifton: Immigration and Settlement Policy, 1896-1905," in *The Settlement of the West* ed. by Howard Palmer (Calgary: Comprint Publishing Co.) pp. 60-85.
7   Canada, *Department of Interior*, letter from William Collyns to Frank Oliver, August 3, 1896, Library and Archives Canada, hereafter LAC, RG15, f. 141376; letter from Fred Burton to Frank Oliver, October 20, 1896, *Pearce Papers*, I-B-8; and letter from J. G. Robb to Hon. David Mills, September 15, 1900, *Pearce Papers*, I-B-8.
8   Canada, *Department of Interior*, memo from R. J. Campbell, to G. W. Ryley citing 280 reserves which were to be cancelled, January 20, 1900; and for consideration of Pearce's role see Alyn E. Mitchener, "William Pearce and Federal Government Activity in Western Canada, 1882-1904," (unpublished Ph.D. thesis, University of Alberta, 1971).
9   Canada, *Department of Interior*, letter from the residents and ranchers of Assiniboia, February 1901; and memo concerning letter from Western Stock Growers Association, May 1, 1901, LAC, RG15, f. 145330.
10   Letter from Alexander Mackay, the company's able secretary, to Sifton, November 9, 1902, quoted by W. M. Pearce, *The Matador Land and Cattle Company* (Norman: University of Oklahoma Press, 1964), p. 88.
11   Canada, *Department of Interior*, letter to M. Mackenzie, June 27, 1904, LAC, RG15, f. 145330 (part 3).
12   Canada, *Department of Interior*, memo by R. H. Campbell on Grazing Regulations, November 6, 1903, LAC, RG15, f. 145330 (part 4); and Canada, *Department of Interior*, Order in Council, December 30, 1904.
13   Canada, *Department of Interior*, memo from Frank Oliver to W. W. Stewart, February 9, 1906, LAC, RG15, f. 145330 (part 4).

14   Paul F. Sharp, "The American Farmer and the 'Last Best West'," *Agricultural History*, 47 (1947), 65-75; and Jack C. Stabler, "Factors Affecting the Development of a New Region: The Great Canadian Plains, 1870-1897," *Annals of the Regional Science Association*, 7 (1973), 75-88.

15   Isaiah Bowman, *The Pioneer Fringe* (New York: American Geographical Association, 1931), p. 70; and the conclusion of studies made by Canadian National Railways that, "Although in exceptional circumstances grain may be hauled distances up to fifty miles, the practical limit is in the vicinity of ten miles." Quoted in W. A. Mackintosh, *Prairie Settlement: The Geographical Setting*, Vol. 1 of *Canadian Frontiers of Settlement* ed. by W. A. Mackintosh and W. L. G. Joerg (9 Vols.; Toronto: MacMillan of Canada, 1938), p. 55.

16   Canada, *Sessional Papers*, 1903, xxxvii, Vol. 12, No. 28, "Report of Superintendent P. C. H. Primrose," p. 51.

17   Canada, *Sessional Papers*, 1904, xxxvii, Vol. 10, No. 25, "Report of J. E. Woods, D. L. S.," p. 117.

18   Karel Denis Bicha, *The American Farmer and the Canadian West, 1896-1914* (Lawrence: Coronado Press, 1968).

19   Canada, *Sessional Papers*, 1906, xl, Vol. 13, No. 28, "Report of Superintendent G. E. Sanders," p. 43.

20   Edward J. Wood, "The Mormon Church and the Cochrane Ranch," *Canadian Cattlemen*, 8 (September, 1945), p. 84.

21   North-West Territories, *Department of Agriculture*, 1903, "Annual Report," p. 159.

22   Alberta, *Department of Agriculture*, 1906, "Annual Report," p. 69.

23   Alberta, *Department of Agriculture*, 1906, "Report of the Cattle Breeders' Association," pp. 203-204.

24   Alberta, *Department of Agriculture*, 1907, "The Beef Commission," pp. 30-46.

25   Canada, *Sessional Papers*, 1903, xxxvii, Vol. 6, No. 15, "Cattle Trade with Europe," p. 143.

26   Canada, *Sessional Papers*, 1903, xxxvii, Vol. 6, No. 15, "Report on Live Stock Cars and Yards," p. 132.

27   For a recent overview of the North Atlantic meat trade see Richard Perren, *The Meat Trade in Britain, 1840-1940* (London: Routledge, Kegan and Paul, 1978); and for a Canadian perspective Simon M. Evans, "Canadian Beef for Victorian Britain," *Agricultural History*, Vol. 53, No. 4, (October, 1979) pp. 748-762.

28   Harry Otterson, "Thirty Years Ago on the Whitemud River," (Unpublished manuscript, Glenbow Alberta Institute hereafter G.A.I.), p. 22.

29   William Murphy, "Ranching, Mining and Homesteading in the Drumheller Valley," (Unpublished manuscript, G.A.I.) p. 94.

30   Letter from A. E. Cross to C. L. Douglas, February I, 1907, *Cross Papers*, G.A.I., B59, F467.

31   *Medicine Hat News*, January 17, 1907.

32   *Medicine Hat News*, February 7, 1907; and L. V. Kelly, *The Rangemen* (Toronto: William Briggs, 1913) p. 377.

33   Wallace Stegner, *Wolf Willow* (New York Viking Press, 1955) p. 137.

34   Cecil H. Stockdale, "Another Hard Winter Story," *Canadian Cattlemen*, xiii (March 1950) p. 58.

35 Saskatchewan Archives, "The 76 Ranch," (unpublished manuscript G.A.I.); and Margaret V. Watt, "McCord's Ranch: Chronicle of Sounding Lake," *Canadian Cattlemen*, 15 (November, 1952) p. 20.

36 Mary Terrill, "Uncle Tony Day and the Turkey Track," *Canadian Cattlemen*, vi (June, 1943) p. 8.

37 Saskatchewan Archives, "The 76 Ranch," p. 3; C. J. Christianson, *Early Rangemen* (Lethbridge: Southern Printing Co. Ltd., 1973) p. 92.

38 Canada, *Sessional Papers*, 1909, xliii, Vol. 7, No. 15a, "The Commercial Livestock Industry in Western Canada," p. 140.

39 J. G. Rutherford, *Cattle Trade of Western Canada*, (Ottawa: King's Printer, 1909) p. 18; and Albert Frederick Sproule, "The Role of Patrick Burns in the Development of Western Canada" (unpublished M.A. Thesis, University of Alberta, 1962) pp. 95-102.

40 Alberta, *Department of Agriculture*, 1907, "Annual Report," p. 121.

41 *Calgary Herald*, August 10, 1909.

42 This point is stressed in a speech by the Honourable W. R. Motherwell to the Agricultural Societies Convention at Regina in December 1907. Saskatchewan, *Department of Agriculture*, 1907, "Annual Report," p. 178.

43 Chester Martin, *Dominion Lands Policy* (Toronto: McClelland and Stewart Ltd., 1973) p. 98.

44 L. S. Curtis, "The Pre-Emption Policy of 1908 Ruined Many Small Ranchers," *Calgary Herald*, August 18, 1949.

45 F. Dunham, "Agriculture in Alberta," *Edmonton Daily Bulletin*, Christmas Edition, 1908.

46 Mary Wilma M. Hargreaves, *Dry Farming in the Northern Great Plains, 1900-1925* (Cambridge: Harvard University Press, 1957) p. 87.

47 Canada, *Sessional Papers*, 1909, xliii, Vol. 16, No. 28, "Report of Superintendent J. O. Wilson, Lethbridge," p. 89; and *Lethbridge Herald*, December 11, 1947, p. 13.

48 Canada, *Sessional Papers*, 1909, xlvii, Vol. 21, No. 28, "Report of Superintendent J. V. Begin, Maple Creek," p. 98.

49 Guy Weadick, "The Origin of the Calgary Stampede," *Alberta Historical Review*, xiv (Autumn, 1966), pp. 20-24.

13. The Matador Pasture: A Pastoral Vision for Prairie Life

Thomas D. Isern

The Perrin boys were lightly rooted in the ranching country at the base of the Cypress Hills in southwestern Saskatchewan. From there they roamed, one of them to ride broncs for the Queen with the 101 Ranch Wild West Show in London, and another, Jess, to find good work somewhat closer to home, on the recently organized Matador Community Pasture. The first pasture manager, the Scot, George L. Valentine, hired him as a rider in 1924 and the next year was in need of additional help. Jess Perrin told Valentine, "I don't know of any men, but I got a kid brother at home. He's not every old, but he's used to horses, and a young lad, if he knows what he's doing, can take the place of a pretty good cowboy." So 16-year-old Peter Perrin also signed on with the Matador pasture. He recalls,

> About the third day Valentine says to me, 'Say, I never asked how old you are.' So I threw my chest out as far as I could and says, 'I'll be seventeen on the 28th of the month [of April].' He took another look at me and says, 'Look, you make sure you don't let a horse kick you until you are seventeen, or there'll be trouble.'[1]

There would be trouble because the Matador pasture was not a private operation but a public institution, employees of which were to be of the age of majority. Cowboys here rode not for rugged individualism but for state initiative and cooperative enterprise. The Matador Community Pasture, established in 1923 by the province of Saskatchewan, was the first community pasture in western Canada.

By and large the decision to convert grasslands into croplands on the Great Plains of North America went unquestioned during the settlement era. Popular sentiment and public policy agreed that the highest and best use for

A family visit to Matador riders in camp, 1920s. (Courtesy of the Saskatchewan Archives Board/S-B9857.)

the prairies was crop production by family farms. Not until the twin disasters of depression and dust in the 1930s were these assumptions rethought and steps taken to restore a degree of pastoralism on the plains, as the Canadian government (largely through the Prairie Farm Rehabilitation Administration) established grazing commons known as community pastures and the American government similar areas known as national grasslands. There was, however, one notable example of a government and its people anticipating these basic reforms in land use and taking action prior to the 1930s: the community pasture program of the province of Saskatchewan, commencing with the historic Matador pasture. The Matador pasture was the grazing lease of the Matador Land and Cattle Company, a Scottish firm headquartered in Colorado. When the province converted this big-business enterprise into the first community pasture it not only exemplified Canadian western ideals of cooperative enterprise but also showed prescience as to land use and social expectations for the prairies.[2]

Generations hence have considered the community pastures of western Canada exemplary state initiatives that have aided prairie dwellers in adapting mixed farming to a semiarid region. Mixed farming—the complementary combination of crops, forages, and livestock in farm operations—was a problem from the outset of prairie farming, as farmers commonly responded to the open landscape with expansive cereal production bordering on monoculture.[3] Community pastures encourage farmers to keep cowherds by providing summer range that is off-farm but local. The grazing lands, as well as the cattle grazing thereon, are under the care of a pasture manager and his hired riders, all of whom are government employees. Mixed farmers thus can deliver their cattle to the community pasture in the spring, devote themselves to field husbandry through the growing season, and retrieve their livestock in the fall to be wintered at home.

The popular conception of community pastures is of those established and maintained by federal initiative, that is, by the Prairie Farm Rehabilitation Administration (PFRA). These pastures came into being under the stress of drought, depression, and depopulation in the late 1930s. The government purchased (or in some cases condemned and purchased) abandoned or eroded croplands; stabilized them by seeding to grass and legumes, mainly crested wheat grass and Siberian alfalfa; made improvements in the form of fences and water supplies; hired managers; and opened the lands as community pastures for local farmers and ranchers, who received allocations to place specified numbers of animals into the pastures. This process commenced in 1937, with pasture development proceeding for a generation thereafter. Today the more than 80 PFRA community pastures are a continuing success story.[4]

The idea of community pastures, however, was not a federal conception, but rather sprang from provincial initiatives, specifically in Saskatchewan. In 1920 the province's Better Farming Commission pointed to the need for common grazing to be made available to farmers. A particular advocate of this in Regina was Deputy Minister of Agriculture F. H. Auld, who urged officials in the federal Department of the Interior to make land available from the public domain. With passage of the provincial Agricultural Co-operative Associations Act in 1922, his department moved swiftly to help farmers organize grazing cooperatives that would use vacant federal lands, which the provincial department leased and then sub-leased to the co-operators. Ranchers opposed this, knowing that it assisted farmers in creating more balanced and tenable operations, but could not stop it. The associations were generally functional, but lacking paid staff, they were not community pastures as the institution subsequently would be defined, and management suffered thereby. On the other hand, Manitoba had no such arrangements, and Alberta would not commence organization of community pastures until 1941 (or, within designated Special Areas, 1939).[5]

While leading the way in the organization of co-operative grazing, Saskatchewan also proceeded to establish the first true community pasture with paid staff. The origin of this historic initiative, the Matador Community Pasture, owed to a congruence of principled opportunism and fortuitous circumstance. In 1904 the Scots-owned Matador Land and Cattle Company, looking beyond its extensive ranching interests in the United States, had taken a lease on six townships of grazing land on the north bank of the South Saskatchewan River. This lease was used for holding and finishing beeves. After the 1921 season, however, due to changes in tariffs as well as to the disastrous postwar beef market, the Scots surrendered their lease. No other cattle company wanted it. The time had come for Deputy Minister Auld and the other advocates of common grazing to act boldly or cease their agitation.[6]

In early 1922 the future of the Matador lands was undecided. The cattle company had driven its remaining stock to South Dakota, leaving its foreman, the legendary Texan "Legs" Lair (so named for his 6'6" height), along with his notorious wolfhounds, in possession of ranch headquarters. A pair of horseback entrepreneurs named Tom Thornton and Lawrence Ohmacht gathered cattle from farmers and ranchers and commenced herding them for pay on the former Matador lease—without government consent. As speculation bubbled, Auld sent a committee of two lieutenants, J. F. Booth and Graham Anderson, to reconnoitre. They brought him back a map and a report saying that the lease—comprising some 117,000 acres, stretching three townships east and west and almost two north from the river—was choice grazing land, with some potentially arable land intermixed. This was a proposition far greater than any co-operative grazing deal the province had considered. It would require resident management and government operation.[7]

Opinion was divided as to the proper disposition of the lands in question. Provincial legislator Wilbert Hagerty wrote Auld "that the ranchers immediately north of the Matador are very anxious to secure leases on portions of this range," but that farmer sentiment "in the districts west of the Ranch especially is very strongly of the 'community pasture' idea."[8] The secretary of one local Grain Growers Association urged to Minister of Agriculture C. M. Hamilton that all the lands be made available for grazing by "surrounding farmers," but the secretary of another local insisted that all arable lands within the block be thrown open to settlement, with the balance—the rough breaks along the river—made into community grazing.[9] Some local business people wanted the entire area thrown open to farmers so as not to be "surrounded by unproductive land."[10] Minister Hamilton declared as to the prospective community pasture, "I have a great deal of hesitation in embarking on a policy of government operation"; but in a frank exchange, he and Hagerty concurred that given the effects of drought on local grain farmers, they had to act to encourage the keeping of livestock.[11]

The deliberate party in the middle, though, was Deputy Minister Auld. He had a vision for the country that was to promote mixed farming even if it meant defying the desires of farmers themselves. "Notwithstanding the fact that a portion of the Matador lease consists of very good land," he wrote already in

April 1922, "it is a very valuable asset for grazing purposes for farmers' stock, and in my judgment should be maintained for that purpose."[12] By July of that year he had concluded an agreement with the federal Interior Department for the province to take over the old Matador lease and organize it as a community pasture—which agreement did not permit opening the lands to settlement. Then there were the freelance pasture men, Ohmacht and Thornton, to deal with. Auld made them pay a fee for each head grazed, allowed them to take a little profit, but let them know they would play no part in the subsequent operations of the community pasture. (Ohmacht was the object of local animosity because of his pro-German sentiments, and Thornton was an alien American.) He ran out the bluff of Legs Lair, who expected to be paid to vacate the old ranch headquarters. Most important, in line with the advice of local farmers and the agricultural press, he enlisted George L. Valentine, a master stockman and an early proponent of the community pasture, to serve as the inaugural pasture manager.[13]

Valentine organized operations forthrightly. His first written communication to the public announced the availability of grazing and declared, "As this is the people's proposition and is of no value except as it serves the farmers and stock raisers of the district, I would like to have you give the matter your best consideration."[14] He hired riders and established the policy that they would go around to farms and collect stock from the various districts, rather than requiring delivery. He alerted the provincial Department of Highways he would need the ferry at Saskatchewan landing open as early in the spring as possible so as to cross cattle to the pasture. Summer grazing rates were set at 50 cents per animal per month, a reasonable charge. All livestock put into the pasture had to be branded and vaccinated for blackleg. Selected patrons would be allowed to place purebred bulls in the pasture in July and would be credited for stud service. Livestock owners were to come to the pasture on announced days in the fall to collect their beasts and trail them home. Auld, meanwhile, placated local municipalities by promising payments in lieu of property taxes.[15]

Valentine's report on the first (1923) summer grazing season showed the extent to which the community pasture was both fulfilling its founding ideals and adapting to circumstances on the ground. More than 100 stockmen placed a total of 2,009 cattle and 150 horses in the pasture. Of these, more than 90 were small operators contributing from 2 to 100 animals, the great majority fewer than 30. Clearly the pasture was serving its mixed-farmer base. On the other hand, C. Janke put in 232 animals, Smith & Knight 207, Legs Lair 130; some substantial ranchers, then, also were using the pasture. The fact that only about 300 of the cattle therein were breeding cows indicates that the larger operators were using the pasture to finish beeves, rather than sustain cowherds. Valentine expressed the hope that as time went on, more small farmers could be induced to use the pasture. Moreover, in the fall the pasture was opened to winter grazing of horses, and the parties who took advantage of this were the larger stockmen.[16]

There was carping from local critics at the outset. As Auld observed of such government enterprises, "everything it does and every person connected with it is a fit subject for criticism by every inhabitant of the province."[17] A local farmer accused Valentine of "willful neglect" and "bad management" for a regular catalogue of alleged offences, the most serious of which involved the mis-sorting of one calf.[18] The following year another croaker disputed at length the selling of a single mare taken up from the pasture as a stray; the fellow, Valentine said, was "more anxious to knock the management of the Matador pasture than he is to recover the mare."[19] It was notable that once operation of the pasture was underway, criticism came only from cranks: there was public acceptance of its general aims and policies. A resolution of the Rural Municipality of Saskatchewan Landing praised the Matador pasture as "the best system of mixed farming that has ever been tried in this district."[20] In early 1925 mixed farmer T. J. Graham of Pennant, in a public address in Swift Current, expressed appreciation for the community pasture. He said he previously was unable to keep stock on his farm, but when the Matador opened, he bought a car of "good young stuff" from Winnipeg, and another the following year. He praised Valentine's management and lauded the community pasture as "one of the greatest contributions to the success of mixed farming in this province."[21]

A financial statement for the 1924 season showed the Matador pasture on sound basis. It took in $9,600 in grazing fees and permits, spent $3,600 on salaries, paid $1,700 in lieu of taxes, covered miscellaneous expenses, and showed a $2,000 balance. Through the 1920s, then, patterns and variations of practice in pasture management emerged. Each year, for instance, local contractors were enlisted to plough six-foot fireguards around the pasture—a perimeter of more than 50 miles at about $10 per mile. Estrays on the pasture were taken up and sold at auction. Fall roundups were adjusted to accommodate delayed harvests. Riders became expert at reading ice for the sake of crossing stock on hard water before or after the ferry season. Old riders from the Matador Ranch days, such as Marvin Swink (first pasture foreman under Valentine, and a notorious hard case), departed; new riders recruited locally, such as Jess Perrin (successor as foreman, much more congenial and popular), settled into long-term service. Cross-fencing with carload lots of cedar posts and barbed wire was commenced, some of it done by the riders, but most by

Corrals on the South Saskatchewan River, 1920s. (Courtesy of the Saskatchewan Archives Board/S-B9853.)

contractors. Permits were issued to farmers to cut hay, and fees were collected from trappers who entered the pasture. In 1928 telephone lines were run into pasture headquarters.[22]

Stability, however, was an illusion. In 1929 farmers on the east side, adjacent to the better arable lands enclosed within the community pasture, requested that these be opened to homesteading. The Department of Agriculture dithered on this question until it was made moot by the onset of depression and drought, creating new sorts of demands.[23] In 1930 came a pile of petitions—six of them, each bearing signatures of from 10 to 50 farmers—denouncing the Matador pasture, which was founded "as a financial aid for farming in this dry and uncertain crop area," for "becoming more and more a course of benefit for Special Privileged stockmen that fill the pasture and have a dominating sway over its operation to a loss for the farm community."[24] Auld convened a public meeting in the Swift Current City Hall at which frustrations were evident and discussion frank. The problem was not that farmers were being excluded from summer grazing—ranchers were not taking this away from them—but ranchers were enjoying winter grazing privileges that farmers coveted. They were having difficulty keeping their workhorses alive without fodder and wanted to turn them onto the Matador. The farmers also wanted more cross-fencing subdividing the pasture, so that it would be easier to sort out their cattle in the fall. These specific demands Auld promised to address, and while the general hostility did not disappear in this climate of stress— protest meetings continued—it did diminish.[25]

Through the mid-1930s the general aims of the pasture were not questioned, but its manager stumbled from one operational crisis to another. Farmers put in their horses, but could not pay their fees in the spring; by various paper ruses Valentine let them have their horses for spring work. New patrons pressured him to take stock, more stock than could be grazed responsibly. Springs and sloughs dried up and had to be dug out. Patrons clamoured for lower rates, and got some relief. The pasture, over-grazed, had insufficient hay and had to buy straw for fodder from farmers. More and more parties sneaked onto pasture lands and illegally cut timber. Farmers continued to agitate about grazing privileges until Auld finally just gave Valentine carte blanche to adjust allotments in whichever way would keep the peace. Riders were forced to collect farmers' starving cattle earlier in the spring, crossing many from south of the river on the ice.[26]

In the late 1930s citizens and officials again discussed the broader issue of land use. In 1937 there were petitions from districts east of the pasture to open its better lands to homesteading, an odd request when no crops had been raised in the vicinity for years. New minister of agriculture J. G. Taggart was not as supportive of the community pasture as Hamilton had been. By 1939 he had come to the conclusion that the arable lands should be opened, the rough lands should be turned over to the PFRA, and the province should get out of the operation. Such talk ceased in 1939 with the advent of World War II—in which year also founding manager Valentine retired and turned over management to his trusted foreman, Jess Perrin.[27]

Following the war came the greatest challenge to the territorial integrity of the Matador pasture: a conflict of values matching the ideals and constituencies of the pasture against public desires to provide benefits for veterans, along with an admixture of socialistic idealism. The Co-operative Commonwealth Federation (CCF) government that came to power in 1944 turned its attention after the war to the creation of cooperative farms. The first cooperative farm founded on lands controlled by the province was the Matador Cooperative Farm, carved from prime acreage in the heart of the community pasture. In this famous example of communal prairie farming, 17 veterans were given three quarters of land each and financial assistance to farm in common. Following an organizational meeting in Regina in April 1946, the veterans—most of them single at the time, a few with their wives—came to the selected site, broke ground with borrowed equipment, commenced farming, and in time constructed a comfortable colony of homes. Nor was this the only such conscription of community pasture acreage, for in 1949 a second cooperative farm, the Beechy Cooperative Farm, was carved from the arable lands in the northeast sector of the pasture (long coveted by prospective settlers). The latter co-op was disbanded in the 1970s when certain tenure arrangements came due, and the land passed to another sort of cooperative farm, the Beechy Hutterite Colony. The Matador Co-op, however, survived the same transition as the reorganized Matador Farming Pool with the help of a New Democratic Party government in Regina and new lease arrangements fashioned by the provincial Land Bank. Of 32 cooperative farms established in Saskatchewan in the late 1940s and early 1950s, the Matador Farming Pool is the sole survivor.[28]

Reduced in size by about one-third, the Matador Community Pasture nevertheless remained an important and evolving institution to farmers and ranchers in the region. Already by the close of the Depression there was evidence it had pivoted toward more careful attention to its mixed-farming mission: there were more breeding cows in the pasture. Perrin reported in July 1939 that he had 659 cows and 29 bulls—"plenty of bull power," he allowed. In 1941 he had 28 bulls servicing 840 cows and was less confident of his bull power.[29] The next pasture manager, Johnny Rempel, acknowledged that during the 1960s the government sought to expand the farmer base further at the expense of larger ranchers:

> The government was cutting out the rancher and kind of trying to look after the farmer a little more. They figured that the ranchers had enough lease land of their own that they could look after their own cattle.[30]

The 1960s, in fact, was a time when the provincial government sought not only to point up the community pasture mission but also to expand the program. For more than two decades the Matador had been the only provincial community pasture. In 1946 the province had opened two more community pastures taking the place of private leases not renewed, and had begun a period of expansion. By the 1950s the province not only was organizing provincial community pastures on federal lands but also buying up abandoned farms, seeding them down, and organizing provincial-owned pastures in the style of the PFRA. A government study in 1962 approved of such initiative and also called for "considerable direct government action"—that is, more community pastures—to encourage farmers to keep livestock. Despite some retrenchment by conservative governments in the 1980s (when the community pasture program was transferred to Saskatchewan Rural Development), the province retained 57 community pastures, including the flagship Matador, in the 1990s.[31]

Another reform of the 1980s was the move to divest government of responsibility for management of community pastures and convert them into patron-operated establishments similar to the old grazing co-operatives. Rumours flew around Beechy and Kyle that the Matador pasture was to be broken into two or three parts, that government managers would be withdrawn, and that the patrons would have to manage affairs on their own. This they did not want, as was made plain at a public meeting in the Kyle Canadian Legion Hall: the patrons did consent to organize the Matador Grazing Co-operative, take over responsibility for providing bulls, and assume greater authority over pasture affairs—but not to provide management.[32]

One reason for the durability of the Matador Community Pasture is that it not only has provided its key economic function but also has assumed an embedded place in community identity—meaning that people have a sentimental attachment to it. The 75th anniversary celebration for the Matador pasture, convened in 1997 at the Kyle Elks Hall, was a gala event declaring that the pasture "ensures environmental sustainability and helps to create success for all involved."[33] Personal recollections are still more telling. A woman who as a farm girl trailed cows to the pasture recalls fondly how she had to open and close the gate from the saddle, for were she to dismount, she could not get back on her horse. Old patrons recollect the easy-going days when, unencumbered by documentation, they just told Jess how many cows they had put in, and he believed them. In cafes and bars, broken-down, old riders reminisce about summers spent at the east camp of the Matador. Johnny and Mary Rempel show albums filled with photos of riders, patrons, friends, horses, and the children they raised at the winter headquarters (which was the old Matador headquarters, on the river) and at the centrally located summer headquarters. During rodeo week in Kyle, visitors stream over from the arena to the Rempel back yard, where Johnny places on display the wooden model of the old Matador headquarters he built, along with the model of the cutter he used to hay the bulls. Over at the Beechy Community Care Home, Peter Perrin, the stripling rider of 1925, muses as to the possible origins of pasture place names like Stud Butte and the Bone Pile Pasture, and recalls meetings with old Legs Lair, a mythic figure.[34]

Sentiment aside, it was a bold and prescient move when Deputy Minister of Agriculture Auld cut through the tangle of divided counsel and petty carping to create the Matador Community Pasture in 1923. It was a decision second-guessed, qualified, and modified over the years, but its fundamental principle has persisted, in fact been refined, namely: common grazing under government management to promote mixed farming by helping farmers sustain livestock. Auld may be considered vindicated by both public policy and popular affections. He himself believed that the Matador was the inspiration and example for the grander PFRA community pasture program; he told a journalist in 1942 the Matador was "a model upon which the operation of the community pastures

"Legs" Lair, date unknown. (Courtesy of the Saskatchewan Archives Board/R-A7532.)

were planned under the PFRA"[35] This was a credit echoed by Saskatchewan Agriculture and Food at the 75th anniversary celebration in Kyle.[36]

The Matador pasture, too, has become a lodestone of local tradition and lore, which is evidence of sincere public acceptance. It has prospered in that middle ground between individualistic capitalism—family farms and ranches—and state socialism. It is a state initiative, state-controlled and state-managed, designed to foster the prosperity of individual enterprises. It is one of the few state initiatives to any degree successful in promoting mixed farming in the face of market and environmental conditions more attuned to extensive field crop production.

A scholar of cooperative farms in 1960 wrote of the "individualistic social climate" that was both context and complement to socialist experimentation on the prairies.[37] As he saw it, farmers drew a sharp distinction between participating in a cooperative enterprise such as a store, marketing pool, or credit union and giving over ownership or control of the means of production. For the latter they, except perhaps the remaining members of the Matador Farming Pool, had little enthusiasm, but for the former they had a persistent fascination. The community pasture falls within this genre of cooperative enterprise that has won the hearts of prairie people: it not only represents a sound adaptation to life on the northern plains but also helps to define the values of those people involved in the process.

NOTES

This article first appeared in *Prairie Forum* 29, no. 1 (2004): 117-128. Presented to the Nordic Association for Canadian Studies in 1999, this paper is an offshoot from a more general study of community pastures in the Prairie provinces of Canada, which in turn is one in a series of essays on topics in the history of western Canadian farming and ranching—my research along these lines going back to 1986. The work has received funding from the Faculty Research Programme of the Canadian Embassy, which I acknowledge with thanks. Thanks also to the Canadian Plains Research Center, University of Regina, with which I have had a long and fruitful relationship as a Research Fellow. Finally, this paper rests mainly on archival and manuscript collections of the Saskatchewan Archives, and so I wish to register heartfelt thanks to the helpful and professional staff in both Regina and Saskatoon.

1 Interview with Peter Perrin, Beechy, Saskatchewan, 28 May 1999.
2 The establishment of PFRA community pastures is treated in my paper, "Community Pastures on the Canadian Prairies," presented to the Western Social Science Association, 1993. A good general history of the Matador Land and Cattle Company is W. M. Pearce, *The Matador Land and Cattle Company* (Norman: University of Oklahoma Press, 1964).
3 Paul Voisey, *Vulcan: The Making of a Prairie Community* (Toronto: University of Toronto Press, 1988), contains a candid discussion of the shortcomings of mixed farming in prairie settlement.
4 This discussion of the origins of PFRA community pastures is abstracted from my paper on the subject cited in note 2. That paper in turn relies heavily on correspondence in Records of the Department of Agriculture, RG 17, Vol. 3287, File 559-13-5, Library and Archives Canada, Ottawa; also Records of the Department of Agriculture, Deputy Minister, 1892-1954, R 261, Saskatchewan Archives, Regina; and annual reports of the PFRA held in the PFRA Library, Regina.
5 Ibid.
6 The activities of the Matador Land and Cattle Company are detailed in my paper, "'We might as well sell out and be through with Canada': How an American Ranch Became Canada's First Community Pasture," presented to the Midwestern Association for Canadian Studies in 1996. That paper in turn relies heavily on three collections at the Saskatchewan Archives Board, Regina (SABR). The first of these is the Letterbook of the Canadian Division, Matador Land and Cattle Company, labelled with the dates 1906-1924 but containing correspondence only from the years 1906–1909, and most of the early blotted correspondence is illegible (hereafter Matador Letterbook). The second is Correspondence of the Matador Land and Cattle Company, 1905–16, loaned for microfilming in 1955 by P. J. Perrin of Beechy, Saskatchewan; this is Microfilm 2.55 (hereafter Micro. 2.55) of the SABR. The third is Correspondence with J. R. Lair, Saskatchewan Landing, 1915–24, Accession R-108, Collection R-822.1, File 2. Also see Pearce, *Matador Land and Cattle Company*, and Stan Graber, *The Last Roundup* (Saskatoon: Fifth House, 1995).

7   Map, "Matador Ranch," by C. G. A., n.d. [1922], "The Report of the Inspectors re
     the Matador Ranch," n.d. [1922], F. H. Auld to C. M. Hamilton, 3 July 1922, Auld
     to Hamilton, 20 October 1922, R-261, Records of the Department of Agriculture,
     Deputy Minister, 1892–1954, File XIV.4, Matador Ranch and Community Pasture,
     1922-42, SABR; George E. Pyne to Hamilton, 1 December 1922, Hamilton to Pyne,
     7 December 1922, M13, Papers of C.M. Hamilton, File 12, "Matador Ranch and
     Grazing Leases," 9 July 1922-28 July 1929, Saskatchewan Archives Board, Saskatoon
     (SABS); and Saskatchewan Department of Agriculture, *Annual Report*, 1925, p. 12.
8   W. Hagerty to Auld, 18 July 1922, R-261, Deputy Minister of Agriculture File XIV.4, SABR.
9   Ed Stephen to Hamilton, 11 February 1922, and M. D. Sherrard to Hamilton, R-261,
     Deputy Minister of Agriculture File XIV.4, SABR.
10   J. W. Phillips to Auld, 1 May 1922, ibid.
11   Hamilton to Hagerty, 8 December 1922, Hagerty to Hamilton, 27 December 1922,
     M13, Hamilton Papers, File 12, SABS.
12   Auld to Sherrard, 19 April 1922, R-261, Deputy Minister of Agriculture File XIV.4,
     SABR.
13   Auld to Lair, 17 July 1922, Auld to Thornton and Ohmacht, 17 July 1922, Auld,
     "Memo for the Matador File," 20 July 1922, clipping from *Saskatchewan Farmer*, n.d.,
     Auld to Thornton and Ohmacht, 14 September 1922, Graham Anderson to Auld, 18
     September 1922, ibid.; B. H. Yorke (Controller, Timber and Grazing Lands Branch,
     Department of Interior) to Auld, 3 May 1922, "Community Grazing lease," 7 August
     1923, R-261, Records of the Department of Agriculture, Deputy Minister, 1892–1954,
     File XIV.5, Dominion Government Correspondence, 1922-29, SABR; George Pyne
     to Auld, [December 1922], Auld to Pyne, 7 December 1922, M13, Hamilton Papers,
     File 12, SABS.
14   Circular [by Valentine], "Matador Community Pasture," n.d., R-261, Records of the
     Department of Agriculture, Deputy Minister, 1892–1954, File XIV.8, Summer Grazing,
     1922–31, SABR.
15   H. Wilson to Auld, 8 January 1923, Auld to Perry D. Lewis, 3 February 1923, H. S.
     Carpenter (Deputy Minister of Highways) to Auld, 13 April 1923, Auld to Deputy
     Minister of Municipal Affairs, 15 December 1923, Auld to A. W. Phillips, 22 January
     1924, R-261, Deputy Minister of Agriculture File XIV.4, SABR; Auld to officials of
     local municipalities, n.d., M13, Hamilton Papers, File 12, SABS.
16   Saskatchewan Department of Agriculture, Annual Report, 1925, p. 13; [Valentine],
     "Stock in the Matador Ranch, June 14, 1923," [Auld], press release announcing avail-
     ability of winter grazing, [1923], R-261, Deputy Minister of Agriculture File XIV.4,
     SABR; "Winter Grazing Permit," [1923], Valentine to Auld, 1 September 1923, R-261,
     Records of the Department of Agriculture, Deputy Minister, 1892–1954, File XIV.9,
     Winter Grazing, 1923–36, SABR.
17   Auld to Ohmacht, 31 January 1923, R-261, Deputy Minister of Agriculture File XIV.4,
     SABR.
18   Ernest Trowell to Minister of Agriculture, 5 November 1923, ibid.
19   Valentine to Auld, 15 December 1924, ibid.
20   Resolution, 6 June 1924, M13, Hamilton Papers, File 12, SABS.
21   Clipping from *Swift Current Sun*, 23 June 1925, R-261, Deputy Minister of Agriculture
     File XIV.4, SABR.

22   Financial Statement RE Matador Ranch (to 15 April 1925), M13, Hamilton Papers, File 12, SABS; Ag.2, Records of the Saskatchewan Department of Agriculture, File 91, "Fireguards in Community Grazing Area," 1924-1930, SABS; Contract for fireguards, Valentine and L. C. Gowin, 20 September 1923, H. S. Carpenter to Auld, 7 May 1924, poster, "Matador Community Pasture, Public Auction," 12 May 1924, Valentine to Auld, 22 August 1924, Valentine to Auld, 27 October 1924, Valentine to Auld, 28 November 1924, Valentine to Auld, 18 March 1926, Valentine to Auld, 21 March 1929, Valentine to Auld, 8 December 1924, Saskatchewan Grain Growers Association, bid for fencing materials, 5 March 1925, Auld to Valentine, 4 March 1927, Hay Permit, n.d., Valentine to Auld, 30 January 1928, Valentine to Auld, 14 February 1928, R-261, Deputy Minister of Agriculture File XIV.4, SABR.

23   W. J. Green to Minister of Agriculture, 29 June 1929, Hamilton to Green, 8 July 1929, M13, Hamilton Papers, File 12, SABS; Malcolm Lenning to Auld, 12 November 1929, R-261, Deputy Minister of Agriculture File XIV.4, SABR.

24   Petitions to Minister of Natural Resources, 1930, R-261, Deputy Minister of Agriculture File XIV.4, SABR.

25   Valentine to Auld, 16 December 1930, Auld to petitioners, 29 December 1930, Minutes of Public Meeting, 6 January 1931, Auld to Valentine, 12 January 1931.

26   A. R. D. Klassen to Valentine, 20 March 1931, Valentine to Klassen, 23 March 1931, Auld to Frank and T. E. Thurslow, 10 April 1931, Klassen to Auld, 15 June 1931, Auld to Valentine, 11 April 1932, Valentine to Auld, 8 June 1931, Valentine to Auld, 23 June 1931, George G. Schmidt to Auld, 21 July 1932, Auld to Minister of Agriculture, 18 January 1933, Auld to R. L. Wright, 24 April 1933, Straw contract with D. Funk, 19 October 1932, Valentine to Auld, 9 March 1933, Auld to John R. Cluff, 3 April 1933, Valentine to Auld, 9 May 1933, Auld to Valentine, 10 May 1933, Valentine to Auld, 1 March 1934, and an abundance of similar correspondence through the mid-1930s reflecting the same issues, all in R-261, Deputy Minister of Agriculture File XIV.4, SABR.

27   Auld to Taggart, 19 February 1937, Auld to C. E. Beveridge, 18 May 1937, E. E. Eisenhauer to Auld, 29 December 1939, Taggart to Auld, 4 November 1939, H. Wilson to Auld, 31 October 1939, Auld to Taggart, 21 October 1939, Auld to Jesse W. Perrin, 11 April 1939, R-261, Deputy Minister of Agriculture File XIV.4, SABR.

28   Henry Cooperstock, *Report to Saskatchewan Federation of Production Co-operatives, March 30, 1960*; George Melnyk, *Matador: The Co-operative Farming Tradition* (Saskatoon: University of Saskatchewan, Center for the Study of Co-operatives Occasional Paper #92-02, 1992); interview with Lorne and Kay Dietrick, Matador Farming Pool, 29 May 1999.

29   Perrin to Auld, 15 July 1939, Perrin to Auld, 13 August 1941, R-261, Records of the Department of Agriculture, Deputy Minister, 1892–1954, File XIV.11, "Jesse W. Perrin, 1939–41," SABR.

30   Interview with Johnny and Mary Rempel, Kyle, Saskatchewan, 30 July 1990.

31   These developments are abstracted from my paper on community pastures cited in note 2. Of particular note in treating the expansion of provincial community pastures in the 1960s (and quoted in this paper) is J. A. Brown, "Community Pasture Development in Saskatchewan," typescript, 1962, Saskatchewan Department of Agriculture Library, Regina.

32 Interview with Britton McCrie, Kyle, Saskatchewan, 29 May 1999.

33 *75 Years*, anniversary program on file at the Kyle Public Library.

34 Interviews with McCrie, 29 May 1999, Hazel McKechney, Kyle, 29 May 1999, Bob Craig, in the Emerald Motel Café, 29 May 1999, Harold Closs, in the Kyle Hotel Bar, 29 May 1999, Johnny and Mary Rempel, 30 July 1990, and Peter Perrin, 28 May 1999. Peter Perrin also is the author of sections in the Kyle district history, *From Basket to Bridge, 1905–1980*, treating the topics "Matador Ranch"(pp. 396–97) and "Matador Community Pasture" (pp. 398–400).

35 Auld to Miriam Green Ellis, 21 May 1942, R-261, Deputy Minister of Agriculture File xiv.4, SABR.

36 *75 Years*, anniversary program on file at the Kyle Public Library.

37 Cooperstock, *Report to Saskatchewan Federation of Production Co-operatives, March 30, 1960*.

Marketing

14. The Politics of Animal Health:
     The British Embargo on Canadian Cattle, 1892-1932

Max Foran

The association of animal health issues with politics has recurred periodically within the Canadian livestock industry. For example, the ninety-day quarantine imposed on American cattle entering Canada in the 1880s was meant to salve British fears over the danger of disease in Canadian export cattle. Similarly, the Canadian reluctance to quarantine the State of Montana during the mange outbreak in the early 1900s was directly related to fears of widespread American reciprocation. However, the most enduring example of the use of animal health to disguise political intent concerned the thirty year (1892-1922) British embargo on Canadian cattle. Canadian livestock officials, convinced that the embargo was unjustly applied, regarded its imposition in 1892 as a British statement of economic rather than animal health protection. Time served to vindicate the Canadian position, for animal health had little to do with the longstanding maintenance of the embargo, and did not figure at all in the heated debate which led to its partial dismantling in 1922. The above notwithstanding, it also appears that a casual attitude towards animal health by Canadian livestock exporters before 1892 gave the British proponents of restriction all the ammunition they needed. The former's apparent obliviousness to British sensitivity over the health of their pedigree herds contributed to the mounting sentiment which resulted in an embargo based on animal health principles. And once having handed the British the excuse they needed, the Canadian livestock industry had to live with the long-term consequences. To the western Canadian cattlemen, these consequences were not as onerous as might have been expected. The embargo had a minimal effect on their British export product of fat cattle. Moreover, in their eyes, the importance of the British live cattle trade, with or without the embargo, was a much diminished factor when compared to the potential offered in the closer and more voracious United States market.

The Canadian live cattle export trade to Great Britain commenced in 1876, and after 1887 was augmented by range fed steers from the great leaseholds in the North-West Territories. The trade involved both finished (fat) animals, and stores (feeders). Shipped by train to Montreal and then by cattle boat to Liverpool or Dundee, the feeder cattle were sold to Scottish graziers, while the fat animals were marketed in Smithfield in competition with British cattle and other finished animals from Europe. With commercial grain growing not yet a major factor, this live cattle trade provided a mainstay in Canadian exports. By 1891, the export of over 100,000 live cattle realized a total value of almost nine million dollars. The British attitude towards the live cattle export trade was ambivalent. For while the availability of foreign beef had its market advantages, the spectre of potential disease generated a perennial nervousness among breeders and owners of pedigree herds.

In late October 1892, the "Huronia" and "Monkseaton," two cattle boats out of Montreal, landed 1,200 head of Canadian store cattle[1] at the Scottish port of Dundee. Within days the animals were sold and relocated to seventy-nine places in several Scottish counties including Aberdeen, Fife, Forfar, Kincardine and Elgin. A week later contagious pleuro pneumonia was detected in one of the animals landed by the "Monkseaton" as well as in some infected home bred animals.[2] All animals that had been on the two steamers were immediately slaughtered, and within a week the British Board of Agriculture announced that Canadian cattle were to be scheduled effective November 21, 1892. Essentially this meant that all Canadian export cattle had to be slaughtered at the port of entry instead of being allowed to move inland for fattening purposes.

More alarmed than surprised, Canadian interests were quick to react, even though the British had attempted initially to ameliorate matters through reference to the embargo's temporary nature. Prime Minister Sir John Abbott expressed his concern a few days before the scheduling notice when he told John Carling, Canada's Minister of Agriculture that "if once scheduled the consequences will be serious and lasting."[3] Accordingly, senior Canadian officials proceeded to exhaust every diplomatic avenue in an effort to persuade the British to change their minds. The Canadian case was based on the absence of the disease in Canada and on the exhaustive measures immediately undertaken to prove same.[4] Certainly the Canadians believed, initially at least, that matters would be righted if the British could be persuaded to delay their decision until completion of a thorough Canadian herd inspection.[5] As a last resort, the Canadians offered to finance a British inspection of Canadian herds in return for delaying the embargo. The offer was refused. Ultimately it did

In 1892 the British Board of Agriculture imposed an embargo on Canadian cattle which lasted for thirty years.

not matter that the resulting inspection detected no evidence of the disease in Canada. The British mind was made up.

Ironically, the other argument used by the Canadians underscored their own culpability. The Canadians claimed misdiagnosis, an argument supported unequivocally by senior veterinarians at the University of Edinburgh who were convinced that what had been mistaken for pleuro pneumonia was actually non-contagious broncho pneumonia.[6] It was put forward that an examination of the viscera of the infected animals by two university veterinarians showed conclusively that the disease had originated in the bronchial tubes which was not the case in pleuro pneumonia. In comparing the severity of this malady with the common cold, the Scottish veterinarians had equated its incidence with the fluctuating hot and cold conditions consistent with those experienced in transporting animals. Robert Wallace of the Agricultural Department of the University of Edinburgh in a letter to the *London Times* went so far as to say that he himself was opposed to the importation of stores cattle because of danger to the pedigree herd, but that "duty will not permit me to stand aside and observe without protesting against what appears to be a meaningless injustice."[7] Despite these assertions, the similarity between the two diseases was undeniable, and in this instance had induced a predictable conservative reaction. It could be argued that the British Board of Agriculture had no choice but to support the opinions of its experts, opinions which in the President's words were "unanimous and unhesitating."[8] Furthermore, the British felt that, given the findings of their own veterinarians, they had no recourse legally.[9]

Despite some rumblings in the British House of Commons, mainly from members representing grazing areas in Scotland,[10] the embargo's status remained unchanged. According to new Board of Agriculture President, Walter Long, in 1895, several instances of pleuro pneumonia had subsequently been found in the lungs of slaughtered Canadian cattle.[11] Confident that they could build up their supply of store cattle and that consumer prices would not be threatened, the British gave permanency to the embargo through The Importation of Animals Act passed in 1896.

By 1903, five years after the last case of pleuro pneumonia had been reported in Britain, it was already clear that the British no longer regarded disease in Canadian herds as a major source of concern. In that year, the President of the Board of Agriculture stated that the embargo was not disease-related but that he was reluctant to remove it due to political pressure.[12] The subject emerged periodically in Great Britain between 1904 and 1914. In the general elections of 1906, several elected members were purportedly pledged to re-admission of Canadian store cattle.[13] Two years later however, Herbert Asquith notified the House of Commons that the government was not prepared to initiate any remedial legislation. Subsequent enquiries in the House met with the same response up to the commencement of the war in 1914 after which time the matter became a dead issue.

The complete absence of pleuro pneumonia in Canada and the British refusal to admit a diagnostic error led Canadian livestock interests to view the embargo as a deliberate attempt to curtail cattle imports. Belgium, which had followed Britain's example in 1892, removed its embargo in 1895 on the grounds that the pleuro pneumonia did not exist in Canada.[14] Matthew Cochrane reiterated the popular perception among cattlemen when he told the Canadian Senate in 1896 that disease containment was only the excuse to protect British agricultural interests,[15] an opinion shared by Canada's widely-respected Veterinary Director General and western Canadian rancher, Dr. Duncan MacEachran.[16] However, despite the conviction with which they were generally held, these contentions were only partially correct. While the embargo did have economic considerations, its imposition was consistent with a heightened sensitivity in the country towards ensuring the health of resident pedigree herds. By ignoring repeated warnings in this regard, the Canadian live cattle exporters were essentially the authors of their own demise.

The preservation of animal health had been a sensitive issue in Great Britain since the mid-1860s. Beset by disease since it had begun live cattle imports following the adoption of free trade in the 1840s, Britain had been brought face to face with a rinderpest plague in 1865. A book on the plague published a year later isolated foreign cattle as a major source of disease and recommended stringent controls, not only for rinderpest but for foot and mouth and pleuro pneumonia, the other two diseases which had established themselves in the country.[17] Britain's seriousness about controlling disease was evidenced in 1877 when it imposed an embargo on several European countries because of pleuro pneumonia. Then, two years later, American cattle were scheduled for the same reason.[18] In fact the retention of Canada as the only country allowed to bring store cattle in to Britain[19] must have seemed a gross oversight to the many groups advocating total exclusion.

The only reason Canada escaped the same embargo as that placed on the Americans in 1879 was her diplomatic success in convincing British officials that the immediate establishment of a ninety-day quarantine against animals coming into the country from United States destinations would protect Canadian and therefore British herds from pleuro pneumonia. Another mitigating element was the fact that Canada was sending a very small number across the Atlantic as compared with the Americans. In 1878, the United States had shipped 169,250 cattle to Britain, while Canada had exported 7,433.

At the official level, Canadian spokesmen, although surprised and delighted in not sharing the fate of their southern neighbours, were acutely mindful of their tenuous situation. In 1880, a federal Department of Agriculture spokesman warned that "it is necessary to be extremely watchful in order to preserve our enviable position.[20] In 1886, for example, when pleuro pneumonia was detected among 200 pure bred Galloway bulls from Scotland, they were all promptly slaughtered plus another 226 head on adjoining farms.[21] In London, the Canadian High Commissioner, Sir Charles Tupper, recognized the British agricultural community's concern over its pedigree herd health, and repeatedly advised Canadians to export only the best healthy stock. As late as 1890, he was warning Canadian producers about the "agitation being again started by some agricultural papers against the introduction of Canadian cattle into this country under the present system."[22]

The agitation referred to by Tupper had been building steadily since the early 1880s. The Royal Agricultural Society and other powerful breeding societies were openly opposed to any importation of foreign cattle on the grounds that home herds were being unnecessarily placed at risk.[23] Canada was no exception, since it was popularly believed that her long border with the United States precluded effective containment particularly in Manitoba and the North-West which had begun exporting cattle in 1887.[24] It was pointed out that quarantine, measures along the Canada-United States border were not enforced with sufficient stringency,[25] a fact reinforced in 1893 in the Canadian Senate by Mackenzie Bowell who noted that "we have been as liberal as possible with our Regulations in the North-West Territories and Manitoba until the present time."[26] It was also widely believed that Canadian cattle shipments contained cattle from the United States, as many as 15 percent according to one source.[27] The escalating risks were obvious in the increased export numbers, augmented after 1887 by the addition of animals from the great leases in western Canada. In the ten years 1880-1890, the number of Canadian live cattle entering Great Britain increased from 32,680 to 66,965. In 1891, the figure stood at an all-time high of 107,689. These enhanced numbers attested to the profitability of the British market and induced many livestock operators to follow liberal marketing policies which included many animals that should never have been exported. Worn-out milk cows and inadequately finished stock, being ill-equipped to endure a long ocean voyage, were prime candidates for disease, particularly pneumonia. A Canadian inquiry in 1891 into the conditions under which cattle were shipped underscored health dangers as a prime issue of concern.[28] Its sweeping recommendations for reform, released in June 1891[29] came too late for them to have any significant impact on alleviating the health problems that prompted the embargo.

The potential problem in shipping deficient animals was realized early and should have served as a warning to Canadian exporters. Cattle boats from Canada were detained nine times in British ports between 1891-93 for suspected disease.[30] Three suspected cases of pleuro pneumonia among Canadian cattle in Dundee, Liverpool and Newcastle in 1890 led to several slaughterings but to no evidence of the disease. In May 1891, a temporary embargo was placed on a Canadian cattle shipment when suspected pleuro pneumonia was detected in the lungs of one slaughtered animal. In this instance, immediate intervention by Sir Charles Tupper led to a more thorough examination of the infected animal by the Board of Agriculture, and resulted in a favourable revision of the original diagnosis. Still, the incident was sufficiently worrisome for the *Canadian Gazette* to report that "the present untoward incident at Birkenhead will undoubtedly tend to revive the suspicion previously entertained, and occurring as it has, its results will be felt throughout the whole year."[31] The validity of the above comment was shown in the subsequent criticism levelled at the President of the Board of Agriculture for finding this evidence of disease too inconclusive to warrant immediate scheduling.[32] The *Edinburgh Scotsman* ran an article which referred to the "cry for prohibition being quickly among English farmers," and for the President of the Board of Agriculture to be "zealous in protecting English cattle."[33]

Another telling factor concerned Britain's own attempt to eradicate pleuro pneumonia. It was felt that the chances of the disease were increasing in line with the expanded cattle numbers and that the present policy of allowing foreign stores into the country belied the existing programme to control pleuro pneumonia, one which had cost over £300,000 by 1892. With the costs for this programme being borne by the national exchequer rather than the local counties,[34] British officialdom had an extra reason to take proactive measures, particularly given the dismal economic circumstances facing agriculture in 1892. Sir Charles Tupper called it "the most disastrous year for the farmer in the present century."[35] The Canadian Government Agent in Liverpool quoted cattle prices as being down between 20 and 30 percent, and wrote of "the apparent hopelessness of farming in this country."[36]

The imposition of the embargo must be set against these several factors. First, the genuine belief among British veterinarians that they had indeed diagnosed pleuro pneumonia in Canadian cattle could not be ignored either practically or legally. Second was the resident sensitivity to herd health and the prevailing belief that no store cattle should be shipped into the country. Third, there was the widespread mistrust of Canadian cattle consignments. Too many shipments had been found to contain animals which exhibited all the signs of chronic pneumonia. Fourth, the practice of importing animals flew in the face of an ongoing expensive programme in Britain to eradicate pleuro pneumonia. Finally, the depressed agricultural conditions prevalent in Britain in 1892 probably stiffened the British official resolve to back the conclusions of its veterinarians. Whether or not British decision makers overreacted, and based their decision on tenuous grounds simply attests to a seriousness of intent, and to an action that was inevitable given the mood of the powerful agricultural lobby. The British simply did in 1892 what they felt they should have done in 1879 when they scheduled American cattle. The embargo represented the last step towards a universal policy of exclusion based on animal health principles. Yet, despite the underlying fear of disease endemic in British conservative rural thinking since the 1870s, the concomitant commercial advantages of excluding Canadian stores also appeared as an indirect benefit. The embargo's maintenance was to be a direct result of this belief.

It cannot be denied that the imposition of the embargo had negative effects on the Canadian cattle industry although here it must also be noted that the most strident calls for lifting the embargo in the late 1890s and early 1900s came from the exporters, stockyard operators, and shipping interests. First, the market in Canadian store cattle had been profitable. According to one source, Canadian cattle destined for inland British farms brought between $5–$25 more than if slaughtered at the port of entry.[37] Second, the loss of weight and condition induced by their long train and boat trip meant that Canadian cattle landed in Britain at a market disadvantage. Indeed, many

Scottish farmers genuinely believed that cattle could not be finished properly in Canada.[38] Finally, the stigma associated with disease continued to rankle in a country that generally boasted a superb animal health record.

While eastern Canadian livestock operators may have felt the burden of the embargo through the loss of a market for feeder cattle too expensive to finish at home, the same cannot be said for the western Canadian cattle industry. Indeed, it seemed little affected by the embargo and may have actually benefited by it. William Pearce, the astute federal Superintendent of Lands and Mines and a strong proponent of ranching, noted amid the embargo debate in 1892 that "scheduling would not be disastrous. In fact it might be of benefit to us."[39] A year later he appeared vindicated. "As far as ranching interests are concerned, the scheduling of cattle has not hurt the same ... more cattle have been shipped in this year than any other."[40] By 1896 he was advocating a policy which would see western cattleman go out of breeding altogether and function instead as range fatteners for eastern bred stockers.[41]

By closing the market for eastern Canadian stores, the embargo enabled the ranches of the west to become a viable replacement. After 1892, western Canadian herds were substantially augmented by eastern Canadian cattle, many of which would have otherwise gone to Britain.[42] It was asserted that the embargo reduced Canada's export potential by 20 percent, a percentage almost precisely matched within four years by the number of store cattle from eastern Canada to the ranches and leases of the North-West Territories.[43] In 1896, the number of cattle shipped to Great Britain from the North-West Territories totalled 17,935. In the same year 16,000 stockers were shipped to the west from eastern farms.

The embargo also removed the need for the ninety-day quarantine on American cattle.[44] Following its lifting in 1897 an increasing number of one- and two-year-old stockers moved across the line to southern Alberta ranches. The movement was two-way and induced, according to the Canadian Department of Agriculture, "a profitable outlet for cattle not suitable in size for exportation to Great Britain."[45] Duncan MacEachran, Canada's Veterinary General with vested interests in ranching in western Canada, wrote in 1898 that "the rapid development of the cattle trade between the United States and Canada especially in the class of feeding purposes has been much beyond expectations."[46] In 1896, live cattle exports to the United States totalled 1,646 head. When the quarantine was lifted in 1897, the number jumped to over 35,000. In spite of the heavy tariff, Canadian cattlemen were able to reap profits of between $5 and $8 a head in the good years around the turn of the century.[47] In 1912, prominent Alberta rancher George Lane received $10.25 per cwt in

Chicago for 300 head shipped from Brooks, a price which eclipsed the old record for grass-fed steers by 500 per cwt.[48]

The embargo did not affect the total number of cattle exported to Great Britain. Before the embargo the highest number shipped was 107,689 in 1891. In the twenty years following the embargo, this number was exceeded fifteen times, reaching a peak of 163,994 in 1906.[49] Of these increasing numbers, the western Canadian share grew from around 10 percent in 1892 to over 33 percent after 1895.[50]

Store cattle did not figure in the western Canadian cattle trade which by its open-range nature was almost exclusively confined to mature animals, usually four-year-old steers finished entirely on grass. Generally, they realized a consistent profit of between $40 and $50 per head, although the vagaries of the market meant diminished returns in some years. The fact that their long trip may have negatively impacted on their sale price was a moot point. For while they doubtless would have sold at a higher price if taken inland to regain their Alberta condition, these great grass-fed steers, after having spent four years on the range, were just too wild to handle and virtually had to be slaughtered at the point of entry. Giving evidence before the 1891 inquiry on the treatment of cattle, one exporter noted that cattle from the North-West had to be loaded directly from the train to the boat since "North-West cattle will not stand any fooling.... If they do not have the proper arrangements made, they will break away and it is no easy task to recover them.[51] In the same inquiry, William F. Cochrane, Manager of the Cochrane Ranche, explained that "we had to kill our cattle when they landed there because they are so wild that we could not take them out of the lairs."[52]

If they were affected negatively by the embargo, Alberta ranchers certainly gave little evidence of it. Rancher D. W. Davis, Alberta's sole representative in the House of Commons, did not raise the subject in the House. The *Calgary Herald* was strangely silent editorially, and seemed more interested in promoting the refrigerated meat trade than in lamenting the negative effects of any embargo.[53] The subject was not mentioned in the meetings of the organized voice of the ranchers in Alberta, The Western Stock Growers Association (WSGA), between 1896 and 1906.[54] In 1906, the newly created Province of Alberta established a committee to investigate the causes behind the languishing livestock industry. In its submission, The Alberta Cattle Breeders Association listed market variables among its seven areas of concern. The embargo was neither mentioned nor even hinted at.[55] Two years later a prominent federal government livestock official referred to the British beef market as "the great maw," and proceeded to tell an audience of Alberta cattlemen that the way to fill it was with finished grain-fed beef.[56] Again, there was no mention of the embargo. In 1914, when the WSGA called a special meeting to coordinate the varying provincial livestock interests in the light of the changing nature of the industry, it detailed several key issues to be addressed. The embargo was not among them. Indeed, after 1904, the only impetus for removal of the embargo resided in Great Britain where the Free Importation of Canadian Cattle Association (FICCA), composed mainly of northeastern farmers and various cooperatives, waged a lone and fruitless battle against the vested agricultural interests.[57]

But if the embargo did not adversely affect the western Canadian cattle industry up to 1906, developments after that date, and the dreadful winter of 1906-07, virtually buried it as an issue. This near obliteration was rooted in three factors in three different countries. The first was the retreat of the ranching industry in western Canada in the face of the large-scale conversion to cash-crop farming.[58] Forestalled by uncertain lease regulations which provided for a two-year cancellation nonce, by indifferent livestock prices, and by the intoxicating appeal of high grain profits,[59] many ranchers began selling off their herds, including breeding stock. A personal glimpse is provided by this rancher east of Calgary: "Some of the ranchers like Shaddock and Strange sold out and left the area... . Billy Bannister ran a few horses but went into grain farming.... Moorhouse started farming too and ceased his range cattle operations."[60] Dismal forecasts for the cattle industry punctuated the reports of government officials in Alberta between 1909 and 1912, typical of which is the following comment issued by the Livestock Branch of the Department of the Interior in 1912:

> There is no doubt that in the western provinces the production of both cattle and swine has been retarded by the unfortunate marketing conditions that have too long prevailed in that part of the country. These conditions ... are largely responsible for their inclination not to indulge in the production of these two classes of stock.[61]

Two years later in 1914, H. S. Arkell, Assistant Livestock Commissioner, noted that livestock liquidation had set in in western Canada, and that high feed prices were "sending more and more people into grain."[62]

Not only was the embargo forgotten, but trade to Great Britain virtually ceased. Only 9,878 head made the trip across the Atlantic in 1913. That the West contributed a negligible number was evidenced by the *Farm and Ranch Review* when it noted in September that it was "the time of year when the big four year old steers came off the ranges and made their way to Liverpool, but we have come to the end of that trade."[63]

The second factor concerns the switch in preference in Great Britain from live cattle to chilled beef imports. Argentina's emergence as a major beef producer plus the declining quality of Canadian export cattle had begun to impact heavily on Canada's place in the British market by 1912. With low production costs, and a highly efficient and scientific chilling process, the South American country virtually redefined the beef export business in the space of a decade. Increasingly, it was becoming more difficult for the British consumer to distinguish between fresh killed and chilled beef. In the ten years between 1900 and 1910, Britain's live cattle imports dropped from 50 percent of the total to 17 percent.[64] In the five years between 1910 and 1914 Canada exported about 310,000 head to Great Britain, a reduction of almost half a million from the five-year period between 1902 and 1906.

However, the most important reason determining the minor place of the embargo among Canadian cattlemen was the availability of a closer and better market. Between 1887 and 1914, Britain was the only viable outlet for cattle exports, and, regardless of the embargo, Canadian cattlemen utilized the British market when the profit margins warranted. At best, it was a capricious market affected by changing prices between shipment and arrival, by unreliable and expensive freight costs, and by fierce competition from Scottish graziers. Thus the sudden emergence of a closer and more lucrative American market in 1914 ushered in an era of unrestrained trade, and, as long as it remained open, it completely negated the British live cattle trade, and reduced the embargo to a non-issue. As a corollary it should be added that it was this need for viable export markets to stabilize domestic consumer prices that affirmed the dependence of the Canadian cattle industry on international variables over which it had no control.

Since the 1880s, Canadians had been paying an average tariff of well over 20 percent on cattle exports to the United States,[65] a factor which, after the quarantine removal in 1897, resulted in a viable but limited trade. The years 1906-13 saw a population increase in the United States of 12 million and a decrease in cattle numbers of 20 million head.[66] The resulting pressure on rising prices forced Washington's hand. Under the Underwood Tariff enacted in 1913, the duties on Canadian cattle entering the United States were removed. For the first time, Canada's ideal export market was open. The impact was immediate. Cattle exports south of the border of both finished animals and feeders increased dramatically from 9,807 in 1912 to 180,383 the following year, and reached an all-time high of 453,606 in 1919. Complemented by favourable prices, the western Canadian cattle industry benefited mightily from its new and seemingly unlimited market, a fact reaffirmed repeatedly in later leaner years by nostalgic references to days that "would never come again."[67] Veteran cattleman George Lane summed up the realities of this halcyon period when he commented in 1916 that "the most dangerous thing that stockmen have to fear is a change of administration on the other side of the line that would slap a duty on everything."[68]

The embargo was thus of no commercial consequence in Canada when Prime Minister Robert Borden and the Minister of Public Works, Robert Rogers, journeyed to London in 1917 for the Imperial War Conference. However, what transpired in London in that spring of 1917 was the beginning of a bitter political struggle in which the embargo was debated openly as an economic and not an animal health issue.

The Canadians' Ministers came to London prepared to lay the issue of the embargo on the table for discussion, but only in the sense of dispelling the stigma it had attached to Canadian cattle. To them, it was a disease-related matter and no more. At one point during the debate on 26 April, Borden said that as long as the slur was removed, "if it is desired to protect the cattle industry in the United Kingdom, then let it be done."[69] Doubtless, he and Rogers were pleasantly surprised when the President of the Board of Agriculture, R. E. Prothero (later Lord Ernle) promised much more. Prothero admitted that the embargo was based on erroneous grounds and that the import of Canadian store cattle was desirable given plans for increased arable acreage in Britain after the war, the need to augment depleted British cattle inventories, and the probable reduction in the Irish role in producing store cattle.[70] A subsequent resolution was passed to "remove the embargo on Canadian cattle as soon as possible." It was also noted that "Mr. Prothero accepts that and that is the end of it." According to the Duke of Devonshire, Canada's Governor General at the time, both Borden and Rogers left London convinced beyond all doubt that the embargo, along with the stigma attached to it, would be totally removed after the war.[71]

Borden might have had reason for private thoughts of disquiet had he been aware that the Colonial Secretary, W. H. Long, who had chaired the meeting, had been the President of the Board of Agriculture when the embargo had been made permanent in 1896. He might have picked up on the way Long had taken issue with Prothero during the meeting for admitting that the embargo had been imposed in error, or the manner in which he neatly turned the discussion back to an argument for removal of the stigma. If Borden had noted the above, he might not have been surprised at Long's turnaround six weeks later when he stated that Canada should be exempt from the Disease of Animals Act but that it would have to be accompanied by a clause that nothing in the excepting act permitted Canada's cattle to be landed as stores.[72] With this statement, animal health ceased to be a pivotal factor, an admittance made outright by the Board of Agriculture in 1919.[73] Reinforced by a resolution in the House of Lords supporting the embargo and by the outspoken support of vested interests like the Royal Ulster Society, the Ulster Farmers Union, the Royal Agricultural Society and other rural organizations worried by the diminished consumer demand for beef,[74] Prothero, now Lord Ernle, issued a public statement in 1920 to the effect that his words at the Imperial Conference in 1917 had been misinterpreted.[75] Inquiries in the House of Common from Scottish MPs respecting the embargo's removal met with flat refusals from two successive Presidents of the Board of Agriculture, Lord Lee of Fareham and Sir Arthur Griffith-Boscawen. The British had clearly drawn their battle lines.

As the year 1919 closed, the debate on the embargo's removal may well have remained a British affair. Given the relative disinterest by Canadian cattlemen, basking as they were in the heady atmosphere provided by still buoyant prices and the unbounded potential of the United States market, the whole issue of the embargo's removal might have gone the road of British political preference. However, matters changed dramatically in 1920, spurred on by two factors. The first was the fear of a return to the dreaded tariff promised by the protectionist lobby prominent in the new Republican Administration in the United States. The second was the sudden and dramatic collapse of livestock prices.

Although there is some evidence to suggest that reciprocal concessions by Canada may have thwarted the return to the tariff in 1921, there can be little doubt that Canada was an unwitting victim of the depressed agricultural market in the United States and by American fears of mounting competition from cheap Argentinean beef and Australian mutton. The fears of Canadian cattlemen were vindicated in 1921 when a 27 percent duty was imposed on Canadian cattle exports to the United States.[76] The embargo re-emerged as an issue only after the failure of the Canadian lobby to defeat the American tariff. In comparison to the loss of the United States market, it posed as a less-than-ideal substitute for a catastrophic setback. To leading ranching spokesman, A. E. Cross, the removal of the embargo was more a palliative than a solution to western Canada's cattle woes "since we can finish here at a good profit."[77] Cross was right. By 1920, the increasing prominence of mixed farming and irrigation had greatly increased Alberta's capacity to "finish" its beef cattle. In all probability, had the Democrats remained in power in 1920, western Canadian cattlemen would have weathered their subsequent economic travail without challenging the embargo.

The economic plight of western Canadian cattlemen was compounded further after 1920 by a dramatic downturn in prices. The high prices reached in 1918 had been maintained artificially by the exceptional economic circumstances following the end of World War I. In 1920, inflated cattle inventories in both Canada and the United States, combined with an unusual coincidence of beef and hog cycles, sent prices in both countries plummeting. Between 1920 and 1921, in the steepest decline in history, cattle prices in Toronto dropped from nearly $13 to $7 cwt.[78] Desperate Canadian cattlemen were anxious to explore any remaining market possibilities including those in traditionally unreceptive European countries, many of which still held the spectre of disease over Canadian cattle. In the winter of 1919-20, entrepreneurial western Canadian stockyard operator, H. P. Kennedy, toured France, Belgium and Switzerland with Toronto's Grand Champion Black Angus and second place Shorthorn in an effort to assuage European nervousness with the quality and potential of Canadian beef.[79] By the summer of 1920, this combination of low prices, surplus inventories, the threat of prohibitive access to their most viable market, and meagre prospects in Europe had left the Canadian cattlemen with the sole option of maximizing its remaining export outlet. However, to optimize the potential of that market, the embargo had to be removed. Thus, in spite of its subsequent intensity, the organized campaign in 1920-21 to have the embargo lifted represented more a desperate response to practical exigency than it did an ongoing desire to remove an historic inequity.

Canadian strategy took two main forms. The first was simply to convince the British that lifting the embargo posed no threat to Britain's herds nor to her beef industry. Visiting Canadian government officials in Great Britain, such as Prime Minister Arthur Meighen who addressed the House of Commons in the summer of 1921, stressed the need for fairness and promised reciprocal tariff concessions should the embargo be lifted. When a group of British editorial writers visited Canada in the summer and fall of 1921, the embargo was the topical issue of the banquet addresses at the many Board of Trade functions held in their honour. Often delivered by high-ranking political figures, these addresses bluntly confronted the visitors with the inequity of using animal disease to mask what was essentially a British fiscal problem.[80] The lobbying campaign was also carried on at unofficial levels and involved high-placed British expatriates to North America. Other agricultural organizations became involved. The powerful United Grain Growers, for example, while declining to become a visible participant informed the Stock Growers Protective Association that it "had people working behind the scenes."[81]

Other practical efforts were made to convince the British to change their mind. In the summer of 1921 the provincial government of Alberta assembled a herd of selected mixed cattle and sent them to Birkenhead in an effort to dissuade the fear of British cattlemen as to either the quality or the threat of the Canadian product. Like many previous cattlemen testing the British market, the government took a loss on the sale, a loss it deemed insignificant to the favourable publicity generated.[82] Efforts had also been initiated in late 1920 by Canadian cattlemen together with vested British interests to establish a feeding association in Great Britain using Canadian stores.[83] In June 1921, attempts were made to float The Livestock and General Brokerage Company. Capitalized at $500,000 this company intended to erect feeding facilities and abattoirs in Great Britain capable of handling 25,000 head annually.[84] Though its grand motto—"Trade Within the Empire"—promised more, this proposed Anglo-Canadian feeding association was really designed to show British beef interests that they had nothing to fear and possibly a lot to gain from lifting the embargo.

The second Canadian strategy was blatantly political. Led by W. F. Stevens, Secretary of the Alberta-based Stock Growers' Protective Association, a concerted campaign was mounted to incite resentment in Britain towards the embargo. Stevens, formerly Alberta's first Livestock Commissioner, combined a wealth of experience and knowledge with an unswerving commitment to his cause. His policy was simple: "Our proper place to hit is in the large industrial centres and show that the embargo is tending to reduce labour in Great Britain and will place the meat trade in the hands of American meat trusts and so reduce the supply of home killed beef to the British consumer."[85] His British ally was long-time friend, James Lennox, a prominent Scottish farmer and spokesman for the National Farmers Union of Scotland. Working through Lennox and E. Watson, Chairman of the Free Importation of Canadian Cattle Association, Stevens launched an advertising campaign in the fall of 1920 designed to convince the urban British consumer that the embargo was directly linked to high beef prices. With Lennox promising to have something in the press every day, articles began to appear in newspapers in Birmingham, Sheffield, Liverpool, London, Manchester and other major cities inciting and fomenting resentment in a hitherto uninformed urban populace. Then in March 1921, the English newspapers of Canadian-born peer, Lord Beaverbrook, infuriated embargo proponents by turning a by-election involving Sir Arthur Griffith-Boscawen, President of the Board of Agriculture, into a bitter fight over the looming spectre of "dear meat."[86] Boscawen's unexpected defeat by the Labour candidate was blamed entirely on the embargo issue, and, more importantly, heralded a significant shift in public opinion. A week later, a meeting sponsored by the Corporation of London and attended by hundreds of representatives from local councils and public bodies across the country went on record as opposing the embargo.[87] With Lord Beaverbrook acting as the Government of Alberta's official representative at subsequent meetings in London's Guildhall,[88] these public forums, representing over 75 percent of the British electorate, ultimately became unanimous in calling for the embargo's repeal.[89] Also, the fact that Britain was currently scheduling Irish cattle for the sixth time since 1907 because of foot-and-mouth disease did little to aid the pro-embargo forces.

In response to this mounting public pressure the British Government was forced to take action. In May 1921, it appointed a four-man Royal Commission with the mandate:

> To inquire into the admission into the United Kingdom of livestock for purposes other than for the immediate slaughter at the ports; whether such action would cheapen or increase the supply of meat in this country, and if so to what extent, and whether it is advisable having regard to the necessity of protecting livestock bred in this country from the introduction of disease and of restoring their losses to their pre-war number.[90]

Over the subsequent six weeks, the Commission called ninety-two witnesses to its twenty-five sittings in the Moses Room in the House of Lords. Its report, released in August, recommended lifting the embargo, but warned that its removal, while increasing the supply of fresh beef, would not significantly alter the consumer price of meat.[91] The Canadians, it appeared, had won their victory even if their political tactics had been obliquely called into question.

Despite the findings of the Commission, the British continued to delay bringing the matter to a vote in the House of Commons. As late as February 1922 the reinstated Arthur Griffith-Boscawen told the House of Commons that "in view of the almost unanimous opinion of all agriculturalists that the removal of the embargo would seriously injure the industry ... and since the commission itself said that it would not affect the price of meat, then we do not propose to introduce legislation to remove it."[92] It was not until July 1922 that the British yielded to the mounting public pressure inspired by the Guildhall meetings and finally agreed to put the embargo issue to a free vote in both Houses of Parliament.

Both debates were long and acrimonious and pitted entrenched agricultural interests in England and Ireland against a mounting swell of popular opinion articulated by the various cooperative societies, Scottish grazing spokesmen and the disenchanted voices of labour. In the House of Lords, W. S. Long, now Lord Wraxall, succeeded in having a motion passed supporting the removal of the embargo, but which imposed restrictions respecting quarantine and eligibility requirements.[93] Despite the outspoken opposition of Boscawen, Britain's highest agricultural spokesman, The House of Commons, on the other hand, voted 247 to 171 for "the removal of the embargo all round." Though Colonial Secretary Winston Churchill delivered a brilliant summation urging his fellow MPs to honour the 1917 pledge, it was obvious that the issue had been carried by urban sentiment. The tally of the "Yeas" vote showed a preponderance of MPs representing cost-conscious consumers in the big British cities.[94] In this sense, Stevens' campaign, masterminded in western Canada, and fairly questioned for its validity, had won the day.[95]

The British agricultural lobby had not yet run its course, however. The ensuing Importation of Animals Act which came into effect in April 1923 had some surprises for Canadian cattlemen in that it restricted import cattle to steers and spayed heifers with all "breeding stock" being excluded to protect home herds. Definitely, the stipulation that female stock had to be spayed militated against the number of heifers shipped as the spaying process was expensive and complex enough to deter some ranchers. Prime Minister Mackenzie King brought the matter up during the Imperial Economic Conference held the same year in London. He castigated the British for their duplicity and pointed out that Canada had granted substantial reductions to its existing preferential tariffs on British items in return for the embargo's removal.[96] The British made no apology for their actions but simply referred to their need to adhere to the strong wishes of their agricultural community. As in 1892, it was a closed matter.

The intent of the agreement was forestalled by the British in other ways. Their veterinarians persisted in declaring cattle designated as stores to be "fat" and therefore subject to immediate slaughter. From 23 April 1923 to 6 March 1924, of the 28,183 head shipped from Canada to Britain as stores, 10,896 were declared "fat" and slaughtered at the ports of entry.[97] Other criticisms charged that pulmonary conditions continued to exist in Canadian cattle, and that shipping practices out of Canadian ports were inhumane.

In the final analysis, the fight over lifting the embargo was scarcely worth the effort. In fact, during his campaign, Stevens had to counter an Ontario-based lobby which saw the embargo's removal as being detrimental to its own promotion of the dressed meat trade.[98] As it was, changing consumer preferences in Great Britain and uncertain prices meant that testing the "Old Country" market remained a risky proposition. Live cattle imports from Canada between 1923 and 1926, averaged about 80,000, well below any year in the period 1891-1911. When prices in Britain collapsed in 1926, trade virtually ceased, picking up only marginally with the passage of the prohibitive Hawley Smoot Tariff in the United States in 1930.

It was Britain's abandonment of her historic free trade policy in the 1930s in favour of a system of imperial preference which brought the embargo issue to the forefront for the last time. At the Imperial Economic Conference held in Ottawa in 1932, Britain signed as agreement with Canada granting her preferential treatment on over 200 items. Under the terms of this agreement, Britain agreed to extend the Importation of Animals Act to include all Canadian cattle. The Free Importation of Canadian Cattle Association celebrated by disbanding. In Canada, however, the whole live cattle business had little future in Great Britain. Any viable meat market in Britain was for pork products with a very small niche for selective top-end cattle. The politics of animal health had run its course.

NOTES

This article first appeared in *Prairie Forum* 23, no. 1 (1998): 1-17.

1   Store cattle refers to animals not classified as "fat," and therefore not subject to immediate slaughter at the port of entry. Generally, they equated to today's feeder or stocker animals, although it was common that animals, ready for slaughter upon leaving Canada, might not be classed as "fat" by the time they landed in Britain. It was estimated that 70 percent of Canadian store cattle were fattened in Scotland.

2   Canada, *Sessional Papers*, No. 50, 1893, "Correspondence Between Canada and the United Kingdom With Respect to the Scheduling of Canadian Cattle." (Hereafter cited as "Scheduling Correspondence.")

3   "Scheduling Correspondence," op. cit.

4   The first action was to trace the infected animals to their source. They were found to have come from Manitoba in the Minnedosa district. All animals in the surrounding area were subject to inspection with no disease being located.

5   The inspection which was continued well into 1893 was both time-consuming and extensive, with up to eleven veterinarians visiting each farm and ranch. In his report one inspecting veterinarian wrote, "In my opinion if Great Britain and the United States were as free from disease as Canada, there would be no need for any inspection or quarantine."

6   "Scheduling Correspondence," op. cit. The examinations were conducted by two veterinarians, one being the Principal of the College. Their conclusions were supported by the Head of the Agricultural Department who had been to North America and had seen firsthand the same post-mortem condition in animals suffering from Cornstalk Disease, a layman's term for broncho pneumonia.

7   "Scheduling Correspondence," op.cit. Doubtless the fact that these findings were by Scottish veterinarians was not lost on the British proponents of the embargo. After all, the chief opposition to the embargo was coming from Scottish graziers and farmers. Most English farmers were in favour of it.

8   Great Britain, House of Commons, *Parliamentary Debates*, Fourth Series, Volume 8, 6 February 1893, p. 1394.

9   This point was emphasized by Tupper who felt that the British might have gone along with the Canadian offer to pay the cost of a British veterinary team to come to Canada to verify any absence of pleuro pneumonia had they been able to do so legally. According to Tupper, government lawyers had advised them that they had no choice but to impose the embargo. See, "Scheduling Correspondence," op.cit.

10 See Great Britain, House of Commons, *Parliamentary* Debates, 6 February, 20 March 1893, 16 August 1895, 9 May 1896.

11 Ibid., 10 August 1895.

12 Quoted in *The Globe* [Toronto], 19 January 1903.

13 See "Royal Commission on the Importation of Store Cattle. Report of His Majesty's Commissioners, 30 August 1921." In the embargo debate in the House of Lords, 1922, the Marquis of Lincolnshire recalled that as a member of that ministry in 1906, he and several other Ministers were indeed favourable to removal but did not do so. The Marquis' somewhat specious argument, probably dimmed with time, suggested that the reason was linked to no one submitting a bill to that effect plus the fact that they had not made a pledge to do so. Great Britain, House of Lords, *Parliamentary Debates*, Fifth Series, Volume 51, 12 July 1922.

14 Canada, Senate, *Debates*, 19 July 1895.

15 See David H. Breen, *The Canadian Prairie West and the Ranching Frontier 1874-1924* (Toronto: University of Toronto Press, 1983), 225-26.

16 Later popular and academic discussions tended to re-affirm the Canadian case that the embargo was imposed on purely economic grounds. For examples, see L. V. Kelly, *The Rangemen* (Toronto: n.p., 1913) and Breen, op. cit.

17 J. R. Fisher, "The Economic Effects of Cattle Disease in Britain and Its Containment, 1850-1900," *Agricultural History* 54 (April 1980): 280.

18 David Zimmerman, "Live Cattle Export Trade Between United States and Great Britain, 1868-1885," *Agricultural History* 36, no. 1 (January 1962).

19 Opinion seems to differ on this. Fisher, op. cit., 28, gives the impression that Canada was the only country allowed to export stores, but there is some evidence to suggest that while several European countries were scheduled including Germany, France, Belgium, Holland and Portugal, the Scandinavian countries were exempt. Of these, Denmark was the only one exporting stores in any quantity. Essentially, Canada was the only country bringing store cattle into Britain in any quantity.

20 Canada, *Sessional Papers*, No. 10, 1880, "Department of Agriculture Annual Report."

21 Canada, *Sessional Papers*, No. 12, 1887, "Department of Agriculture Annual Report."

22 Quoted in Canada, Senate, *Debates*, 15 April 1890.

23 "Scheduling Correspondence," op. cit.

24 Ibid.

25 The British seemed smugly secure in this argument stressing their utmost confidence in the health of Canadian cattle while hinting at the impossibility, despite every best effort, of preventing American cattle from coming into the country, the ninety-day quarantine notwithstanding.

26 Canada, Senate, *Debates*, 2 February 1893. Bowell was responding to a suggestion from a Senator from British Columbia that quarantine regulations should not be extended to British Columbia. Three years earlier another Senator had tried to convince the House that American cattle should be allowed to be shipped to Britain

from Canadian ports. See 14 February 1890 in Canada, Senates, *Debates*. Both were rejected by the Upper House.

27 "Scheduling Correspondence," op. cit. Quoted in English press. It was an oft-repeated comment, and one doubtless reinforced by misleading impressions about the severity of Canadian winters.

28 Canada, *Sessional Papers*, No. 7B, 1891, "Export Cattle Trade in Canada." This inquiry came as a result of British reformer Samuel Plimsoll's efforts to have the live cattle trade discontinued altogether on humane grounds.

29 For information, see *Macleod Gazette*, 4 June 1891.

30 Canada, *Sessional Papers*, No. 8d, 1894, "Appendix to the Minister of Agriculture Report, 'The Scheduling of Canadian Cattle, the Canadian Case'."

31 See *Macleod Gazette*, 18 June 1891 quoting *Canadian Gazette*, 28 May 1891.

32 "Scheduling Correspondence," op. cit.

33 *Edinburgh Scotsman*, 5 November 1891.

34 Canada, Senate, *Debates*, 15 April 1890.

35 A statement made by Sir Charles Tupper in his early correspondence to Canadian officials following the identification of pleuro pneumonia. Tupper appeared far more upset than surprised by the action taken by the British.

36 Canada, *Sessional Papers*, No. 13, 1893, "Department of Agriculture Annual Report."

37 Zimmerman, op. cit., p. 52. Zimmerman was quoting a statement made by the Canadian Minister of Agriculture in 1881. Other figures mentioned in 1890 were between $10 and $20, and a differential of 20 percent. See, Canada, Senate, *Debates*, 14 February, 15 April 1890.

38 Kelly, op. cit., p. 391. This assertion was made by the Dominion Veterinary General, Dr. J. G. Rutherford. Kelly had inserted the entire text of a report Rutherford had written in 1909 entitled, "Cattle Trade in Canada."

39 Canada, *Sessional Papers*, No. 13, 1893, Department of the Interior Annual Report, 1892. "Report of the Dominion Superintendent of Lands and Mines."

40 Ibid., "Report of the Dominion Superintendent of Lands and Mines, 1894."

41 Ibid., "Report of the Dominion Superintendent of Lands and Mines, 1896."

42 This view is shared by Simon Evans who has done extensive work in the history of ranching in western Canada. See Simon Evans, "Canadian Beef for Victorian Britain," *Agricultural History* 53, no. 4 (October, 1979): 757.

43 Canada, *Sessional Papers*, No. 13, 1896. Department of the Interior Annual Report. "Report of the Superintendent of Lands and Mines, 1895."

44 William Pearce had argued strenuously for this removal in his annual reports after the embargo was imposed, feeling that the freer movement of cattle across the line would augment the stocking of the Canadian ranges. Interestingly, however, the first resolution passed by the newly formed Western Stock Growers Association in 1896 was one in favour of maintaining the present quarantine regulations. Clearly, the southern Alberta ranchers were holding a narrower view than Pearce in that they feared American cattle would encroach on their leases.

45 Canada, *Sessional Papers*, No. 8, 1898, "Department of Agriculture Annual Report."

46 Canada, *Sessional Papers*, No. 8, 1898, "Report on Cattle Quarantine in Canada from November, 1896 to October, 1897."

47 Kelly, op. cit., p. 341.

48   Alberta, *Department* of Agriculture *Annual Report, 1912*, "Report of Livestock Commissioner."

49   For full export numbers on cattle exports to Britain, 1874-1937, see *Canadian Cattlemen*, June 1938.

50   Evans, "Canadian Beef," op. cit., p. 757.

51   "Export Cattle Trade of Canada," op. cit.

52   Ibid.

53   See *Calgary Herald*, 29 October, 11 November 1892.

54   Glenbow, Western Stock Growers Association Papers (hereafter WSGA Papers), "Minute Book, 1896-1914," Box 1, ff. 3.

55   Alberta, *Department of Agriculture Annual Report, 1906*, Appendix C, "Report of Alberta Cattle Breeders' Association."

56   Alberta, Department of Agriculture Annual Report, 1908, Appendix B, "Alberta Cattle Breeders' Association Annual Report." The remarks were made by Duncan Anderson, Special Representative of the Livestock Branch of the Dominion Department of Agriculture, at the annual meeting of the above association.

57   So much so that when asked by Canadian interests for assistance when the embargo debate was heating up in 1920, the FICCA chairman, E. Watson informed W. F. Stevens, Secretary of the Stock Growers Protective Association, that he felt there was no hope for the cause and that the fruitlessness of his long tenure as chairman had left him drained and exhausted. WSGA Papers, Box 8, ff. 45. Correspondence dated 15 November 1920.

58   For a good account, see Simon Evans, "The End of the Open Range Era in Western Canada," *Prairie* Forum 8, no. 1 (1983): 71-87.

59   In 1911, it was reported that cropped land around Fort Macleod had increased by 20 percent in one year. In the same year grain shipments from Stavely and Claresholm in the heart of former ranching country totalled over 1.5 m bushels. *Sessional Papers*, No. 28, 1911, North-West Mounted Police Annual Report, "Report of Supt. P. C. H. Primrose, D Division, Fort Macleod."

60   C. H. McKinnon, *Events at LK Ranch* (Calgary: Phoenix Press, 1979), 59. This is an excellent account of the history of one of Alberta's most enduring and successful livestock enterprises. Consisting mainly of detailed reminiscences, particularly those of the family patriarch, Lachlin McKinnon, the narrative chronicles the evolutionary nature of ranching in Alberta.

61   Canada, *Sessional Papers*, 15C, 1912. Department of the Interior Annual Report, "Report of the Livestock Commissioner." Other comments referred to the "ranching industry dying as quickly and as decently as it was able"; to "the rancher's day being a thing of the past"; to "many ranchers ... disposing of as much stock as possible"; and as a "general tendency being to market anything and everything."

62   *Agricultural Gazette*, December 1914.

63   *Farm and Ranch Review*, 20 September 1913.

64   See Evans. "End of...," op. cit.

65   For information on the United States tariff structure up to 1935 see F. Albert Rudd, "Production and Marketing of Beef Cattle from the Short Grass Plains Area of Canada" (Master's thesis, University of Alberta, 1935), 105.

66   *Farm and Ranch Review*, 20 September, 4 November 1913. The magazine also noted that United States beef exports had declined by 96 percent during the same period.

67 This was the period when Alberta rancher Alfred Ernest Cross consistently topped the Chicago market with his range fed steers. For example, see *Calgary Herald*, 22 December 1915. On 5 December 1918, a 1,700-pound Alberta Shorthorn Hereford cross brought the highest price ever paid for a range steer on the Chicago market. The price of $18.75 cwt led the *Canadian Cattlemen*, which reported the incident in December 1938, to reflect nostalgically, "Will those days ever return?"

68 *Calgary Herald*, 3 October 1916. Lane had also noted in 1913 the enormous potential of the emerging market in the northwest United States. He stressed Alberta's decided locational advantage over American suppliers and felt that profits would increase by as much as $7 a head over the average. Time was to vindicate Lane's prophesy, a prophesy more accurate today than when Lane made it in 1913. See *Farm and Ranch Review*, 20 September 1913.

69 See "Extracts from the Discussion at the Imperial War Conference, April 26, 1917"; Canada, *Sessional Papers*, No. 42A, 1917. "Minutes of the Proceedings of the Imperial War Conference, 1917." During the debate, Rogers also commented that the embargo was a matter of policy for the development of British live cattle and that "we have no objection to a policy of protection."

70 "Extracts from Discussion," op. cit., pp. 3-4. Doubtless, the British gratitude towards Canada for her significant contribution to the war effort contributed to their feelings of benevolence at the time.

71 Great Britain, House of Lords, *Parliamentary Debates*, 12 July 1922. Remarks by the Duke of Devonshire.

72 Great Britain, House of Commons, *Parliamentary Debates*, Vol. 157. Quoted in embargo debate, 22 July 1922.

73 Ibid., Vol. 119, 14 August 1919.

74 Ibid., Vol. 119, 6 August 1919; Vol. 121, 12 November 1919.

75 Ibid., Great Britain, Vol. 157. Quoted in embargo debate. 24 July 1922.

76 The Young Emergency Tariff passed in 1921 was given permanence by the Fordney-McCumber Tariff in 1922.

77 Glenbow, Cross Fonds, Box 113, ff.908, "A. E. Cross to James A. Lougheed, Minister of the Interior, March 24, 1921." Cross rightly believed that the embargo had far greater implications for Ontario producers.

78 For a good contemporary article, see E. C. Hope, "Livestock Cycles in Canada," *Scientific Agriculture* 11, no. 2 (October 1930).

79 "A Half Century of Service," *Canadian Cattlemen* 1, no.2 (September 1938).

80 For example, see address given by R. B. Bennett, 20 August 1921, in Western Stock Growers Association Papers, Box 8, ff. 45.

81 WSGA Papers, Box 8, ff. 45. "UGC Livestock Superintendent to W. F. Stevens, September 16, 1921."

82 *Farm and Ranch* Review, 20 May, 5 October 1921.

83 WSGA Papers, Box 8, ff. 45. "W.F. Stevens to E. L. Richardson, Secretary, Western Canadian Livestock Union, October 12, 1920."

84 Cross Fonds, Box 113, ff. 913, Correspondence dated 23 July 1921.

85 WSGA Papers, Box 8, ff. 45. "Stevens to E. Watson, November 30, 1920."

86 The incident in the Dudley by-election was discussed at great length in the House of Commons debate on the embargo in July 1922. Apparently, Beaverbrook's *Daily*

Express flooded the electorate with leaflets. One MP was quoted as saying that "we were threatened with consequences should we dare oppose Lord Beaverbrook's will."

87 See "Royal Commission on the Importation of Live Cattle. Report of His Majesty's Commissioners, August 30, 1921."

88 *Calgary Albertan*, 24 July 1921.

89 See *London Times*, 15 June 1922.

90 "Royal Commission...," op. cit.

91 Ibid.

92 Great Britain, House of Commons, *Parliamentary Debates*, Vol. 150, 9 February 1922.

93 Great Britain, House of Lords, *Parliamentary Debates*, Vol. 53, 12, 26 July 1922.

94 Great Britain, House of Commons, *Parliamentary Debates*, Vol. 157, 24 July 1922. Prominent cabinet ministers voting against the removal of the embargo included J. Austin Chamberlain, Lord Privy Seal and Leader of the House, Stanley Baldwin, President of the Boards of Trade, and of course Arthur Griffith-Boscawen, President of the Board of Agriculture.

95 The Royal Commission was right in its conclusions. Meat prices did not change appreciably with the lifting of the embargo.

96 Canada, *Sessional Papers*, No. 36, 1924, "Record of Proceedings and Documents, Imperial Economic Conference, October-November, 1923."

97 Great Britain, House of Commons, *Parliamentary Debates*, Vol. 171, 17 March 1924.

98 See *Calgary Herald*, 7 January, 15 February 1921.

15. The Manitoba Grain Act: An "Agrarian Magna Charta"?

D. J. Hall

" **A** gigantic combination," fumed the *Winnipeg Daily Tribune* on September 14, 1897, had been forged "for the purpose of cinching the farmers on their wheat this season." The combine allegedly involved "practically all the grain dealers, big millers and grain syndicates," creating a monopoly sufficiently powerful to force the price of wheat downward by six cents per bushel at many shipping points, and able to pressure the banks "to shut down on small traders and drive them out of business."

The revelation of this monopoly confirmed the worst suspicions of the prairie farmer. Heated demands were instantly raised across the Canadian West for government control of the grain trade, and for abandonment by the railways of policies which encouraged grain elevator monopolies at most shipping points. Increasing pressure over the next five years eventually forced a reluctant Dominion government to concede the first effective regulatory legislation governing the grain trade. The issue also contributed to the transformation of western agrarian organizations, and a dramatic shift in the relationship between politicians and farmers.

Agrarian grievances had always germinated as quickly as the wheat itself in the rich prairie soil. As early as 1883 the Manitoba and North West Farmers' Union, the first effective western Canadian farm protest organization, was denouncing the National Policy protective tariff, the monopoly of the Canadian Pacific Railway, and federal control of public lands.[1] Often overlooked in the farmers' catalogue of demands in 1883 was the call for an end to the developing grain elevator monopoly. The owners of the CPR were familiar with the problems of transporting grain both in the United States and in central Canada.[2] They were determined that they would create from the beginning in the West the most efficient, modern grain handling system possible. Anyone who cared to

For many years, western settlers held the view that eastern businessmen milked the profits from the farmer, as is illustrated in this cartoon that appeared in the *Grain Growers' Guide* December 15, 1915. (Courtesy of the Saskatchewan Archives Board/R-A19422.)

build a mechanical elevator with a minimum capacity of 25,000 bushels would be guaranteed a monopoly at a given shipping point. The railway would service only elevators (including competing elevators at those points, rare before 1900, where more than one was built); but flat warehouses and loading platforms would not be serviced once an elevator was built.

From the railway point of view this system was most practical. Elevators required a sizeable capital investment, and few people would risk their capital without a guaranteed volume of trade. The elevators had facilities for cleaning and drying grain, and bins for storing the grain of individual farmers. Most important, the elevators enabled the railway to handle the grain crop with maximum despatch: cars could be loaded in minutes, compared to the one or two days necessary to load by hand from flat warehouses or loading platforms. At a time when there were almost no facilities for farm-stored grain, and few large storage elevators at the Lakehead, it was vital to employ the limited available rolling stock as efficiently as possible to move the crop before the Great Lakes shipping season closed.

The farmers were unimpressed with the benefits of the monopoly. In Ontario and parts of the United States, farmers had often had the advantage of several buyers competing for their grain at the local shipping points. Alternatively, if the price was not considered high enough, the farmer could order up his own railway car, load it (from a flat warehouse or loading platform) and attempt to market the grain himself in a larger centre. This was not possible in Manitoba wherever elevators were built. There the farmer had either to sell his grain to the local elevator operator at whatever price was offered, or to pay the elevator 1 and a half to 2 cents per bushel for cleaning and storing while a railway car was ordered to take his grain to market. Inevitably the farmer believed that he was being cheated: that his grain was being graded too low; that there was excessive dockage; that the elevator operator engaged in mixing grades; and that prices were ruinously low. Many of the elevators were built by large milling companies which many farmers believed to be in league with one another and with the CPR to defraud the producer.

There is no question that the Manitoba farmer of this period was in a vulnerable position. World prices were generally low and freight rates high until the late 1890s, leaving a precariously narrow profit margin. Thousands of prairie farmers failed to survive these testing early years, and it became an article of agrarian faith that the entire grain-handling system would have to be brought under government regulation or control before stability in the economics of farming could be anticipated.

The farmers achieved some small successes in this sphere in the 1890s,[3] but not until the early twentieth century were they able to produce permanent agrarian organizations to further their demands. The Manitoba and North West Farmers' Union had failed in part because it was the tool of aspiring Provincial Rights or Liberal politicians.[4] Its successor, the Patrons of Industry movement of the 1890s, also failed because its leaders were inept and naive, and because it was easily outmanoeuvred by the brilliant young Attorney General of Manitoba, Clifford Sifton.[5]

Sifton's attitude to the farmers was of crucial importance. As Minister of the Interior, he would be the western representative in the Dominion cabinet after the victory of Wilfrid Laurier's Liberals in the Dominion election of 1896. For six years, between 1882 and 1888, Sifton had been engaged in stirring up agrarian discontent to the advantage of the Provincial Rights-Liberal Party. After the provincial Liberal victory of 1888 he set about soothing agrarian tempers, outmanoeuvring aspiring Patron politicians, conceding the minimum legislation necessary to silence agrarian demands, and trying to keep Manitoba a safe and attractive place for the businessman, investor and capitalist. Sifton was a man of political genius and great energy, and was also extremely successful. By manipulating the press and packing public meetings of farmers with his supporters, by alternately attacking and compromising with the Patrons, he managed to deflect or destroy opposition before it got out of hand. He had come to the conclusion, after nearly a decade and a half of political activity, that what he privately termed "the simple-minded farmer" was easily manipulated.[6] Unscrupulous demagogues, in his view, could all too easily whip up agitation over any number of imagined grievances.

Such was Sifton's analysis of the renewal of agrarian agitation in the years after 1896. Yet, while there were superficial signs to support his claims, there were underlying changes in the public mood that would shortly render his analysis invalid. To western Liberals the Laurier victory of 1896 portended drastic cuts in the protective tariff, firm restraints upon and competition for the CPR monopoly, and a government generally more responsive to the needs of the West than the Conservatives had been. In each of these areas there would be disappointment followed by disillusionment. Those western Liberals who felt betrayed had nowhere to turn. The Patron movement was in a shambles after 1896, and independent political action for the time being was discredited. An anti-Sifton cabal soon developed within the Liberal Party in Winnipeg in an effort either to make the government responsive to western demands, or to drive the Minister of the Interior out of office. It was strongly supported by the *Winnipeg Tribune,* whose owner, Robert Lorne Richardson, was also Member of Parliament for the rural constituency of Lisgar. Richardson's paper, which reflected the somewhat populist-Liberal views of its owner, was not a sufficiently reliable vehicle for the political machinery which Sifton was trying to construct. When Sifton purchased the *Manitoba Free Press* in 1897-98 to serve as the main cog in the government's propaganda network in the West, the aggrieved Richardson seized the opportunity to use the *Tribune* to voice the complaints both of the dissentient Winnipeg Liberals and of the angry farmers.[7]

The period between the demise of the Patrons and the rise of the permanent agrarian organizations of the early twentieth century, then, does not mark any sort of lull in agrarian agitation. On the contrary, in the columns of the *Tribune* (and of the *Edmonton Bulletin,* organ of Frank Oliver, MP for Alberta) the farmers' complaints were regularly agitated, and demands set forth for government control of the grain trade and railways. Oliver and Richardson were two-thirds of the trio of MPs who constantly pressed radical western demands upon the government; the other was the Reverend James Moffat Douglas, Patron-Liberal MP for Assiniboia East.[8]

It was in the midst of these developing rivalries and tensions in western politics that the *Tribune* exposed the "gigantic combination" of grain dealers and other interests in September 1897. The paper claimed that while there were four major syndicates in the grain business which "all operate separately as regards their financial interests, they have formed a mutual arrangement as regards the prices they will pay for wheat." This could have resulted in effective syndicate control of Manitoba wheat prices. At first the various interests denied that there was any such arrangement, but shortly it was revealed that in fact the grain interests had agreed to cooperate, though not so fully and formally as at first alleged by the *Tribune*. They had suffered significant losses by excessive competition in the previous crop year, and were cooperating in reducing and sharing staff at various shipping points. The result, of course, was that, even at points with two or more elevators, farmers would find no price differential.[9] Moreover, there was evidence that at some points elevator operators were refusing to accept for storage grain which had not been sold to them, at least until the elevator's own backlog had been shipped. Because of the CPR-supported monopoly enjoyed by the elevators, this effectively deprived farmers of any alternative way to sell their wheat.[10]

This situation led to renewed attacks on the railway. One irate independent grain dealer, J. K. McLennan, attempted to have the Winnipeg Grain Exchange pass a motion denouncing the CPR for not responding to farmers' needs as quickly as had the Manitoba and North Western, and Northern Pacific and Manitoba Companies. The CPR did not maintain sufficient rolling stock, declared McLennan, and "should not attempt to move a mountain with a mole-hill equipment." It was "the lack of cars" that "really creates the monopoly."[11] The motion was defeated 28 to 9, clear evidence to the angry farmers of the influence of the combine, supposedly allied with the CPR. The railway protested that a significant factor in the delay in moving the grain was the rapid increase in the Manitoba crop; in 1895 some 10.6 million bushels of wheat were shipped from Fort William, in 1896 some 12.7 million, and in 1897 an estimated 17:6 million.[12] Unimpressed by the railway's difficulties and explanations, the *Tribune* charged that the combine had defrauded the Manitoba farmer of at least $750,000 during the 1897 season. Parliamentary action was demanded to crush the combine "before another season comes around. It will not do to leave our farmers in the hands of the Philistines any longer."[13]

Speaking in Winnipeg a few days after the story about the combine had broken, J. M. Douglas contended that it was time for action to secure western Canadian farmers the same privileges possessed by their American brethren in Minnesota and the Dakotas.[14] There, he stated, by law, the farmer had "the right to demand cars of the railway company and ship his grain to market himself." The advent of elevators at Wapella and Fleming in his constituency had markedly depressed prices received by the farmer compared to when there had been competitive small buyers. Douglas's views were echoed by R. L. Richardson, who claimed that he also had concrete evidence of the misgrading of wheat. The *Tribune* repeatedly urged farmers' groups across the province to pass resolutions on the combine issue, and on the right to sufficient railway cars. The farmers responded enthusiastically, and the resulting resolutions were printed with much fanfare.[15]

The case against the railway and the syndicate had seemingly been amply documented. So had the rapid spread of agrarian indignation. According to the *Tribune*, the Dominion government had indicated that it would lend a sympathetic ear.[16] When the 1898 session of the House of Commons commenced, therefore, both Douglas and Richardson were optimistic about their chances of success. Each introduced a private member's bill into the House, Richardson's in the form of an amendment to the general Railway Act, both bills having the object of forcing the railways "to accept shipments of grain from farmers' wagons or sleighs, or from flat warehouses at all way stations where Grain is shipped." In addition they were seeking changes in the system of grain inspection. The prospects for success looked even brighter when Clifford Sifton, already recognized as one of the most powerful men in the Cabinet, himself attacked the syndicate and indicated that he would pilot Douglas's bill through the legislative shoals.[17]

Unhappily the westerners had miscalculated the strength of the lobby which was beginning to weigh in against the bill. A furious Rodmond P. Roblin, a leading grain man and Opposition politician in Manitoba, accused Richardson of "villifying [*sic*] men whose shoe latchets he was not worthy to unloose."[18] The grain men had invested heavily in elevators, operated on the narrowest of profit margins, and had no desire to be ruined by competition from flat warehouses and loading platforms. Roblin claimed that "the grain men did not make as much as their clerks last year [1896]." He admitted the existence of both a CPR monopoly and a grain syndicate, but viewed them as crucial to the survival of the industry. From Manitoba and the Territories grain men began to descend upon Ottawa to press their views on the government and to testify before the Railway Committee of the House, to which Douglas's bill had been referred. There the CPR contended that while it did not object to the principle of the bill, the real problem lay with the elevators which had refused to accept farmers' grain. "If the elevator company did their duty," argued the railway representatives, "there would be no necessity for the bill." More important, it was argued, if competition forced the closure of elevators, transportation costs would increase as the result of inevitable delays, further reducing the farmers' return on their wheat.[19]

So effective had the lobby proven that Douglas was forced to agree to a compromise with the CPR. There would be no platform loading; flat warehouses would be permitted; and farmers would be allowed two hours to load each

A farmer with a horse-drawn grain wagon using a loading platform to shovel grain directly into railcars. Note this practice was carried on for some years after the passing of the Manitoba Grain Act, as is evidenced by the Grand Trunk and Canadian National boxcars pictured. (Courtesy of the Saskatchewan Archives Board/R-A15067-1.)

car, after which fifty cents per hour demurrage would be charged, or $5 for twelve hours. On this basis the Railway Committee passed the bill, with the support of most western members.[20]

Staunchly opposed, however, were Richardson and Oliver, the former declaring that the compromise was "a piece of damn nonsense and of no use to anyone."[21] The difficulty was that farmers could not load the cars in two hours. The farmers would be worse off than ever, because as things stood they at least had twenty-four hours to load cars at points where such loading was permitted. Outraged protests from western farmers and independent grain men encouraged Oliver and Richardson to approach the renegade Conservative, D'Alton McCarthy, an able lawyer and a longtime sympathizer with farmers' organizations in Ontario, to draft a series of amendments which would substantially alter the bill when it reached third reading. The proposed amendments would have required the railways to maintain flat warehouses wherever they were demanded in writing by an individual; railway cars would have had to be supplied on demand to farmers for loading from vehicles, flat warehouses or elevators; while demurrage and other charges would have been effectively eliminated or greatly reduced.[22]

D'Alton McCarthy worked out these amendments on Friday, May 6, 1898. He went to Toronto for the weekend, and on the Sunday was thrown from a moving vehicle, sustaining serious injuries. On May 11 he died, and with

Loading grain from a horse-drawn wagon by means of a portable grain loader run by a tractor, Radville district of Saskatchewan, 1928. (Courtesy of the Saskatchewan Archives Board/R-A15067-3.)

him died all hope for effective legislation in 1898. On the day of his death the government announced that it would not permit the bill to proceed.

Belatedly and rather weakly Douglas explained that the bill which had emerged from the Railway Committee did not reflect the compromise to which he had agreed, and he promised to renew his battle in the next session of Parliament.[23] To keep the fires burning Richardson and the *Tribune* were active in founding The Anti-Elevator Monopoly Association in June 1898, at a meeting which enjoyed Manitoba-wide representation. Supporters of the Association were convinced that the tide of public opinion would force the government to concede legislation. Even the CPR seemed to be bowing to the popular will when it announced in midsummer that it was abandoning its monopoly policy and would allow farmers to load grain directly on to cars from their vehicles or from platforms.[24]

Events showed, however, that this was no more than a paper concession. In 1898 there was a large crop once again, and the CPR was scarcely able to meet the demand for cars from elevators, let alone from individual farmers. The elevator monopoly therefore remained "a burning question throughout the province [of Manitoba] and territories."[25] When Douglas reintroduced his bill in 1899, it provided for practically all the demands made by Richardson and Oliver in 1898, and added a requirement for a government grain inspector. On this occasion Clifford Sifton again gave the measure his general support, stating that the government agreed with its objectives, but not with its every provision. Upon his motion the measure was referred to a special committee of the House, where once again it would be buried.

Sifton could not afford to let the bill pass as it stood. Douglas had been strongly supported by Patrons in his constituency, and passage of his bill would have lent renewed credibility to the dying movement. Even more important, as one of Sifton's supporters pointed out to him, he could not afford to let Richardson "pose as a real Legislator."[26] A victory on the elevator issue would strengthen Richardson's pretensions to being the real spokesman for the Manitoba farmer.

Nor had the lobby against the legislation weakened since 1898. When the special committee met in May 1899 it was confronted by at least thirty opponents of the bill, whose names read like a *Who's Who* of Canadian grain trade interests. Among them were President Shaughnessy of the CPR, John Mather of Lake of the Woods Milling Company, W. W. Ogilvie of Ogilvie Milling, and L. M. Jones of Massey Harris. When several leading bankers joined the crowd, Frank Oliver declared that "it was evident a more gigantic combine existed than the people dreamed of." After one of the grain men protested that each of these groups was present to protect the interest of the farmers, Richardson retorted in a sarcastic despatch to the *Tribune:*

> The conclusion, therefore, was irresistible that the farmer was a lucky dog when he has the elevator owners, railways, grain dealers and bankers all represented and bound to see that his interests are protected. No wonder that he should be described as pampered, puffed and spoiled, and that those who claim that he suffers ... should be characterized as professional agitators and dangerous demagogues. Great is Humbug![27]

The grain and transportation representatives nevertheless claimed that it was the long-term interest of the grain trade that they wished to protect. Western Canada, asserted Sir William Van Horne of the CPR, had developed an efficient system for moving grain which contributed significantly to the relatively high prices paid to the farmers for their wheat. They also were looking into the immediate future; Sifton anticipated that within a decade the wheat crop would triple, and that it could not possibly be moved without an efficient elevator system. "The proposition to encourage the construction of flat warehouses," he snorted, "is practically on a par with the proposition to encourage ox-carts as compared with railroads." He gave the opinion that "some of the farmers have been loaded up with the windy clap-trap and nonsense which appears in the *Tribune* and have got wrong notions about this question. ... It is an artificial agitation raised by three or four scamps for the purpose of making themselves popular, totally regardless of what harm they may do to the permanent development and interests of the country.... "[28]

Angry as Sifton might have been, the reality of western frustration with the second burial of the Douglas bill had to be considered. A reply had to be made to Douglas's bitter charge that "it was the cold steel of Clifford Sifton" that had defeated the bill.[29] The government therefore responded by appointing in the fall of 1899 a Royal Commission on the Shipment and Transportation of Grain in Manitoba and the North-West Territories. Comprising the Commission were Justice E. J. Senkler of St. Catharines, as chairman; three western farmers, W. F. Sirrett, W. Lothian, and C. C. Castle; and C. N. Bell, a Winnipeg grain merchant, who served as secretary.[30] It was anticipated that the Commission would provide ample grounds for the government to take some action and that, if carefully chosen, it would come to the conclusions the government wished.[31]

The Minister of the Interior expected that a good deal of political hay could be made from the Commission. The *Tribune* charged that it would be heavily influenced by Bell, the grain merchant. Sifton told the editor of the *Free Press*, A. J. Magurn, that it should be made clear that Bell was only "Secretary, and has no voice in coming to conclusions or framing a report, and three out of four members of the Commission are farmers from the Province of Manitoba." "This is the first time in Canada that an important commission on a business matter has had a majority of farmers upon it," added Sifton, "and you should roast the *Tribune* for belittling them and treating them with contempt."[32]

Despite Sifton's protestations, however, much of the work of the Commission indeed fell to Bell, who planned its hearings, reported frequently to Sifton, and undertook a trip to Minnesota to inspect the system of elevators and grain handling adopted there in 1897. On March 19, 1900 the report of the Commission was presented to the House of Commons. Included in its schedules was recommended legislation to govern the operating of terminal and country elevators, weighing, cleaning and docking procedures, and the like. It suggested the appointment of a warehouse commissioner, and the granting of the right of "any ten farmers residing within twenty miles of a shipping point" to erect a flat warehouse or loading platform.[33]

Sifton worked closely with Bell in preparing legislation to be presented to Parliament in 1900, when he would be absent in Europe in a futile search for a cure for his growing deafness. When he sent the draft bill to the Minister of Inland Revenue, Sir Henri Joly de Lotbiniere, under whose jurisdiction the legislation would fall, Sifton adjured him,

it is an absolute necessity to the success of the Liberal party in
the west that a Bill should be carried through and the question
dealt with at the present Session. ... Our people in the North
West are very much excited about the question and they would
regard the failure of the Bill to become law, especially during
my absence, as a very serious reflection upon me.[34]

Initially the western members were pleased with the legislation, the government evidently having been willing to adopt many of their suggestions. The
bill provided for licensing of warehouses and elevators, for an inspector, and
for specified procedures to be used in the weighing, cleaning and storage of
grain. It required elevators and warehouses to accept the grain of any farmer
when space was available, prohibited the mixing of grain while in storage, and
generally attempted to meet the demands of farmers with respect to grain
handling. It also provided for the right of ten farmers within twenty miles of
a given shipping point to erect a flat warehouse of a minimum 6,000-bushel
capacity, with the railway company to provide land and a siding. The bill did
not, however, strengthen existing legislation under which farmers could individually order grain cars.[35] Indeed, the chief concern of the western members
was to broaden the flat warehouse clause to include any ten farmers within
forty miles of a given shipping point—a significant commentary on how far
many farmers had to haul their grain—and to reduce the minimum size to
3,000 bushels. Reluctantly the government accepted the changes.[36]

Having thought that they had secured a good if not a perfect bill, the
western members were stunned when, on third reading several days later, the
government proposed an amendment which in their judgment negated the
whole thrust of the bill. Once a warehouse had been provided at a given point,
the amendment ran, those desiring to erect additional warehouse facilities
would have to compensate the railway for the land and the cost of the spur line.
At once a fundamental philosophical cleavage emerged between the majority
of members and the western representatives of the farmer. What they had
wanted, the westerners declared, was "absolute free trade" in the shipment
of grain, the right of the farmer to ship by whatever means he chose, and to
ensure that adequate facilities would be available to permit him to do so. The
railways had objected that there might be no limit to the construction of flat
warehouses, that the elevator companies might be ruined, and that there would
be substantial losses to the railway in terms of land and construction of branch
lines. From the point of view of the farmer, the railways were there to serve
the producer, not to control him, and land granted to railways was not private
land in the usual sense. It was granted for a public purpose, to facilitate trade.

The issue had suddenly become of sufficient interest to engage the Leader of the Opposition, Sir Charles Tupper, and the Prime Minister. For once they agreed with each other as they bore down on the western members. The proposals of the farmers "would mean confiscation," declared Laurier. "It is not in the spirit of British legislation to give power over one man's land to another man unless adequate compensation is provided for it." It was only reasonable that those benefiting from additional facilities should have to provide "adequate compensation." Farmers would have to understand that the railway and elevator companies were not there only to serve them, but that efficient handling of grain required a partnership of all interested parties, and some concessions on all sides. Meanwhile the western proposals were simply "socialistic legislation; it is not British legislation as we understand it." The government amendment passed by 93 votes to 10; every available representative of the western grain farmer opposed it, irrespective of party.[37]

The *Tribune* attributed the change of policy to the malign influence of the CPR, and added:

> ...the people will continue to be robbed and enslaved; politicians will continue to stand in with the corporations; the electorate will continue to be corrupted; the press will continue to be operated by the corporate politicians until the people wake up and cease to support and applaud their betrayers because of a party name, and until they make up their minds to stand by those who uphold the public welfare.[38]

Such indignation was, perhaps, justified. The regulations granted in The Manitoba Grain Act, 1900 provided nothing more than the grain interests and the CPR were willing to concede. The railway company was already allowing individual loading of cars when they were available, and was required to do nothing more. It was as interested as the grain men in having elevators and warehouses licensed and inspected, and in securing more permanent grain standards and prohibitions against mixing. Abuses of the system by some grain dealers resulted in public attacks on everyone connected with the system, including the CPR. Even the grain men themselves had earlier appealed to Sifton for changes in the system of grading, but he had replied:

> It would be very unwise for the grain men themselves to take up the question of having the method of grading changed. The farmers and some demagogues who make their living by talking nonsense to the farmers about things that they do not know

any-thing about would immediately set up a cry that the steps which were suggested were being taken in the interests of the grain dealers and for the purpose of enabling them to beat the farmers. I know that you beat the farmers as much as you can, but still I am in sympathy with your movement to make the grades more permanent. When anything is done it should be by representative and general movement.[39]

Hence the Royal Commission, and hence the acceptability of the report to the grain men. If neither the elevator owners nor the railways were ecstatic about the provisions for loading platforms and flat warehouses, these at least would be limited in number and, it was hoped, would undermine criticism of the system. There seemed in any case to be little to prevent the usual favouring of elevators by the railway.

All this was well understood by the farmer. Welcome as the standards, inspection, licensing and so forth were, they would not occasion dramatic changes because the worst abuses of the industry had been largely internally corrected or controlled in the previous few years. But the cards still remained stacked in favour of "the interests." If a flat warehouse was full, for example, most farmers would still have to sell to or ship through elevators. Loading platforms, it was estimated, would benefit only those farmers within five miles because of the need to haul enough grain to fill a railway car within twenty-four hours—when a car was available. Furthermore, the specified size of the loading platform meant that at most shipping points only one car could be ordered and loaded at a time. True independence and freedom of trade still seemed remote.

The Liberal government naturally played up The Manitoba Grain Act as a great achievement in guaranteeing farmers' rights. Southern Manitoba in particular was flooded with propaganda depicting the benefits to be derived from the legislation.[40] Insofar as the 1900 election reflected opinion on the grain issue, however, the verdict among farmers in Manitoba can only be considered negative. Only two Liberals (Sifton and W. F. McCreary) were returned from Manitoba's seven seats. Richardson was re-elected despite the most desperate attempts to defeat him. In the Territories the Liberals did better, but both J. M. Douglas and Frank Oliver had been vigorous opponents of the government's policy. Plainly contemporary farmers would hardly have concurred with the historian who in 1928 praised the Act "as a veritable agrarian Magna Charta."[41]

Proof that the farmers' concerns were well founded occurred in the famous grain blockage of 1901. Almost perfect weather conditions and the beginning

The grain blockade of 1901-02. Scenes such as this at Wolseley, North-West Territories, in February 1902, were not uncommon in many of the towns and villages along the Canadian Pacific Railway that winter. The farmers in outlying areas were unable to market their grain because of congestion in elevators and insufficient railway accommodation. As a result, many farmers found it necessary to build their own more or less temporary granaries wherever they could find room in the towns or the villages, in order to have their grain handy to market through elevators or to load directly onto railway cars as space became available. Many of the farmers in this picture probably had travelled distances of 40 to 50 kilometres (25 to 30 miles), perhaps across the Qu'Appelle Valley, and over primitive roads to say the least. Often, they would have travelled in convoys, so as to help one another over the more difficult stretches. (Courtesy of the Saskatchewan Archives Board/R-B2969.)

of the rapid settlement of the West combined to produce a staggering record crop of 60 million bushels. By the time the shipping season closed on the Great Lakes, only one-third of the crop had been moved, while half remained in the farmers' hands for lack of storage facilities.[42] The rolling stock of the CPR had been shown to be utterly inadequate, while farmers found that the pressure on the railway made a mockery of their hopes to order cars for individual shipment. Indignation and frustration led directly to the formation of the Territorial Grain Growers' Association late that fall, and when Sifton came west towards the end of 1901 to prepare the ground for a by-election,[43] he was left in no doubt that the blockade was the political question of the day. At once he shifted his ground and developed a strategy to secure the farmers' votes.

First, when the CPR applied for increased capitalization, Sifton was behind the government's imposition of certain important conditions: some $8 million would have to be spent on locomotives and cars for the western trade, and $4 or $5 million for improving and doubling trackage west of the Lakehead, and increasing terminal elevator facilities.[44] Second, Sifton moved in May 1902 to amend The Manitoba Grain Act. The government accepted almost *verbatim* a resolution of the TGGA calling for compulsory supplying of railway cars without discrimination; the local railway agents were to keep an order book, and supply cars in the order in which applications were made, whether to elevators, flat warehouses, loading platforms, or otherwise. To avoid block bookings by elevators, it was stipulated that when insufficient cars were supplied to cover outstanding orders, they were to be distributed in rounds at the rate of one car per applicant in the order on the list. Limits on the construction of flat warehouses were lifted, and any person within forty miles of a shipping point could apply to erect one, the railway being compelled to provide land and a siding.[45] This was the very demand that Laurier, two years earlier, had dismissed as "socialistic." In 1903 The Manitoba Grain Act was completely recast to further entrench farmers' rights; among the additional changes was a provision for much larger loading platforms than in the past to accommodate more cars.[46]

A third part of the government's strategy was to go further than the grain men had ever intended in establishing permanent grades for grain by statute, and in making provisions to preserve the identity of individual farmers' grain. Each year it granted a little more to the farmers, finally consolidating the legislation in 1904 in The Grain Inspection Act.[47] The fourth part of the government's strategy—perhaps the most popular of all—was a decision to support a court case against the CPR for discriminating in the distribution of cars. The success of the Sintaluta test case, launched late in 1902, was a political triumph.[48]

The significance of the changes in the Dominion government's attitude was demonstrated early in 1903 when the grain dealers, who previously had been able to rely on Sifton, waited upon the Minister of the Interior to oppose the changes in The Manitoba Grain Act. The Act, they told Sifton, "is detrimental to the best interests of the country in restricting and interfering in trade and commerce, and is manifestly unfair to those in the grain trade, who have capital invested in elevators." It was nothing more than "class legislation." Complex as the question was, responded Sifton:

> when the grain producer comes to parliament, and says: "I have produced a commodity which is in universal demand, and I object to its going through the hands of middlemen who will take an undue toll on the product of my labor," then I want to say plainly, that this is a complaint and an objection that parliament is bound to recognize. And I tell you, moreover, gentlemen, that no parliament will ever be elected in Canada that can afford to disregard this protest.

These remarks, Sifton was later told, were worth several thousand votes to the Liberal Party.[49]

Significantly, Sifton directed the editor of the *Manitoba Free Press* to give a good deal of attention to farmers' movements:

> I have acquired in the last three years the opinion that the farmers['] vote in the west is going to be influenced along lines somewhat different to what we have been accustomed in the past. I think that with a reasonable amount of effort the Liberal party may be kept in line with the trend of thought amongst the farmers. What I mean is that such organizations as the Grain Growers' Association demanding emancipation from disabilities which are in themselves unjust and unfair ought to receive straightforward and emphatic support from the Liberal party.... .
>
> There are, no doubt, in the resolutions passed by some of the farmers['] meetings fantastic and impracticable propositions, but this always happens in such cases. Nevertheless, if the reasonable and practicable demands are met the others are generally lost sight of. In fact as a rule they are passed to placate cranks who are in the meetings.
>
> I think you should take strong grounds in favour of retaining and perfecting the privileges given to the farmers by the Grain Act, and resisting any kind of legislation which will ever again compel the farmers to ship through elevators against their will.[50]

It is difficult to believe that this was the same man who, four years earlier, had declared that the farmers' demands were utterly impossible, stirred up solely by self-seeking demagogues, and could be defused by skilful political manipulation of the meetings. The formation of the TGGA, and in 1903 of the Manitoba Grain Growers' Association, combined with the willingness of farmers to vote according to their perceived class interests, had convinced Sifton that the Liberal Party could no longer easily control "the simple-minded farmer." Now the party was to be "kept in line with the trend of thought amongst the farmers."

Sifton certainly was influenced by the fact that in 1900 the CPR had swung its support to the Conservative Party, and was thereafter regarded by the Liberal government as an enemy. The grain men could no longer compete in influence with the organized farmers. Despite his apparent conversion, however, Sifton never became a populist. He did believe that the western agrarian vote would play a significant role in the future, and that the Liberal government could capture that vote by responding to legitimate demands. It was the duty of the government, he said in 1902, "to put the farmer in as independent a position as possible, so far as the wheat buyer is concerned," to enable the farmers "to say to the grain buyers: We are independent and we can deal with it ourselves."[51] Between 1900 and 1904, then, there took place a revolution in the relationship between the Dominion government and the farmers of western Canada. The events described occurred in the context of, and contributed to, a growing western regional consciousness. The rapid expansion of the prairie West and the increasing importance of western grain in Canada's export economy lent a weight to farmers' demands which the government no longer could afford to ignore.

If any of the resulting legislation deserves to be called "an agrarian Magna Charta" it is indubitably The Manitoba Grain Act of 1903 (and not that of 1900) combined with The Grain Inspection Act of 1904. These Acts by no means solved all agrarian problems, but they were the first serious recognition by Ottawa of western agrarian rights. Neither did the Acts result in the demise of the elevator system as the grain men in 1903, and Sifton in 1899, had dolefully predicted. Indeed the number of flat warehouses declined;[52] but the number of loading platforms increased dramatically, more than 15 million bushels of wheat being shipped from them in 1908. They became "real competitors of the elevators," helping to protect the interests both of farmers who used them, and of those who did not.[53] This legislation, then, marked one of the first substantive victories for the organized farmers, and signaled the coming of age of the agrarian movement in western Canada.

NOTES

This article first appeared in *Prairie Forum* 4, no. 1 (1979): 105-20. An earlier version of this paper had been presented at the Canadian Historical Association in Saskatoon in June 1979.

1   Brian Robert McCutcheon, "The Economic and Social Structure of Political Agrarianism in Manitoba: 1870-1900," Ph.D. thesis, University of British Columbia, 1974, pp. 97-107, 111, and "The Birth of Agrarianism in the Prairie West," *Prairie Forum*, 1, 1976, pp. 79-94; *Brandon Blade*, 8, 22, 29 November 1883.

2   Library and Archives Canada (LAC), Canadian Pacific Railway Records, Van Horne letter-book 41, pp. 912-14, Sir William Van Home to W. F. Luxton, 28 November 1892; letterbook 49, pp. 859-63, 921-22, Van Horne to Charles Braithwaite, 3, 14, September 1895.

3   Constant popular pressure during the 1880s and 1890s forced the provincial government of Manitoba to negotiate with provincially aided railways not only lower freight
rates, but freedom from monopoly at shipping points. The Lake Manitoba Railway
and Canal Company (the forerunner of the Canadian Northern), for example, agreed
in its contract in 1896 to permit the loading of grain from farmers' vehicles and flat
warehouses. Archives of Manitoba (AM), Manitoba, Railway Commissioner, Mortgages, Leases, Agreements between the Railway Commissioner of the Province of
Manitoba and the Canadian Northern Railway Company Lines, 1896-1916, Indenture
dated 1 August 1896 between Lake Manitoba Railway and Canal Company and the
Government of Manitoba-Mortgage to secure bonds of Lake Manitoba Railway
and Canal Company; and see T. D. Regehr, *The Canadian Northern Railway: Pioneer
Road of the Northern Prairies 1895-1918* (Toronto: Macmillan, 1976), ch. I.

4   B. R. McCutcheon, Ph.D. thesis, op. cit., pp. 140-46, 181-99.

5   Ibid., pp. 278ff.; *Winnipeg Daily Tribune*, 17-19 January 1894.

6   LAC, Clifford Sifton Papers, vol. 238, pp. 35-36, Sifton to E. H. Macklin, 27 July 1900.

7   On the development of Liberal divisions in Winnipeg, see A. R. McCormack, "Arthur
Puttee and the Liberal Party: 1899-1904," *Canadian Historical Review*, LI, no. 2, June
1970, pp. 144-48.

8   On Oliver's early career, see W. S. Waddell, "The Honorable Frank Oliver," M.A.
thesis, University of Alberta, 1950; and on Douglas, see Gilbert Johnson, "James
Moffat Douglas," *Saskatchewan History*, VII, 1954, pp. 47-50.

9   *Tribune*, 15, 16, 17 September 1897.

10   Ibid., 19 October 1897.

11   Ibid., 20 October 1897.

12   Ibid., 12 January 1898. The total wheat crop for Manitoba in 1897 was 18.3 million
bushels, and for the North-West Territories was 2.5 million bushels.

13   Ibid., 3 November 1897.

14   Ibid., 17 September 1897.

15   See, for example, ibid., 10 December 1897, 8 January 1898.

16   Ibid., 29 January 1898.

17   Ibid., 11 February, 17, 18, 22 March 1898; LAC, Sifton Papers, vol. 226, pp. 199-200,
Sifton to A. McBride, 30 March 1898; Canada, House of Commons, *Debates*, 1898,
col. 450 (14 February 1898); col. 671 (17 February 1898); cols. 2059-83 (17 March 1898).

18   *Tribune*, 19 March 1898.

19   Ibid., 29 April 1898.

20   Ibid., 3 May 1898.

21   AM, J. J. Moncrieff Papers, file 4, Richardson to Moncrieff, 6 May 1898.

22   *Tribune*, 5, 6 May 1898.

23   Douglas's explanation is printed in full in ibid., 16 May 1898.

24   Ibid., 14, 16, 19 July 1898.

25   Ibid., 24 December 1898.

26   LAC, Sifton Papers, vol. 65, 47301-4, A. J. Magurn to Sifton, 22 April 1899.

27   *Tribune*, 9, 10, 12 May 1899.

28   Ibid., 19 May 1899; LAC, Sifton Papers, vol. 233, pp. 513-19, Sifton to G. D. Wilson,
13 July 1899.

29   *Tribune*, 3 June 1899.

30　Canada, Parliament, *Sessional Papers*, 1900, No. 81.
31　The government was inclined to use Royal Commissions for political ends, not to determine policy but to collect information, to educate electors and legislators, and to perform "the safety valve function for the explosive agrarian mentality" which then existed. Claims V. C. Fowke, "so sure was the Dominion government of what it wanted to be forced to do that it would entrust to no one but farmers the task of manning its early agricultural commissions." (Fowke, "Royal Commissions and Canadian Agricultural Policy," *Canadian Journal of Economics and Political Science,* xii, 1948, pp. 168-75.) The government wanted to *appear* to be sympathetic to the agricultural point of view, but it was careful in selecting farmers who were not supporters of government ownership, nor likely to come to conclusions which seriously conflicted with Sifton's views.
32　lac, Sifton Papers, vol. 234, pp. 737-38, Sifton to Magurn, 9 October 1899.
33　*Sessional Papers*, 1900, No. 81a, schedules D and E.
34　lac, Sifton Papers, vol. 236, pp. 781-83, Sifton to Sir Henri Joly, 7 March 1900.
35　63-64 V., c. 39 (Manitoba Grain Act). Under the general Railway Act, 1888, railway companies were prohibited from discriminating against any shipper, but the provision had never been strongly enforced with respect to the grain trade. See 51 V., c. 29, s. 240.
36　*Debates*, 1900, cols. 5757-5805 (21 May 1900); cols. 5809-26 (22 May 1900).
37　Ibid., cols. 6258-6308 (30 May 1900). The westerners included N. F. Davin (Assiniboia West), J. M. Douglas (Assiniboia East), F. Oliver (Alberta), A. Puttee (Winnipeg), R. L. Richardson (Lisgar), W. J. Roche (Marquette), and J. G. Rutherford (Macdonald). T. O. Davis (Saskatchewan) was paired, but indicated that he opposed the amendment. Absent were A. A. C. La-Riviere (Provencher) and J. A. MacDonnell (Selkirk), and of course Sifton (Brandon) who was in Europe.
38　*Tribune*, 31 May 1900.
39　lac, Sifton Papers, vol. 69, 51043-46, W. L. Parrish to Sifton, 2 March 1899; and reply, vol. 231, pp. 392-93, 14 March 1899. Until this time grain standards were established in each year by a Grain Standards Board, appointed by the Dominion government, and having representatives from central as well as western Canada. Since standards could not be established until a substantial part of the harvest was in, the situation led to instability for the purchaser as well as the producer of grain.
40　Ibid., vol. 237, pp. 793-94, Sifton to A. J. Magurn, 19 July 1900.
41　H. S. Patton, *Grain Growers' Cooperation in Western Canada* (Cambridge: Harvard University Press, 1928), p. 30.
42　L. A. Wood, *A History of Farmers' Movements in Canada* (Toronto: Ryerson, 1924), pp. 171-72.
43　He had been instrumental in unseating R. L. Richardson for corrupt practices, and was preparing to crush him in a by-election held in February 1902. For details, see D. J. Hall, "The Political Career of Clifford Sifton 1896-1905," Ph.D. thesis, University of Toronto, 1973, pp. 551-54, 769ff.
44　lac, Sifton Papers, vol. 247, pp. 233-37, Sifton to J. W. Dafoe, 21 January 1902; *Debates*, 1902, cols. 2938-40 (17 April 1902); cols. 4583-89 (10 May 1902); *Manitoba Free Press*, 27 January 1902; *Sessional Papers*, 1902, No. 48.
45　2 Edw. 7, c. 19 (1902), An Act to Amend the Manitoba Grain Act, 1900.

46 3 Edw. 7, c. 33 (1903).

47 4 Edw. 7, c. 15 (1904).

48 See L. A. Wood, op. cit., pp. 179-80; V. C. Fowke, *The National Policy and the Wheat Economy* (Toronto: University of Toronto Press, 1957), pp. 125, 158.

49 *Manitoba Free Press*, 14, 15 January 1903; LAC, Sifton Papers, vol. 251, pp. 371-73, Sifton to J. W. Dafoe, 14 March 1903; vol. 250, p. 595, Sifton to J. McLaren, 12 February 1903.

50 Ibid., vol. 251, pp. 371-73, Sifton to Dafoe, 19 March 1903.

51 *Debates*, 1902, cols. 4354-56, 4486-515 (7, 9, 10 May 1902).

52 H. S. Patton, op. cit., p. 30, fn. 2.

53 V. C. Fowke, *The National Policy and the Wheat Economy*, op. cit., p. 106, fn. 5. This was precisely what the *Tribune* had predicted when agitating for reforms prior to 1900.

16. "The Better Sense of the Farm Population":
The Partridge Plan and Grain Marketing in Saskatchewan

Robert Irwin

What was the ideology of the early western Canadian farm movement? The question has intrigued historians for a generation. It has special relevance because of the connection so often made between the farm movement and later political developments in the Prairies. In Saskatchewan, the crusade for government ownership of the elevator system provides an excellent background for discussion of farmer ideology. The most radical of the proposals for change, the Partridge Plan of 1908, attacked the entire grain-marketing system. Named after the originator of the scheme, Sintaluta farmer E. A. Partridge, the plan called for a radical realignment of the grain trade away from the principles of the competitive market towards an alliance of producers and government. In Saskatchewan, Partridge's proposals for a government-owned elevator system gained momentum in 1908 and 1909, but never reached fruition. Instead, the Saskatchewan Grain Growers' Association (SGGA) accepted the government-sponsored alternative—the farmer-owned Saskatchewan Co-operative Elevator Company (SCEC).

Historians writing of the farmers' attempt to reform the elevator system have suggested that farmers believed the elevator business was effectively controlled by monopolistic interests. According to this interpretation, the farmers sought fairness within the competitive market system. They believed that government ownership of facilities was one solution to the monopoly problem, but when the government presented viable alternatives, farmers quickly acquiesced.[1] Political scientist Duff Spafford even suggests that many farmer leaders never accepted government ownership at all and used it only as a threat to win reform from the government.[2]

The farm movement's shift from the Partridge Plan to the SCEC scheme, however, is more complex. Earlier work emphasized the homogeneity of agrarian society and accepted a monolithic farm movement. These writers consequently found the source of agrarian ideology in the common sense of economic dislocation, and reforms were interpreted as a basic drive to improve financial returns in the agricultural industry.[3] Some more recent work still accepts the homogeneous class perspective despite greater methodological sophistication.[4] A few writers, however, have challenged the concept.[5] Alvin Finkel, for example, has outlined ideological splits within the populist ideology of Social Credit.[6] Several writers have highlighted contrasting left-wing and right-wing tendencies within populism generally, and in one recent study four variations of populist ideology were delineated.[7]

Some works on the agrarian movement de-emphasize political activity. Taking their lead from Lawrence Goodwyn's study of Texas populists, Ian MacPherson and David Laycock emphasize the agrarian interest in the philosophy of cooperation. Both of these studies stress diversity within the organized farm movement, and competition for the support of everyday farmers. They have found that some farmers rejected the capitalist philosophy of competition promoting instead a utopian philosophy of cooperation.[8] These farmers defined society in organic terms, and believed cooperation was a method of achieving a new economic and social order. In some circumstances they associated the achievement of this goal with a combination of government ownership of sectors of the economy and cooperative enterprise.[9] Laycock and MacPherson identified another group within the farm movement. Laycock, using the terminology of W. L. Morton, called them crypto-liberals. This group accepted the capitalist philosophy, but recognized the need to protect themselves from corruption within the system. Cooperation in this philosophy became a defence mechanism against abuses of capitalism. This group, furthermore, did not accept an active role for the state in the marketplace.[10]

This perspective is helpful in reassessing the significance of the Partridge Plan. In fact, it represented much more than a scheme for nationalization of the elevator system. It was a panacea for grain handling, grain marketing, grain transportation, and farm credit problems. It reflected the ongoing enthusiasm of some farmers for a utopian form of cooperation. Although the plan contained several contradictions, in many respects its terms foreshadowed demands made by the Wheat Pool movement and changes implemented by the government Wheat Board. The acceptance of the Partridge Plan in 1908, and its rejection in 1911 demonstrate that a homogeneous ideological perspective based upon class position did not exist. At least two groups existed in Saskatchewan. One believed that the capitalist system could be changed through cooperatives; the other that cooperatives were simply a useful mechanism for operation within the system. A variety of factors including government manipulation, the report of the Saskatchewan royal commission, and the rise within the Saskatchewan Grain Growers' Association (SGGA) of a new leadership group, under-mined the hopes of the idealists in 1911. The struggle for influence in the SGGA during this period affords insight into the complex ideology of the farm movement, and into the difficulty of making significant radical changes in a regulatory environment.

In order to fully understand the implications of the Partridge Plan, some well-known facts must be reviewed. There is no doubt that the primary motivation behind the grain growers' demand for changes in the grain elevator system was improved profitability of farm operations. Furthermore, monopolies were a central issue for the farmers. They claimed that:

> the price of nearly every article which they consume has been artificially raised by combinations among the manufacturers or the dealers, while the price of their own product (grain) ... has been artificially reduced by a combination of the large milling and elevator interests.[11]

The first benefit the movement claimed would come from elevator reform was improved prices for grain.[12]

It is also important to remember that E. A. Partridge was one of the most remarkable figures in the western Canadian farm movement.[13] A splendid orator and tireless agitator, he was especially concerned about the marketing system. A visit to the Winnipeg Grain Exchange in January 1905 convinced him that a combine composed of the elevator companies, millers, and exporters controlled the exchange. He concluded that the corruption and speculation he witnessed at the exchange was damaging to farmers, and he worked throughout 1905 to convince farmers of the need for reform. At the 1906 convention of the SGGA, he delivered a scathing attack on the terminal market and the combine.[14] Farmers, he stated, were forced to sell their grain immediately after the fall harvest to obtain cash to meet their credit obligations. The elevator companies, millers, and the exporters, he argued, manipulated the market to ensure that prices remained low in this period. Having obtained farmers' grain at a bargain price, they would then contract to deliver grain to the English millers at a future date at a much higher price.

Partridge's accusations touched a nerve amongst farmers, and many found his arguments convincing. He won broad support when, on his own initiative, he made efforts to solve the problem. First he helped to organize the Grain Growers' Grain Company (GGGC) as a commission firm. This company, it was hoped, would provide competition for the local buyers and act as a watchdog on

Edward Alexander (E.A.) Partridge, circa 1910. One of fourteen children, Partridge was born on November 5, 1861, on a farm near the small village of Dalston, Ontario. He came west in 1883 and died in Victoria in 1931. (Courtesy of the Saskatchewan Archives Board/R-A15253.)

the grain exchange. Second, Partridge encouraged the federal government to participate in the terminal market as a terminal elevator operator. Partridge was successful in getting the grain growers' organizations to press his agenda forward, and the government caved in to the pressure of the farm lobby, appointing a royal commission in 1906. The Millar Commission (named for its dominant member, SGGA secretary John Millar) discovered abuses of the market by the grain dealers, but recommended that improved regulation of the industry, rather than government intervention, was the best solution to these problems.[15]

The farm community reacted with indignation to the Millar Commission report despite the popularity of its chair. Partridge became the champion of reform in the farm movement, and thus enjoyed a high stature in the grain growers' associations as the issue of grain elevators came to prominence. In 1907, primarily due to Partridge's influence, the SGGA convention considered another resolution calling for the nationalization of the terminal elevators. For the first time, Partridge focussed his attack on the initial delivery system as well. He linked problems at the local elevator to the grain-marketing issue, and demanded the provinces take a role in improving conditions:

It is proposed to construct and operate an elevator or elevators at every shipping point throughout the province and grade on a uniform plane. These elevators would be equipped with an up-to-date cleaner. At points equipped with elevators the owners should be given an opportunity to dispose of the elevators at a fair valuation. I think this plan should be put into effect by the local Government because the Dominion Government is too far away.[16]

Partridge's plan was not a coherent, structured proposal for reform. Rather, it was a collection of objectives with government elevators at its centre. The SGGA was leery of the proposal, but when presented in Manitoba it received a favourable response. The executive of the MGGA gave its conditional support to the concept and following the Manitoba election campaign in early 1907, the MGGA requested that the provincial government explore the viability of such a system.[17]

Partridge presented his ideas again at the 1908 SGGA convention. Resolution Fourteen, sponsored by Partridge, read:

> Resolved that this convention places itself on record as being strongly in favour of Dominion owned and operated terminal elevators and also a system of provincially owned and operated internal storage elevators at internal points where grain would be weighed and graded through a government agency with provision for a sample market in Winnipeg.[18]

The convention report termed it the "most important resolution" discussed. The disappointment in Saskatchewan following the release of the Millar Commission report, and the belief that the Winnipeg Grain Exchange had persecuted the GGGC unfairly added to the agitation at the convention. Although Partridge confused several issues in his presentation, a majority of SGGA members greeted the Partridge Plan with enthusiasm, and Resolution Fourteen was carried. In a vein similar to American populists identified by Lawrence Goodwyn, the Partridge Plan of 1908 proposed to deal with far more than grain-handling reform.[19] It addressed Partridge's concerns regarding grain handling, the grain blockade, farm credit problems, and the speculative market system. While Resolution Fourteen was intended to clean up the corruption in the internal elevator system, it also made provision for introducing better cleaning facilities and major changes to grading and inspection procedures. The resolution called for total reform of the grain trade through government ownership of the local elevators.

One of the major issues addressed by the Partridge Plan was the corrupt practices of the elevator companies such as charging excessive dockage, giving light weights, refusing to special bin grain or replacing specially binned grain with lesser quality grain, and refusing to allow farmers to deal with non-company buyers once they had purchased storage space in the elevator.[20] These problems all arose because the farmers had no alternative but to deal with the large grain-handling firms. By limiting access to storage space at the elevator, the grain-handling firms forced the farmers to sell their grain to the local market.

Given this "tyranny of the elevator monopolists,"[21] accusations of improper grading and maintaining an overly large spread between street and spot prices were inevitable. Unable to withhold their grain due to their responsibility to creditors, farmers were forced to accept the conditions imposed upon them by elevator agents. Partridge argued that the participation; of the government in the elevator business would reduce the problems caused corruption.

The monopoly issue and farmers' perception of corruption have been the primary focus of most accounts of the Partridge Plan.[22] The plan, however, did not stop there. Partridge believed that the impotence of street sellers, indeed of all grain sellers, in the market place would come to an end if the monopoly on grain purchases by exchange members was addressed. The control of the local elevators by the grain buyers was, he argued, the key to the monopolies which existed in the grain trade.[23] Under the existing system, the line elevator companies were both grain handlers and grain marketers. By refusing storage space to farmers, agents forced farmers to sell grain at low prices and substandard grades at the local elevator. The larger the spread the grain buyers maintained between the price for grain at the elevator and the price at the Winnipeg Grain Exchange, the larger their profits. By removing grain buying from the activities of the elevator agents, and turning grain elevators into handling and storage facilities, Partridge believed that these problems would disappear. To handle street wheat, he proposed to combine wagon lots of equal quality, advance up to 50 percent of the total value to the farmer, and then sell them as car lots with the price received being divided amongst the sellers.[24] In this sense the Partridge Plan was visionary. Its proposals for dealing with wagon lots of grain from farmers unable to produce a car lot resemble the practice later used by the Wheat Pools and the Wheat Board.

The Partridge Plan also visualized an end to the autumn grain blockade, and the competitive disadvantage of the massive fall selling which caused it. He believed that the blockade would be resolved if farmers were not forced to ship their grain immediately following the harvest. In his opinion, the major obstacle to any solution to the problem was the lack of adequate storage facilities on the prairies:

> If you build granaries on your farms you would have to finance yourselves. You will have to build adequate storage facilities to carry your wheat for well on into spring. Now are you going to do this or will you have the government do it? Which would be the cheaper way? You may have to leave it on your farms until well into the summer season. Why not provide the storage facilities by government intervention.[25]

The grain blockade would also be alleviated by changes government elevators would engender in the agricultural credit system. When discussing the elevator issue, farmers often referred to the problems they encountered when obtaining credit from the banks. Most often they referred to the banks' refusal to take grain stored in the initial elevator system as collateral. Section 86 of The Bank Act specified that only warehouse receipts or bills of lading could be accepted as collateral by a bank.[26] A farmer received a bill of lading when he shipped his grain to the terminal market and consigned it to a commission agent to be sold. A warehouse receipt was issued when a farmer placed his grain in general storage at either a terminal elevator or an initial storage facility provided that the warehouse operator and the farmer agreed upon the weight and grade of the grain. If the farmer desired to preserve the identity of his grain, on the other hand, he was said to have special binned the grain and a warehouse receipt was not issued; instead, he received a special bin receipt.[27]

The nature of the farmers' attack on the banks makes it difficult to determine if the terms of repayment were unreasonable, or if they felt the banks should take special binned grain as collateral, or if they felt the banks were refusing to accept a warehouse receipt issued by an initial elevator.[28] The rhetoric of the farm movement captured all of these concepts. The one issue on which the farmers and the banks agreed was that, under the current law, grain with an undetermined grade and weight was not accepted as collateral. Partridge believed that under a government storage system banks would accept stored grain as collateral for loans, thus allowing farmers to market their grain leisurely as prices warranted.[29] As their ideas for credit reform evolved, the SGGA became convinced that the government itself should become involved in grain marketing by advancing money to farmers on grain in storage.[30]

Finally, Partridge argued, a government-owned system would make it possible to establish a sample market thereby eliminating grading problems and the speculative market. Grain growers complained that the grading system did not correspond to the milling value of the grain.[31] Its emphasis upon colour and weight of the kernel was unfair to bleached and frosted grain. Moreover, the grading system promoted mixing at the terminal elevators.[32] A sample market, or a market where grain buyers purchase grain from the producer following a visual inspection of a sample of the grain for sale, eliminated the need for a grading system and provided for direct contact between the producer and the terminal buyers. At the same time, the speculative trade in grain futures would be eliminated. A grain exporter would purchase the grain on the basis of a sample when he was prepared to export it.

The Partridge Plan should thus be considered a farmers' panacea for problems in grain handling, grain transportation, credit, and especially grain marketing. In many respects it resembles the petit-bourgeois confusion over capitalism described by J. F. Conway.[33] It attacked the instruments of the capitalist grain trade, yet it did not completely reject the capitalist system. As the Interprovincial Council of Grain Growers' and Farmers' Associations (IPC)[34] attempted to build a coherent proposal out of Partridge's concepts, they summarized their objectives:

> The backbone of the grain combine would be broken. The general level of prices would be raised. The creation of a co-operative agency for the disposal of farmers' grain at cost would be made easy of accomplishment. A scientific classification of grain according to its intrinsic value or the requirements of the millers, by the operation of a sample market under the most favorable circumstances would be made possible. The creation of storage in the interior where weight and grade certificates could be obtained would permit the borrowing of money by the farmer to discharge his pressing liabilities at an early date, benefitting all who have business relations with him, except the grain dealer who formerly 'cinched' him.
>
> His ability to finance on the security of his grain would permit the farmer to market gradually, so that his offerings kept step with the milling and export demands, making the price received higher for the farmer, though not necessarily for the consumer, since the farmer would only obtain the benefit formerly absorbed by the speculator.[35]

Throughout 1909 farmers began to discover that the Partridge scheme came with problems. The premiers of the prairie provinces concluded that the Partridge Plan's emphasis on marketing made it too radical for easy implementation. They informed the IPC that major constitutional amendments would be necessary to put the Partridge Plan into effect.[36] In order to eliminate marketing from the internal elevator system, the provincial government would require new authority to control, regulate, and govern the storage of grain. Secondly, they required power to regulate weights and grades of grain. Finally, the provinces needed power over trade and commerce in grain since the plan had extraterritorial implications. Hence, the Partridge Plan required vast constitutional changes.

Grain growers in Saskatchewan and Manitoba, however, continued to voice their support for the plan despite the premiers' position. The proposals for modest changes in the operation of grain elevators made by the three prairie premiers in 1908 were "totally inadequate to safeguard the interests of the farmers in marketing their grain.... ."The IPC requested that "the government *acquire and operate the interior storage facilities along the lines previously stated."* [37] It acknowledged that there were limitations to provincial authority in the grain trade, but suggested that they were limited to the issue of monopoly and that the advantages of government ownership were too important to ignore. [38] Even if the Dominion refused to cooperate with the provinces, a government-owned system ensured farmers protection from corrupt practices such as inferior grading, light weights, excessive dockage, inadequate cleaning facilities, and the lack of special binning. Moreover, a sample market, regarded by farmers as a great improvement over the grading system, would be created by the increased special binning within government elevators. If the Dominion agreed to coordinate its activities with the provinces—the grain growers assured the premiers that it would—then simple amendments to the Inspection Act would make the elevator agents official weighmen and samplers. The agents would then be able to issue storage receipts with the grade and weight determined. As the banks demanded that these two factors be assessed before they would accept grain in storage as collateral, this amendment would remove a major obstacle to farm credit. Farmers able to borrow capital to meet their immediate requirements could market their grain leisurely as prices warranted alleviating the grain blockade. [39]

In a return to the days of the "patrons of industry," the rhetoric and tactics used by the grain growers and Partridge were confrontational. [40] The IPC claimed that their demand for elevator reform was "part of the world-wide protest of the WORKERS against the wrongs inflicted upon them by the SCHEMERS." [41] Partridge referred to a set of counter-proposals prepared by the premiers in 1908 as being in the "best tradition of diplomacy, statecraft, and the game of flim-flam." [42] He also advocated political action on the part of farmers. [43] He actively campaigned against the Scott government in 1908, and was so persistent that Scott informed Liberal party members within the SGGA that "such conduct is a poor way to win favour for the Grain Growers." [44]

A monolithic class acceptance of the Partridge Plan never materialized. Many farmers expressed concern that the proposals were too radical and that Partridge acted like a demagogue. One report to Scott suggested that only one-third of the SGGA opposed the Partridge Plan as early as 1909. [45] Indeed the convention's refusal to support Partridge's political tactics in 1909 suggested that his hold on the delegates was tenuous. By 1910 more reports were received that farmers were divided by the plan. [46] Party men in the association, such as Levi Thompson, A. G. Hawkes and George Langley, were finding their position difficult.

Throughout the negotiations of 1908 and 1909, the farmers' official proposals remained firm. In Manitoba, Premier Roblin, after first appearing to lack interest, acceded to the MGGA demands in time for a December 1909 election campaign.[47] In Saskatchewan, however, Scott, a Laurier Liberal with a philosophical opposition to government ownership of industry, refused to accept the SGGA proposals. Instead, he appointed a royal commission to make "a searching enquiry into the proposals looking to the creation and operation of a system of elevators to effect the objects outlined by the Grain Growers' Association."[48] He then carefully selected the members of the commission to ensure that they were not wedded to a scheme of government ownership.

Selecting a chair for the commission proved to be the most difficult task. Scott needed someone amiable to the concept of government assistance to the grain growers rather than government ownership, and at the same time acceptable to the farmers. He decided upon an academic and requested President Walter Murray at the University of Saskatchewan make a recommendation.[49] Murray's first choice, Professor Adam Shortt, was unavailable. Undeterred, Scott interviewed several other candidates and chose Dr. Robert Magill, a respected political economist, as the chairman.[50] Educated at Queen's University in Belfast, and at the University of Jena, Magill had no experience with the grain trade. A professor at Dalhousie University, Magill had conducted the Nova Scotia Royal Commission into the Hours of Labour 1908-09 (Eight-hour Day Commission), and had won universal acclaim for his handling of the affair. His work on labour organization in Britain demonstrated an understanding of the problems of working people, while subtly attacking the socialist perspective.[51] Most important, his positions on matters such as government ownership were largely unknown.

Once he had selected the chair, Scott found it much easier to fill out the commission membership. In response to SGGA demands that their organization have a majority of members on the commission, Scott found two men on the SGGA executive without a commitment to government ownership.[52] Rather than allow the SGGA to nominate members of the organization whose views the premier did not know, he had the names of Fred Green and George Langley submitted to the executive for their approval.[53] Scott's appointments were not pawns, and Green especially proved difficult on the commission.[54] Scott, nevertheless, was able to authorize a commission of enquiry sympathetic to his views against government ownership yet endorsed by the SGGA.

A critical moment in the campaign for government elevators occurred at the 1910 SGGA convention. Premier Scott had not yet appointed any members to the royal commission, and sent a key cabinet minister—former grain growers' president W. R. Motherwell—to discuss the matter with the SGGA.[55]

Delegates attending the Saskatchewan Grain Growers' Association convention of February 9-11, 1910, pose in front of Prince Albert City Hall. (Courtesy of the Saskatchewan Archives Board/R-B1595-2.)

Motherwell informed the convention that the commission would be composed of an academic, an elevator industry representative, two SGGA members and an independent farmer. It appeared that the SGGA might lose control of the elevator reform process. At this key moment, Partridge was absent from the convention due to a serious illness. George Langley and Fred Green took on the role of chief proponents of elevator reform and refused to accept Motherwell's proposal. Hence Langley and Green gained new respectability in the movement. Partridge's absence and the discovery of new champions of elevator reform proved key developments.

The SGGA, however, remained committed in principle to government elevators. At the first public hearing conducted by the Saskatchewan Royal Commission on Grain Elevators, 17-23 May 1910, at Moose Jaw, the SGGA presented its plan and its objectives in detail. The SGGA directors at their meeting 19 April 1910, decided to recommend that the government create a system of grading, storing, transportation, and marketing of grain by building government-owned elevators.[56] Their plan was a compilation of the ideas discussed during the 1909 negotiations and went far beyond mere public ownership and control of elevators.[57] It called for an elevator system which would stop malpractice within the industry and give farmers a 'fair' deal. The elevators would have the proper equipment for cleaning grain, as well as ample storage space for special binning. The operators, meanwhile, would be qualified to sample and weigh grain and provide graded and weighed storage certificates. On the basis of these certificates, the government-owned system would be secured to advance to the farmer 65 to 80 percent of the price of his grain upon delivery of wagon lots, and large-scale farmers would be able to secure bank loans on special binned grain. The wagon lots received at the elevator were to be combined on the judgement of the operator and stored for sale as car lots. The storage of grain at internal points instead of the terminals would allow farmers to sell grain directly to other farmers for seed purposes, to sell to the local millers, or to sell to the terminal market at their leisure. Such a system, the SGGA argued, would alleviate the grain blockade, provide opportunity to establish a sample market, and free farmers from the responsibility for damages or loss of grain during shipment.

Nevertheless, the position of the grain growers demonstrated that they did not fully understand or were unwilling to accept the implications of Partridge's proposals. The premiers had argued that the system demanded by Partridge entailed the establishment of a provincial monopoly in the elevator business. Given the radical changes in the marketing system demanded by Partridge, this was certainly the case. In Manitoba, the experience of the Manitoba Elevator Commission clearly demonstrated this point.[58] Yet the SGGA proposal to the royal commission suggested that the farmers' organizations were unsure of their position.[59] A monopoly was unnecessary, and thus, they concluded, any constitutional problems which a monopoly would produce were avoided:

> A system which offers such solid advantages over a private system, namely, security against fraud, opportunity to sell on sample, to raise money on grain before shipment without pledging it to dealers, to save screenings, and to give small farmers equal prices to those obtained by car lot shippers, and which can only be attacked by a method which must bring private owners under suspicion of making up losses by robbery will more than hold its own in a competitive struggle from the first.[60]

Such a position was not in line with Partridge's plan.

Completed in late October 1910, the *Report of the Royal Commission on Elevators* was a thorough investigation of the entire grain-handling industry. It not only criticized the SGGA proposal, but also largely substantiated the claims made by the premiers in 1909. Magill, Langley, and Green concluded that none of the proposals they had examined would ensure a successful solution to the elevator problem, and that the SGGA proposal and Manitoba Elevator System were especially inadequate. Both of these systems required that the elevators' income be derived solely from handling and storage charges. The evidence presented by the independent farmers' elevator companies, according to the commissioners, proved that this was unrealistic unless a total monopoly was achieved.[61]

The commissioners believed that the problems with corruption were being addressed by legislation as well. C. C. Castle, the warehouse commissioner, reported that many of the farmers' grievances could be remedied by utilizing the provisions of the Manitoba Grain Act.[62] For example, the act legislated against the provision of light weights by elevator agents, and by providing municipal scales, farmers could keep watch over the honesty of these operators. The act also provided for taking samples of the grain for inspection at Winnipeg to prevent undergrading or charging excessive dockage. Castle also pointed out that section 61 contained a method for preserving the identity of grain through samples to prevent substitution. Castle thus concluded that government ownership of elevators was unnecessary to prevent corruption.

The commission also found the SGGA proposals too broad to be carried out effectively.[63] Echoing the earlier objections of the three prairie premiers, the commission argued that the province should avoid involvement in the terminal elevator and marketing systems. The terminals were under the jurisdiction of the Dominion government, and any tampering by the provinces would raise constitutional questions. The commissioners also questioned the SGGA demand for a sample market where grain purchases were made on the basis of visual inspection rather than grading. Mixing grain was a requirement of a sample market, yet the SGGA argued that government ownership of the terminal elevators would end the practice of mixing grain.[64] Moreover, the commission, supported by T. A. Crerar of the GGGC, concluded that the exchange was a competitive market. "There is at all events the appearance of competition in the Exchange. If it is only appearance it is well affected."[65] The commissioners had also discovered during their visits to the American markets that a sample market could be developed without government elevators; therefore, if the SGGA desired this type of market, despite the evidence in the report, it could be created without the elevator system that they proposed.

Given the evidence, the commissioners concluded that a much less radical and risky plan would meet the farmers' expectations. Looking to the examples of the Saskatchewan creamery and telephone systems, they recommended that a farmers' cooperative elevator company be created.[66] They suggested that farmers take a financial interest in the company and that the elevators not only store and handle, but also buy and sell grain. Financed by low interest loans

Directors and officers of the Grain Growers' Grain Company, 1908-09. Standing (left to right): W. H. Bewell, M. C. McCuaig, David Railton, John Spencer, Robert Elsom. Seated (left to right): John Allan, A. M. Blackburn, I. T. Lennox, T. A. Crerar, D. K. Mills, John Kennedy, E. A. Partridge. (Courtesy of the Saskatchewan Archives Board/R-B7456.)

from the government, such a company would take advantage of the system as it existed, while providing farmers with an interest in its success. Organized on the principle of maximum local control consistent with central management, a cooperative organization would provide farmers with the necessary control of the grain-handling system.

The reaction to a synopsis of the commission report, released in the *Grain Growers' Guide*, appeared mixed. It demonstrated that a homogeneous class ideology did not exist in the organized farm movement. The *Guide* criticized the commission for leaving the burden of responsibility for the system on the farmers and was displeased with its dismissal of the Manitoba Elevator System.[67] One SGGA local was even more critical, claiming that the commission had caved in to the government.[68] The response from C. A. Dunning, a SGGA director first elected in 1910, was indicative of a new attitude in Saskatchewan, however. Dunning, although he considered a "co-op" solution premature, agreed that there were serious disadvantages to government ownership.[69] It was a significant break in class solidarity.

At the 1911 convention of the SGGA, the Whitewood delegates, because
their local association unanimously supported government-owned elevators,
moved "that in the opinion of this convention the finding of the Elevator
Commission is not in accordance with the expressed wishes of the farmers of
the province and that this convention is in favor of a system of government
owned interior elevators."[70] Partridge spoke in favour of the resolution. His
opposition to the commissioners' recommendations was rooted in his desire
for a grand scheme of economic and social change. Partridge argued that the
cooperative elevator system proposed by the commission failed to address
many of the issues. Specifically, grading and credit problems were left unsolved
by the scheme. The cooperative company made no provision for the creation
of a sample market nor for government advances on stored grain. These two
reforms, key to Partridge's original conception of government-owned elevators
as the means to total grain trade reform, were, in his opinion, plausible only
if the government owned the elevator network. Partridge did acknowledge,
however, that certain aspects of the report were favourable. He applauded the
commission's recommendation of an independent system of management for
the elevators and supported the concept of having farmers take a financial
interest in the elevators by raising 15 percent of the capital.[71] Partridge could
not influence the delegates, however. When the vote was called, a substantial
majority favoured the commission report.

No simple explanation can suggest why this happened. There were certainly
members of the SGGA who had never been totally committed to the Partridge
Plan.[72] Even the premier, well aware of the existence of these individuals, did
not quite understand the swing away from Partridge. He wrote:

> the reports coming from the convention were by no means
> cheerful for us. Langley was not at all sanguine. I daresay it was
> one of the very frequent cases where the kickers make all the
> noise and appear to be the whole show. Everybody was simply
> amazed when on taking the vote that night anywhere from five
> to one to ten to one stood up in favour of our scheme.[73]

Certainly the premier had gone out of his way to influence the debate at
the convention. On 2 February, four days before the SGGA convention began,
he introduced a bill in the Legislature to implement the commission's recommendations.[74] The *Guide* argued that the Scott government's introduction of the
elevator bill, while the convention debated the issue, influenced the delegates.
Many SGGA delegates, it argued, concluded that the "co-op" was possible while
government ownership was not. Rather than have the elevators remain in the

hands of the elevator tyranny, they pragmatically chose to support the reforms which were available.[75] Statements made during the debate suggest the *Guide's* hypothesis may have some merit. A delegate from Saltcoats stated that the government intended to hand over the elevator system to the farmers; to refuse to accept it was, in his opinion, "like looking a gift horse in the mouth."[76] Dr. T. Hill, a SGGA director and a supporter of government ownership, also urged the convention to accept the reforms which were available.[77]

Still, other factors must be taken into account. First, Scott's manipulation of the royal commission had paid off. Both Green and Langley had demonstrated their independence at the 1910 SGGA convention and were prominent members of the SGGA. Many delegates expressed faith in their abilities. G. H. McKague, for example, proclaimed, "Mr. Langley and Mr. Green are the best friends the farmers of this province ever had. I recommend, gentlemen, that we support the commission because we appointed them in honour."[78] Second, the grain growers admired cooperative ideals. The *Guide*, by explaining the merits of cooperation and promoting the GGGC as a shining example of its success, contributed to farmers' acceptance of a cooperative solution. Moreover, the provincial cooperative creamery program, similar in many respects to the elevator proposal, was a respected system within the farm movement.[79] It is also important to make reference to a new leadership group which emerged at the 1911 convention. The men whom L. D. Courville called the "co-op" elite—men such as J. A. Maharg, C. A. Dunning, A. G. Hawkes, and J. B. Musselman—obtained key positions on the SGGA directorate.[80] From their new positions of strength, they could promote their commitment to cooperatives as a tool for reforming the system rather than transforming it.[81] And finally, the cooperative scheme appeared to ensure farmers that they would receive a fair deal at the local elevator. All of these factors played a role in the final decision to support the commission report.

With the support of the SGGA for the elevator bill now assured, Scott moved forward with the legislation. The bill calling for the creation of the Saskatchewan Co-operative Elevator Company (SCEC) received third reading in the provincial Legislature on 2 March 1911.[82] Opposition member and SGGA director, F. C. Tate, although he had joined the SGGA convention in its approval of the royal commission report, moved to block the legislation on the grounds that the cooperative scheme did not provide adequate relief from grain-marketing problems.[83] Following the longest debate in the history of the Legislature,[84] the legislation passed on 14 March 1911. The "Co-op," as the company would be known, with the SGGA executive acting as provisional directorate, was created. The SCEC was chartered as an elevator company and given the authority to act within the existing system.[85] The Co-op, however, did not change the existing marketing system. In fact, it became quite adept at maneuvering within the system and was often accused of being no different than the line companies with which it competed.[86]

Partridge and elements of the farm movement never truly accepted the scec. Following the 1911 convention, he argued that the scec was simply competition for the gggc; it solved none of the major marketing problems. "Perhaps ... the better sense of the farm population," he wrote, "will revert to the original idea of government ownership of storage facilities as being an essential part of the wider program for the establishment of an ideal market at Winnipeg."[87] His ninety-minute speech at the 1911 convention had led one delegate to remark, "Mr. Partridge has spoken more of selling wheat than the elevator question."[88] The comment was intended to be derogatory. Yet, it better than anything else sums up E. A. Partridge's objectives. The Partridge Plan called for government ownership of the elevator system to break the power of a perceived combine in order to ensure a "fair deal" for farmers. But it was more. His plan was a rejection of the competitive market system in the grain trade. It called for a realignment of the grain trade along the lines of a government-producer alliance. He had made reform of the elevator system into a panacea for all the major grain-marketing problems.

Contradictions in his plan existed. The demand for a sample market for example was not compatible with either the desire to combine the loads of street sellers or the opposition to mixing. The plan was also risky. It required the government to become involved in advancing money to farmers on grain in storage and it was based on weak financial prospects for the operation of elevators as storage and handling facilities. But it was visionary. Indeed, in many respects his plan called for changes which would be adopted later in the wheat pool movement and the wheat board.

The emphasis upon government ownership versus government assistance has caused historians to overlook these key concepts in Partridge's proposal. Moreover, the belief in a monolithic class ideology kept historians from scrutinizing the competing philosophies within the farm movement. By 1909 at least two distinctive ideologies can be discerned. The first perceived the capitalist market system as flawed and sought to trans-form it through a major overhaul of the internal elevator network. The second sought to improve the existing system rather than realign the system itself. Examined through this perspective, the Scott government's ability to get the sgga to accept the scec was a dramatic victory for a crypto-liberal leadership group and a defeat for Partridge and his utopian cooperative ideals. The decision to support the cooperative scheme represented a remarkable shift in the sgga. E. A. Partridge, and those who supported him, such as William Noble and

F. C. Tate, lost influence in the organization throughout the 1910-11 period. A new group had risen to dominate the meetings. Key individuals in this group were J. A. Maharg, C. A. Dunning, George Langley, J. B. Musselman, and A. G. Hawkes. These new leaders rejected the confrontational tactics of the utopian cooperators. They represent the group which Courville called the "Co-op" elite, and the decision to accept the SCEC marked their ascension in the ranks of the grain growers.

NOTES

This article first appeared in *Prairie Forum* 18, no. 1 (1993): 35-52. The author acknowledged Dr. W. A. Waiser of the University of Saskatchewan under whose supervision much of the original research for this paper was undertaken, and Dr. Paul Voisey and Dr. David J. Hall of the University of Alberta who read and commented on earlier versions of this paper.

1   Two early studies which set the tone for future interpretation are W. A. Mackintosh, *Agricultural Cooperation in Western Canada* (Toronto: Ryerson, 1924), 34-35 and H. S. Patton, *Grain Growers' Cooperation in Western Canada* (Cambridge, MA: Harvard University Press, 1928), 80-81. Vernon Fowke; *The National Policy and the Wheat Economy* (1957; Toronto: University of Toronto Press, 1983), chapter 7, flushed out this interpretation. More recent works dealing with the elevator issue in the context of government ownership versus government assistance include: J. W. G. Brennan, "A Political History of Saskatchewan, 1905-29" (Ph.D. dissertation, University of Alberta, 1976); K. Murray Knuttila, "The Impact of the Western Canadian Agrarian Movement on the Federal Government, 1900-1930" (Ph.D. dissertation, University of Toronto, 1982); Knuttila, "E. A. Partridge: The Farmers' Intellectual," *Prairie Forum* 14, no. 1 (Spring 1989): 59-74, especially p. 63; and John Everitt, "A Tragic Muddle and a Cooperative Success: An Account of Two Elevator Experiments in Manitoba, 1906-28," *Manitoba History* 18 (1989): 12-24.

2   D. S. Spafford, "The Elevator Issue, the Organized Fainters and the Government, 1908-11," *Saskatchewan History*, 15 (1962), 92.

3   C. B. Macpherson, *Democracy in Alberta* (Toronto: University of Toronto Press, 1951) is the key Canadian study in this regard. For his influence on the interpretation of the elevator issue see Fowke, *The National Policy*, 122-24.

4   J. F. Conway, "The Nature of Populism: A Clarification," *Studies in Political Economy*, no. 13 (Spring 1984): 137-44; Conway, "The Prairie Populist Resistance to the National Policy: Some Reconsiderations," *Journal of Canadian Studies* 14, no. 3 (1979): 77-91; Conway, "Populism in the United States, Russia, and Canada: Explaining the Roots of Canada's Third Parties," *Canadian Journal of Political Science* 11, no. 1, (1978): 99-124; Peter Sinclair, "Class Structure and Populist Protest: The Case of Western Canada," *Canadian Journal of Sociology* 1 (1975): 115; and Jeffery Taylor, "The Language of Agrarianism in Manitoba, 1890-1925," *Labour/Le Travail* 23 (1989): 91-118.

5   L. D. Courville, "The Saskatchewan Progressives" (M.A. Thesis, University of Sas-
    katchewan, Regina Campus, 1971); Courville, "The Conservatism of the Saskatchewan
    Progressives," *CHA Historical Papers/Communications Historique*, (1974): 157-181.
6   Alvin Finkel, *The Social Credit Phenomenon in Alberta* (Toronto: University of Toronto
    Press, 1989).
7   David Laycock, *Populism and Democratic Thought in the Canadian Prairies, 1910 to
    1945* (Toronto: University of Toronto Press, 1990).
8   See especially Ian MacPherson, *Each for All: A History of the Cooperative Movement
    in English Canada 1900-45* (Toronto: Macmillan, 1979), chapter 13, especially pp. 13
    and 467; and David Laycock, *Prairie Populists and the Idea of Cooperation 1910-45*
    (Saskatoon: Centre for the Study of Co-operatives, University of Saskatchewan,
    1985), 11 and 20-28. Laycock breaks this group into two sections which he refers to
    as radical democratic populists and social democratic populists. The distinguishing
    characteristics revolve around the use of the state in the economy and group govern-
    ment.
9   Laycock, *Prairie Populists*, 24; Laycock, *Populism and Democratic Thought*, 290; and
    MacPherson, *Each for All*, 46.
10 Laycock, *Prairie Populists*, 7.
11 Saskatchewan Archives Board (hereafter SAB), Interprovincial Council of Grain
    Growers' and Farmers' Associations (hereafter IPC), Pamphlet G361.1, A Provincial
    Elevator System, p. 1.
12 Ibid., 5.
13 Knuttila, "E. A. Partridge," is the best examination of Partridge's philosophy and the
    impact of his tragic life on the farm movement.
14 SAB, Saskatchewan Grain Growers Association, Annual Convention Reports (hereafter
    SGGA Reports), 1906, 7-16.
15 Canada, *Sessional Papers*, 1908, no. 59, "Report of the Royal Commission on the Grain
    Trade of Canada (1906)," p. 18.
16 SAB, SGGA Reports, 1907, 39.
17 SAB, IPC, Pamphlet G361.2, "The Struggle for Government Elevators with Reasons
    Therefor," 1. The difference in reception for Partridge in Manitoba and Saskatchewan
    can be explained by looking at their stage of development. Obtaining an elevator was
    the most important issue to many Saskatchewan farmers, while Manitoba farmers
    were more concerned with making the system work. The MGGA decision was also
    influenced by the opposition to the GGGC at the Winnipeg Grain Exchange.
18 SAB, SGGA Reports, 1908, 18.
19 The best synopsis of the Partridge Plan is found in the Archives of Manitoba (AM),
    United Farmers of Manitoba, MGGA Convention Reports (hereafter MGGA Reports),
    1908, Appendix. See Lawrence Goodwyn, *Democratic Promise: The Populist Move-
    ment in America* (New York: Oxford University Press, 1976), Appendix "The Omaha
    Platform" for a review of American populist ideology.
20 SAB, IPC, G361.1, "A Provincial Elevator System," 2.
21 SAB, Walter Scott Papers (WSP), IV.87, F. M. Gates, SGGA vice-president to Scott, 20
    April 1908, 40111.
22 For example, Fowke, *The National Policy*, 121-25 and 138-40.
23 SAB, MGGA Pamphlet, G315.1, p. 1.

24   SAB, IPC, G361.1, "A Provincial Elevator System," 7.

25   SAB, SGGA Reports, 1907, 41.

26   The Bank Act, *Revised Statues of Canada*, 1906, c. 29, s. 86.

27   The Manitoba Grain Act, *Statutes of Canada*, 1900, c. 39, s. 19 and 34.

28   *Report of the Saskatchewan Royal Commission on Elevators* (Regina: n.p., 1910), 20; and SAB, WSP, IV.87, Gates to Scott, 20 April 1908, 40111-40115. Gates wrote that the plan would "induce the banks to treat the farmer in a more liberal manner in the matter of advances."

29   Partridge, like many SGGA members, often confused the inability of banks to loan money on grain in storage under the Bank Act with reluctance.

30   *Report of the Elevator*, 24-26.

31   Ibid., 20.

32   Mixing meant that grain of lower grades was combined with grain from higher grades to produce an overall sample which was at the lowest possible level of the highest grade.

33   Conway, "The Nature of Populism," 137-38.

34   The IPC was formed by the SGGA and its sister organizations in Manitoba and Alberta in 1908 specifically for the purpose of lobbying the provincial governments for government elevators. The IPC was replaced in 1912 by the Canadian Council of Agriculture, SAB, WSP, IV.87, M. D. Geddes (secretary IPC) to Scott, 9 March 1908, 40102.

35   SAB, IPC, G361.1, "A Provincial Elevator System," 8.

36   SAB, WSP, IV.87, Premiers' reply to the IPC, January 1909, 40191-8.

37   Ibid., Minutes of the Interprovincial Council Meeting, 19 May 1908, 40121.

38   Ibid., Roderick McKenzie to Scott, 25 February 1909, 40212.

39   Ibid., 40124.

40   Taylor, "Language of Agrarianism," 104-06.

41   SAB, IPC, G361.1, "A Provincial Elevator System," 1.

42   *Grain Growers' Guide* (GGG), June 1908, 10.

43   SAB, SGGA Reports, 1909, 14-16.

44   SAB, WSP, IV.87, Scott to Langley, 27 November 1908, 40177.

45   Ibid., IV.86, Frank Moffat to Scott, 20 February 1909, 39836; T. M. Bryce to Scott, 18 February 1909, 40208; Levi Thompson to Scott, 22 November 1909, 39878. Scott had also been informed in confidence by some SGGA executive members that other schemes besides the Partridge Plan could be considered. Scott to Roblin, 13 December 1909, 40252.

46   SAB, W. R. Motherwell Papers, T. Storrar to Motherwell, 17 January 1910, 7133; and A. H. Shaw to Motherwell, 14 February 1910, 7142.

47   SAB, WSP, IV.86, Scott to Bulyea, 20 December 1909, 39939. Scott believed that Roblin had astutely saved the announcement for the campaign.

48   *Report of the Elevator*, 10

49   Barry Ferguson, "The New Political Economy and Canadian Liberal Democratic Thought: Queen's University 1890-1925" (Ph.D. Dissertation, York University, 1982), 404-14. Ferguson's work suggests an academic would fulfill Scott's expectations admirably. Adam Shortt, for example, thought monopolies were the inevitable outcome of efficient production, and that the state should be used to alleviate the problems

of monopoly rather than break them up or become involved as a producer. See pp. 157-63 and 170-72.

50 University of Saskatchewan president Walter Murray had recommended several men for the position of chairman. After Shore, he believed Magill best suited Scott's needs. SAB, WSP, IV .88, Murray to Scott, 6 January 1910, 40386. Magill was concerned about his lack of experience with the grain trade. Ibid., Magill to Scott, 16 January 1911, 40558.

51 R. Magill, "Trades Unionism in Britain." *Queen's Quarterly* 14 (July 1906): 4-15.

52 SAB, W. R. Motherwell Papers, II.55, Motherwell to C. Lunn, 8 January 1910, 7131; and WSP, IV.86, Scott to Bulyea, 29 December 1909, 39938. The process of stacking the royal commission is discussed at length in Brennan, "A Political History of Saskatchewan," chapter 1.

53 SAB, WSP, IV.88, Scott to Langley, 28 February 1910, 40451.

54 SAB, WSP, IV.88, Magill to Scott, 7 March 1911, and Scott to Magill, 18 March 1911.

55 SAB, SGGA Reports, 1910, 17.

56 *GGG,* 4 May 1910, 20.

57 *Report of the Elevator,* 24-26.

58 D. M. McCuaig, chairman of the MEC, cited in Everitt, "A Tragic Muddle," 18.

59 The legal counsel for the IPC concluded that the "constitutional difficulty [Scott] set forth only applies to matters outside and unnecessary for carrying out our requests." SAB, WSP, IV.87, McKenzie (secretary IPC) to Scott, 25 February 1909, 40212.

60 *Report of the Elevator,* 34.

61 Ibid., 40. SAB, WSP, IV.86, Survey of Farmers Elevators 1908, 39801-39821.

62 *Report of the Elevator,* 75-76.

63 Ibid., 26-30.

64 Ibid., 30. The grading system allowed for certain variations in the quality within each grade. Mixing grain was the practice by which grain of a lower grade was mixed into high quality grain to reduce the overall quality of the grain to the minimum standard of the higher grade. In this manner a terminal could increase the total amount of high grade grain it exported, A sample market required an exporter to purchase grain in small quantities on sample and then mix these small samples together.

65 Ibid., 65.

66 Ibid., 96.

67 *GGG,* editorial, 30 November 1910, p. 2.

68 *GGG,* 4 January 1911, p. 12.

69 *GGG,* Dunning to editor, 21 December 1910, p. 11.

70 SAB, SGGA Reports, 1911, p. 19.

71 Ibid., 20.

72 Spafford, "The Elevator Issue," 92. Spafford argues that the SGGA leaders used public ownership as a club to hold over the government's head. He states that while a few directors such as E. A. Partridge were deeply committed to nationalization of utilities, most SGGA directors used it as a threat. While this hypothesis certainly must be considered, it trivializes the deeply held beliefs of many farm leaders. The work which the IPC and the SGGA put into the preparation of the Partridge Plan suggests that it was more than a simple threat. Moreover, MGGA president McCuaig defended the system won by his association with determination and heart. Given the close

coordination of activities between the SGGA and the MGGA, it is hard to believe that many Saskatchewan farmers were not as deeply committed to public ownership. Indeed, while Spafford's explanation may be relevant to the question of elevators, the Partridge Plan was far more than just an elevator proposal.

73 SAB, WSP, IV.88, Scott to Magill, 23 February 1911, 40569.
74 *Saskatchewan Legislative Assembly Journals 1910-1911.* The bill was introduced on 2 February 1911 and commenced second reading on 7 February 1911, see pp. 49 and 66.
75 *GGG*, editorial, 15 February 1911.
76 SAB, SGGA Reports, 1911, 28.
77 Ibid., 29.
78 Ibid., 22.
79 *Canadian Annual Review*, 1911, 561. The Co-operative Creamery system boomed during this period. The number of cream suppliers in the program rose from 553 in 1908 to 1596 in 1911.
80 Courville, "Saskatchewan Progressives," 107.
81 Laycock, *Populism and Democratic Thought*, 61.
82 *Saskatchewan Legislative Assembly Journals 1910-1911*, 49. The name of the company in the original bill was the Grain Grower's Elevator Company of Saskatchewan. It was changed to Saskatchewan Co-operative Elevator Company before the final reading to avoid confusion with the GGGC.
83 *Saskatchewan Legislative Assembly Journals 1910-1911*, 80; *Canadian Annual Review*, 1911, 569. Haultain also maintained his position in favour of government elevators.
84 SAB, WSP, IV. 88. Scott to Magill, 18 March 1911, 40583; *Canadian Annual Review*, 1911, 569.
85 *Statutes of Saskatchewan, 1910-1911*, c. 39, The Saskatchewan Elevator Act, sec. 2.
86 Robert Irwin, "Farmers' Expectations and the Saskatchewan Co-operative Elevator Company" (M.A. thesis, University of Saskatchewan, 1988), chapters 4 and 5.
87 *GGG*, "Difficulties Multiplied," 8 March 1911, p. 32.
88 SAB, SGGA Reports, 1911, p. 22.

17. Farmers and "Orderly Marketing": The Making of the Canadian Wheat Board

Robert Irwin

INTRODUCTION

C ontroversial is the best word to describe the Canadian grain trade and the role of the Canadian Wheat Board within that trade. Complaints by farmers led to the establishment of several federal Royal Commissions and numerous provincial enquiries.[1] Given the grain trade's important economic place in Canada, historians, economists, and publicists have also written extensively on this topic. The recent attacks by some Canadian farmers on the mandatory pooling structure of the Canadian Wheat Board have similarly led to numerous new publications.[2] Few of these works, however, systematically address the process by which Canada adopted single-desk marketing. Those that do offer a simplistic interpretation which fails to address the economic complexities of the grain trade or focus extensively on the political aspirations of the farm movement.[3] This article examines the origins of the Wheat Board in the context of the economic structure of the Canadian grain trade and the perceptions of Canadian farmers. Many farmers demanded a compulsory Wheat Board because they believed it would improve their returns and solve long-standing problems they had regarding the economic structure of the grain trade. They were neither anti-capitalist nor politically motivated. They simply sought to order the capitalist market place in an international context.

This present analysis is divided into three parts. Part one examines the Canadian grain trade and discusses the nature of farmers' complaints against the system. Grain marketing was composed of four components: grain handling, grain transportation, grain inspection, and grain marketing. Each of these components, while nominally distinct, functioned in combination to determine the price a farmer received for grain, and affected a farmers' ability to sell grain in the open market. This information is essential to an understanding of the complexity of the problems farmers faced.

In part two, the article analyses various efforts to improve the market system, considering both the solutions envisaged by the farm movement and by the government. The article demonstrates that the farm movement made efforts to regulate and reform all the various components of the grain trade system and eventually concluded that a pool was a necessary aspect of grain marketing.

In part three, the emergence of the Wheat Board is examined. The crisis of the Depression led the government to experiment in different avenues of wheat marketing, first as an effort to improve farm returns and subsidize Canadian agriculture, and later as a method of organizing the marketing of Canadian wheat. World War II provided the impetus for the government to act on issues and concerns already identified.

In their recent examination of the Wheat Board, historian David Bercuson and political scientist Barry Cooper argued that the government's desire to control the prairie grain crop and fight inflation during World War II provided the rationale for making the Wheat Board a compulsory pool. They maintain that the dual marketing system established in 1935 answered farmers' concerns about the grain trade and did not require modification. In an even more gratuitous attack on the Wheat Board, publicist Don Barron has concluded that the Wheat Board was foisted upon western Canadian grain farmers by left-wing, anti-capitalist propagandists within the Social Gospel movement. Neither of these two interpretations takes into consideration the structure of the Canadian grain trade nor the long-standing grievances of many prairie farmers as expressed through the farm organizations of the period. The campaign for a compulsory pool originated in a rational critique of the prairie grain marketing system as farmers sought to improve the return they received for their grain. Their complaints focused on all elements of the system including grain handling, grain transportation, grain marketing, grain inspection and the terminal market; addressed areas of concern related to corruption in the grain buying and handling business, and perceived problems related to concentration and monopoly in the grain trade.

Farmers, unlike other capitalist enterprises, could not respond to falling prices by withholding produce, reducing supply, and thus influencing the supply/demand relationship. They made seeding decisions in April and the crop had to be sold in September-November to meet credit commitments and avoid carrying charges caused by the close of navigation on the Great Lakes. The farm organizations were especially concerned about the spread between street and track prices for grain. Street sellers—the approximately 50 percent of farmers who sold grain in small wagon lots to the local elevator agent—received a lower price for their grain than other sellers in the market. These farmers, furthermore, delivered large volumes of grain in the fall when supply overwhelmed demand in the market. The farm movement thus demanded "orderly marketing" whereby the supply/demand relationship could be stabilized. Both regulatory measures and cooperative grain handling and marketing agencies failed to establish "orderly marketing." Indeed, the experiments demonstrated that "orderly marketing" required involvement in all aspects of the grain trade including grain handling, marketing, transportation, and terminals. The experiments in government marketing during and immediately following World War I suggested a new direction might be of benefit to those farmers unable to take advantage of the track market. During the Depression the government established a dual market whereby a voluntary Wheat Board co-existed with the open market. This, however, was a price support system, rather than a system of "orderly marketing," destined to last only until conditions on the international wheat market returned to normal.

Rather than normal conditions, however, the Liberal party was forced to deal with World War II and the complete disruption of international commerce. More than ever, the government needed to foster an "orderly market" to meet its international commitments to Britain. A subsidy system no longer met the requirements of the federal government, and a compulsory pool was established in 1943. Following the war, the government had to assess its position. If the government did not desire to re-establish a price support system, it had to abandon the Wheat Board and restore the competitive market or continue the compulsory pool. The wheat pool proponents, including the most active of the farmers' organizations, lobbied for the continuation of orderly marketing; and faced with continued international commitments and unable to finance a farm subsidy program, the Liberal party reluctantly continued the Wheat Board.

THE PRE-WHEAT BOARD GRAIN TRADE AND FARMERS COMPLAINTS
The Open Market System

The grain trade was a complex system of interdependent components in the grain handling, grain transportation, grain inspection, grain marketing, and export sectors. Each of these sectors influenced the price a farmer received for grain. Grain was purchased by weight, subject to quality and dockage. The owner of the grain, moreover, was responsible for all charges assessed as grain passed through each of the sectors. Until delivered to the terminal market, the owner was responsible for any damage or shrinkage of the grain shipment. Final payment and the transfer of the grain from the Canadian grain trade to the export market occurred at the terminal market. This transaction occurred at the Winnipeg Grain Exchange, and until that time, the owner of the grain was responsible for carrying charges. The price of grain at the terminal elevator prepared for shipment to exporters was the "spot" price. Consequently, grain which arrived at the Lakehead terminal market and could be shipped immediately received a higher price than grain which had to be carried by the owner until the navigation system opened in the spring. Besides selling actual grain, the organization of the Winnipeg Grain and Produce Clearing Association in 1904 provided for the trading of future delivery contracts on margin. These contracts speculated upon the price of grain, ready for shipment at the terminal market, at a certain date in the future. The contracts were for October, December, May or July delivery, in 1,000-bushel allotments of a certain contract grade. Futures contracts, at one and the same time, followed trends in the market and projected prices into the future. Although futures contracts could be speculative, they also allowed grain owners to hedge their supply and thereby helped to stabilize the market place by reducing the risk to the owner of the grain.[4] The price a farmer received, therefore, depended upon the grain's position within the trade, the timing of the sale, and the price on the futures market.

The first opportunity for farmers to sell their grain came at the local delivery point. Farmers could sell grain to the local elevator in wagon lots for the "street price."[5] This price was the lowest available to farmers and reflected the transfer of all risk for the product from the farmers to the purchasers at the earliest possible moment. In this sale, the quantity, quality, and dockage of the grain shipment had to be agreed upon between the buyer and the seller. The buyer, furthermore, had to make allowances for carrying charges, insurance, and shrinkage. To protect themselves from the risks they assumed in a street purchase, the buyers hedged the grain by selling a futures contract. In this manner, any change in price between the time of purchase at the local delivery market and final sale at the terminal market was foregone. At the local delivery point, farmers could also sell their grain for the "track price." The track price was paid for wheat loaded into a rail car and awaiting transportation to the market. Track sellers maintained risk related to quality and weight until the car lot reached the terminal market for official inspection. Track sellers were also responsible for inspection and weighing charges, cleaning charges, freight (including freight on dockage), commission on sale, and interest on any advance made prior to sale at the terminal market. Finally, track sellers were also responsible for any handling and storage costs incurred getting the grain into the rail car prior to the track sale.

Farmers able to sell track grain had other avenues of sale available to them. If the price offered for the track delivery was too low, or if farmers believed a better price could be obtained at a later date, they shipped the grain to inspection points themselves. It could be sold following inspection or delivered to the terminal elevator, where it could be cleaned and treated for sale at the spot price. The sale required farmers to purchase the services of a member of the Grain Exchange on commission of one cent per bushel. In this manner a farmer assumed the risk of price changes during the shipment period and paid any charges incurred. This mechanism of sale improved the bargaining position of the track seller. A farmer delivering grain in this manner could hedge his delivery, like the grain buyers, by selling a futures contract, and then buying it back upon the sale of his grain.

Complaints by Farmers

Corruption

Farmers complained vigorously about corruption in the grain trade. Excess dockage, light weights, improper grades, refusing to special bin grain or replacing special binned grain with lesser quality grain, replacing grain in storage with grain of inferior quality, and refusing to allow farmers to deal with alternative buyers once a farmer placed grain in storage at an elevator were amongst the more serious charges.[6] They also complained about the corruption of railway agents in the disbursement of rail cars during the peak marketing system.[7] Primarily these complaints were made because of prairie farmers' undue reliance upon the local grain handling facility as a primary market for their produce. They felt that they were forced into the hands of the "tyranny of the elevator monopolists."[8]

Monopoly

The charge that a monopoly controlled the grain trade in Canada is somewhat misleading: the grain trade contained a number of corporate enterprises. In reality, farmers were more concerned about combinations to restrict competition in the grain trade. In this regard, they noted the attitude of the Canadian Pacific Railway in signing restrictive deals on car disbursement with elevator operators and, more importantly, to the activities of the North West Grain Dealers' Association in Manitoba and Saskatchewan and its counterpart, the Western Grain Dealers' and Millers' Association in Alberta. The price of grain changed daily at the Winnipeg Grain Exchange and in order to communicate this information to grain buyers at local delivery points in the countryside, these associations—the membership included virtually all of the corporations involved in grain handling and grain buying at local delivery points—sent notifications of price offerings for track and street purchases to their local buyers. Usually one telegram was sent to the delivery point, and the agent would distribute the information to his competitors. All grain buyers at a local delivery point, consequently, were instructed to offer the same price for the grain.[9] Few farmers complained about the track prices offered. These prices, set daily on the basis of the closing spot price in Winnipeg, reflected the producers' ability to deliver grain to the spot market themselves if they were unsatisfied with the track price offered. Farmers, however, believed that the street price was unduly influenced by this combination of the grain buying interests.

Marketing

The railways, a second important component of the grain trade, also influenced the price a farmer received for grain. All grain in Canada moved to the terminal market on the railway. The closing of navigation at Thunder Bay in December placed a tremendous burden upon the railway system: any grain not reaching Thunder Bay in time for shipment had to be carried over the winter and thus had extra carrying charges placed upon the owner. Nearly 75 percent of the prairie grain crop was shipped between the end of August and the end of November.[10] In November 1925, the railways delivered a total of 51,852 grain cars to the Lakehead; in May 1926, the largest shipping month in the off-season period, the railways delivered 11,069 cars; in August 1926, the lowest month of the shipping year, the railways delivered 1,867 cars.[11] The railways organized their car delivery system quite effectively, each car making approximately three trips to the Lakehead in the three-month shipping season; but given the short-term nature of the demand for rolling stock, the railways were reluctant to provide sufficient rolling stock to adequately move the crop, and each year the availability of cars reached crisis proportions. The massive movement of grain in September to November, furthermore, often congested the rail lines beyond their capacity. The grain blockade had a significant influence on prices paid at the local delivery point. At times of rail congestion, track prices could be reduced because of the length of delivery. More importantly, farmers delivering street grain in November faced additional discounts on the price of street grain, as local grain buyers factored in the risk of not delivering the grain to the Lakehead in time for shipment and faced the possibility of additional carrying charges for winter storage.

The spread between track and street prices, a spread emphasized during the grain blockade, was a constant concern for farmers who sold street wheat. Farmers argued that the spread was often as high as five or six cents per bushel once all charges were considered. The Royal Grain Inquiry Commission (1925) concluded that the extra risk associated with street purchases, after all charges were assessed and hidden costs considered, averaged only two cents per bushel. But D. A. MacGibbon, a member of the Commission, wrote

Railcars (left) transferred grain to the Lakehead terminals at Fort William. From there, prairie crops were loaded onto lake freighters and eventually transported around the world. (Courtesy of the Saskatchewan Archives Board/R-B211.)

that the spread averaged three to four cents per bushel.[12] As already noted, many farmers believed the excessive spreads occurred because a combine controlled competition. Some observers, however, remained unconvinced. Based on testimony from the Saskatchewan Co-operative Elevator Company, the Royal Commission concluded that the excess spread between street and track prices occurred because the elevator companies needed to recoup losses incurred in the storage and handling business. The tariff set by the Board of Grain Commissioners for elevating, handling, storing for fifteen days, and insuring a farmers' grain in graded storage was 1.75 cents per bushel, and the Commission noted it was insufficient to cover the costs of this service to the elevator companies.[13] More likely, the local grain buyers discounted the price of street wheat simply because they could: farmers delivering into the street market were a captive market for them. On the whole, these farmers were either too small to produce a car lot of wheat, did not have a car lot of grain of similar quality and kind, or lived too far from the local delivery point to make it cost-effective to deliver a car lot to the elevator or flat warehouse within the fifteen-day free storage period.[14] Since they could not turn to the track market, nor ship themselves, they had to deliver in the street market no matter what the discount. Improved competition would be necessary to reduce spread between street and track prices; but even after the farmers' corporations entered the market, complaints regarding the spread between street and track prices continued. The only real solution lay in technological improvements on the farm such as the introduction of motorized trucks and mechanization with the resulting increased farm size.

The most significant complaint made by farmers against the open market system was the injurious fluctuation of the price of grain on a seasonal basis. Farmers noted that the price of grain in October rarely compared with the price in May and July. While seasonal variation in the price of a seasonally produced commodity is quite normal and reflects the carrying and financing costs of withholding the product from the market, farmers blamed the price variation on the Winnipeg Grain Exchange. Most importantly, they noted that the low point of the marketing system seemed to coincide with the period when most farmers were forced by the financial commitments of operating loans to liquidate their crops.[15] Farmers informed the Royal Grain Inquiry Commission that speculation led to the unduly low autumn prices and that speculators could influence the market downward through the sale of "wind bushels."[16] The president of the Saskatchewan Wheat Pool reflected this attitude in 1931:

> The organized farmers for many years, and as strongly today as at any time in the past, feel that the present system of futures trading does not work out in their best interests. They feel the price they receive for their wheat from day to day is largely influenced by the attitude of mind of the uninformed speculating public, and that such a method of determining or influencing the price level is too insecure and unstable a foundation upon which to build any industry. They feel that the effects of uncontrolled speculation results in much wider fluctuations in the market price than would otherwise be the case.[17]

A small number of farmers tried to take advantage of the seasonal price variation by purchasing May futures following the marketing of their crop.[18] These farmers, therefore, entered the market as speculators themselves, and by purchasing futures continued to maintain their risk position regarding price variations in a pattern similar to withholding the grain from the market.

REGULATION AND REFORM
Canada Grain Act

The passage of the Manitoba Grain Act (1900) and the consolidation of legislation affecting the grain trade into the Canada Grain Act (1912) solved many of the grievances of the farm movement regarding corruption in the grain trade.[19] As warehouse commissioner C. C. Castle later reported, most of the farmers' complaints about grain handling could be dealt with through better enforcement of the regulations.[20] The Canada Grain Act had provisions for the prevention of light weights and contained an appeal mechanism for disputes regarding grades and dockage. Furthermore, the Act contained mechanisms for the prevention of substitution for special binned grain and provisions for rail car allotment. Indeed, one of the first activities of the Territorial Grain Growers' Association had been the successful prosecution of the Sintaluta railway agent for failure to maintain a proper car order book. The Canada Grain Act also provided the Board of Grain Commissioners with powers to regulate grain elevators and examine other aspects of the trade. Through regulation, consequently, the government dealt with one aspect of farmers' concerns.

Farmers' Organizations

Farmers dealt with their concerns regarding combines and the lack of competition themselves by organizing producer-controlled grain handling and commission-selling cooperatives. In this project, they received generous support from the prairie provincial governments.[21] The United Grain Growers, Ltd. (UGG) and the Saskatchewan Co-operative Elevator Company (Co-op) both emphasized special binning in the construction of their elevators, created commission sales departments in Winnipeg to act on behalf of farmers, and offered competitive street prices in the markets they served. According to the Royal Grain Inquiry Commission (1925) these two farmers' companies operated at 41 percent of the prairie delivery points. At these points, the general manager of the Co-op asserted the line elevator companies had improved their street prices to meet the competition.[22] Both UGG and the Co-op operated terminals and export subsidiaries, but both remained dependent upon the basic marketing structure of the Winnipeg Grain Exchange. Although they certainly improved competition in the local delivery market and were active participants in the trading at the Exchange, the two farmers' grain handling companies did not address farmers' concerns with the disordered market place.

Marketing Solutions

Sample Market

The concept of a sample market became one of the most problematic, yet the most open market consideration, promoted by the farm movement. E. A. Partridge, a leader in the Saskatchewan Grain Growers' Association (SGGA) and one of the founders of the Grain Growers' Grain Company, introduced this idea in conjunction with his proposal for a government-owned grain handling system. In essence, a sample market was a grocery store for wheat. Farmers would send samples of their crop to the market, and grain buyers would buy directly from the farmer following their inspection of the sample. The sample market, consequently, did away with grading and government inspection as well as the speculative trading in wheat products by bringing the buyer and producer into direct contact.[23] A sample market, however, was a cumbersome marketing activity and led directly to the mixing and reducing of the overall quality of Canadian grain about which farmers complained. Partridge's radical plan, furthermore, never received unanimous support even within the Grain Growers' Associations. Although the idea of a sample market emerged periodically after World War I, other more efficient methods of improving the marketing system were already under consideration.

Orderly Marketing

Orderly marketing is the term used by the wheat pools and the Wheat Board to define the new marketing technique. In his classic study of the grain trade, C. F. Wilson suggests the term had two different meanings. He notes that orderly marketing at times appeared to mean that farmers desired to sell an equal amount of grain each month of the year and thus equalize the supply into world markets and avoid seasonal price variations. At other times, it appeared to mean selling grain when prices were high and holding while prices were low, in order to secure the best possible return.[24] In some ways farmers saw these two different concepts as synonymous. In his testimony to the 1925 Royal Commission, Henry Wise Wood noted that the consumption of wheat occurred on a yearly supply scale rather than a seasonal pattern. The objective of a centralized selling agency would be to sell wheat so that supply and demand were in balance. Since farmers understood that Canadian wheat was primarily a high quality blending wheat used to mix with other wheat in the milling process, the demand for it was rather steady in international markets. Thus equal marketing across the year would accomplish the goal.[25] Implementing such a system, however, would prove difficult. During the debate over the Partridge plan at the SGGA convention in 1908, the desirability of withholding grain from the market in the fall and marketing it across the year to improve the average return to farmers was discussed. In his comments F. W. Green, the secretary of the SGGA, provides one of the earliest indications of how organized farmers believed orderly marketing could be implemented:

> It would be necessary to bring the whole trade under the management of one agency. This agency must be created by the farmers who owned the grain and it must be subject to their control. They must be willing for this agency to market the

entire product in the best interests of the whole. This could be done by an independent commission nominated by the Grain Growers' Association, and appointed by the government. A sum of money sufficient to handle the whole trade could be borrowed on the credit of the country, and should be kept entirely separate from other channels of trade. This would render them independent of the banks or grain exchange, and would enable them to sell their wheat in the world's best markets, though to do this they must be able to order it forward independently of individuals in the best interests of the whole.[26]

The idea of a government-sponsored pool, therefore, surfaced at least as early as 1908.

The members of the SGGA, however, recognized the difficulty for most farmers of withholding grain from the market. Credit obligations, the SGGA noted, meant most farmers needed to secure cash income in the fall. Indeed, the SGGA presentation to the Saskatchewan Elevator Commission in 1910 blamed the credit policy of the banks for forcing farmers into the hands of the elevator monopolists. They recommended, consequently, that the government advance money on grain delivered into the country elevator system and pay farmers a final payment once the season's grain was marketed.[27] Thus by 1910 the basic parameters of the Wheat Board scheme had been advanced by the organized farm movement.

Wheat Board, 1919-20

The closure of the Winnipeg Grain Exchange in 1917, and the creation of the Board of Grain Supervisors in June 1917, changed the marketing system dramatically. The system, although it did not create a market along the lines suggested by producers, did address their concerns with seasonal price variations and the spread between street and track wheat. The Board of Grain Supervisors had the power to fix prices paid to producers and by overseas purchasers.[28] It used this power to make arrangements for purchasers to cover the carrying charges on grain in storage at the terminal market, and to fix the spread between street and track prices. The temporary measures of the Board of Grain Supervisors prevented wheat prices from inflating beyond the capacity of the British market's ability to purchase the commodity, and eliminated the seasonal variation in prices. When the Winnipeg Grain Exchange reopened on July 21, 1919, prices quickly rose on speculation, and then on July 29 the government decided to reintroduce a measure of control.[29]

In certain respects, the Wheat Board created for the 1919-20 crop year operated on the basis of the SGGA proposals of 1910 and made an effort to match
supply with demand in international markets. A review of the condition of
international markets concluded:

> Under these abnormal conditions, resulting in uncertainty of
> price and instability of market, it would appear that in order to
> secure that the early movement of the Canadian crop which is
> so essential, and that fair distribution among our wheat produc-
> ers of the actual value of their product as determined by the
> world demand for the same throughout the entire season of
> marketing, which is equally desirable, action should be taken
> by the Government, looking to purchase, storage, movement,
> financing and marketing of the wheat grown in Canada in 1919.[30]

In designing the new control system, the Canadian government turned to the
wheat marketing controls used in Australia during World War I.

In 1910, the Saskatchewan Elevator Commission was critical of farmers'
proposals that the government serve as a bank to allow farmers to market
their grain in an "orderly fashion." Yet in the aftermath of World War I that is
precisely what occurred. Under the Wheat Board, farmers received an initial
payment following their delivery of grain into the elevator system, thereby
allowing them to meet their credit obligations. The Wheat Board then took
control of the grain and marketed it to overseas purchasers in an orderly
fashion. At the end of the crop year, after all expenses and interest charges on
the advances had been considered, the government made a final payment to
producers based upon their deliveries. Producers, therefore, received an average
payment from the prices obtained throughout the year. The Board, meanwhile,
sold grain to domestic interests at prices approximating world prices, and
into international markets at the best obtainable price; but it existed for only
a single year, and as prices fell following the government's abandonment of
single-desk marketing, the farm movement demanded its return.

Return to the Open Market and the Wheat Pools

The campaign to restore the Wheat Board in 1920-22 produced frustration
amongst farmers, who did not necessarily believe that its reinstatement would
restore pre-war prices.[31] Rather they believed that the Wheat Board would
provide them with an average yearly price and provide for better marketing of
Canadian wheat. As one farmer told Saskatchewan Premier Charles Dunning:

> Our plan is this, we want an average price of the season's wheat price. The big percentage of farmers are forced to sell at thresh- ing time irrespective as to the price they get.[32]

This demand for orderly marketing reflected similar ideas within the Lib- eral government. The minister of Agriculture, W. R. Motherwell—a former president of the SGGA—noted the primary benefit of a Wheat Board would

> be due to the ability of such a Board to finance advance pay- ments and hold our wheat off the market at a time when it is otherwise glutted—in short, feed the market in accordance with its ability to consume and pay for our produce and thus avoid this annual glut and depression of our market that has occurred almost continuously since there was a west.[33]

Farmers also demanded that the new Wheat Board be compulsory. Both Dunning and Motherwell favoured the creation of a Wheat Board, but both were hesitant about the concept of compulsion. Dunning informed a member of the Grain Exchange:

> In the course of my connection with public life I have never known an issue upon which the people were so absolutely of one mind as is the case in connection with the present demand for a compulsory Wheat Board. There are objectors of course, and there are many, like myself, with an instinctive distrust of the principle of compulsion as applied to trade.[34]

There was ample evidence, however, that a voluntary board or pool was prob- lematic.

James Stewart and Fred Riddell had gone to great lengths to explain the problems of a voluntary pool. Stewart had participated in all the war-related marketing schemes and had served as chair of the 1919-20 Wheat Board. Riddell was the general manager of the Saskatchewan Co-operative Elevator Company. In 1921, Premier Martin, Dunning's predecessor, requested Stewart and Riddell prepare a report answering several questions about the grain trade. The first question asked if it was

> possible for any kind of pool comprising less than the whole of the western wheat crop to market the crop to the same advan-

tage from the producers' point of view as a system of national
marketing of the whole crop by a Canadian Wheat Board.

Stewart and Riddell answered it was not.[35] These two experienced grain traders, furthermore, concurred with farmers' opinions that the primary benefits
of a pool involved marketing grain over the entire season and the subsequent
reduction in price volatility and seasonal slumps.[36] Thus, Stewart and Riddell's report suggested that without 100 percent control of the crop, orderly
marketing could not be properly implemented.

The collapse of the Wheat Board proposals in 1922 led farmers sympathetic
to the orderly marketing ideal to pursue cooperative wheat pool ventures.[37]
These pools functioned on a contract basis—rather than a strictly voluntary
system—and offered farmers an initial payment for grain delivered into the
elevator system and consigned to the pool. Pool elevator customers, since their
grain was combined with other members, avoided the differential between
street and track deliveries. The pool then marketed this grain through a central selling agency in an orderly fashion. They obtained bank financing for
the purchase of grain based upon the margin between the initial price and
the open market price, and thus did not hedge grain on the futures market.
At the end of the crop year, farmers received final payment based upon the
price obtained, less costs and carrying charges, for the year's wheat sales. By
1925, the wheat pools had also entered the elevator business, building elevators in all three prairie provinces; and in Saskatchewan the pool purchased
the Saskatchewan Co-operative Elevator Company. At their peak, the three
prairie wheat pools marketed 52 percent of the prairie wheat crop.

The problems of a non-compulsory pool, however, were quickly apparent
to the wheat pool directors. Problems had emerged as line elevator companies
refused to ship farmers' grain to the pool terminals and kept the revenue from
storage and handling charges for themselves. Similarly, at locations where
pool elevators did not operate, small street sellers had difficulty pooling grain
for shipment as car lots and continued to be forced to deliver into the street
market.[38] In April 1925 another problem emerged: the wheat pools nearly
collapsed as prices on the open market fell dramatically and approached the
pools' initial price offering. Although rumours circulated that the April price
collapse had been a deliberate attempt to break the wheat pools, the realiza-
tion amongst pool officials that the large amount of grain controlled by the
pools influenced the market place was the more important lesson of the 1925
experience. On April 6, 1925, as the price collapse approached catastrophic
levels for the pools, the Central Selling Agency began to purchase futures

"Clearing the Right of Way" appeared in the *Grain Growers' Guide* January 12, 1921. The cartoon illustrates how farmers imagined wheat pools through co-operative and orderly marketing would bypass "Speculators, Profiteers, Gamblers & Middlemen" in getting their crops to world markets. (Courtesy of the Saskatchewan Archives Board/R-A19419.)

in Winnipeg to support the price of wheat on the Exchange. A. J. McPhail noted in his diary:

> We are simply forced to take these measures to fight the bears on the market. Apparently a few strong grain interests can bear down the market if there is no bull resistance. All the bullish interests are afraid to buy for they do not know when the pool may be forced to unload. The pool appears to be the only organization that can go in and change the trend in the market and to do it we must take steps which we would not under ordinary circumstances take. But we must fight the devil with his own weapons.[39]

The pool, controlling approximately half the Canadian wheat crop and operating outside of the open market system, could by its very presence influence a downward slide in Canadian wheat prices. Since traders did not know what price the pool would deliver grain for, or how much it would place on the market, they would tend to watch from the sidelines at times of price collapse. The experience led to the demand for the 100 percent pool. As historian Allan

Levine notes, "the Pools could not eliminate speculation, nor could they raise the price of wheat—as some of their more radical spokesmen claimed. It was clear these two objectives would never be realized unless the Pools had total control of the Canadian wheat crop."[40]

The cooperative pools, like many corporations and agencies, collapsed in the Great Depression. The decline in prices during the 1929-30 crop year forced the pool to sell wheat into international markets below the initial price offered to pool farmers. Its efforts to stabilize the price of wheat by withholding it simply opened room for open market grain sellers to enter the market. Without a complete monopoly, the pools' efforts at orderly marketing in the Depression crisis simply produced a large surplus of unsold grain. The pools, meanwhile, faced the impossible task of recouping from their patrons some of the initial advance made on delivery. The impossibility of this task led the pools to demand that the government guarantee the initial payment at a generous level in all future debates.[41] The Depression, therefore, introduced a new element to the marketing debates: the desire for a government guarantee on the initial price. In 1930, the federal government bailed out the pools; it guaranteed the pool loans, took control of the pool wheat surplus, and a new era of wheat marketing began in Canada.

THE EMERGENCE OF THE WHEAT BOARD
Price Support System
Grain Price Stabilization

Prime Minister R. B. Bennett placed J. I. McFarland, former president of the Alberta Pacific Grain Company, in charge of the pools' wheat with the purpose of liquidating the surplus and repaying the loans. McFarland, however, with the support of the government, pursued a very different goal in wheat marketing. For the next five years, he held the pool wheat off the international market and purchased even greater amounts of wheat through the Wheat Pool Central Selling Agency in order to support wheat prices.[42] The government meanwhile sought to negotiate wheat marketing deals with other wheat exporting countries in order to reduce international wheat supplies, thereby following one of the basic parameters of orderly marketing: matching supply to demand.

Voluntary Wheat Board

McFarland's wheat price stabilization policy, however, could not be maintained indefinitely. By 1935, the surplus in the pool accounts had grown to unwieldy proportions, and its presence alone may have sustained low prices.[43] Charles Wilson notes that in 1934-35 "speculators had returned to the market on the short side in the expectations that prices could only decline in response to the size of the government-held surplus.[44] Given these circumstances, McFarland informed the Prime Minister, "I am sure you think I am against the Board, and it is true I have been, but from here on I have no argument against it." He continued, "It might however be a good time to turn the whole thing over to a Government monopoly. I put this up for your consideration."[45] Bennett understood such legislation would be popular with many farmers: in 1933, when the Conservatives had first floated the idea of a national marketing board, the Saskatchewan Association of Rural Municipalities had orchestrated a letter-writing campaign in support of such legislation.[46] In 1935 Bennett placed legislation before Parliament creating a Wheat Board with a monopoly on grain trading. A special committee reviewed the bill and heard testimony of support from the prairie wheat pool representatives.

Liberal members of the committee, nevertheless, succeeded in having the bill amended before it passed third reading. The draft Wheat Board legislation created a Board which fulfilled virtually all of the parameters of the orderly marketing vision of wheat pool supporters. In its original form it would have had responsibility for all grains (rather than just wheat), taken control of the local elevator network, controlled interprovincial and international movement of grain, and through its monopoly powers eliminated the futures market. The Canadian Wheat Board Act, however, fell well short of this. The monopoly had disappeared, and consequently the futures market continued to function. The Board's activities, furthermore, were restricted to trading in wheat. Rather than replacing the open market it offered a marketing option.[47] In reality, the 1935 Wheat Board addressed only one issue for prairie farmers: by providing an initial guaranteed price to farmers who delivered wheat to the Board, it set a floor price on wheat.[48] This floor price effectively became a subsidy mechanism. Thus the 1935 Wheat Board, rather than being a vehicle for orderly marketing, was a subsidy to prairie farmers.

The actions of Mackenzie King's newly elected Liberal government in 1935 clearly demonstrate that it perceived the Wheat Board as a temporary price support measure. The Wheat Board was to make use of the existing marketing system, and use its own agencies only if necessary. As noted in the official publication *The Canadian Wheat Board*, the real purpose of the legislation "was to protect the producer against untimely developments in the international wheat situation."[49] The government made it clear that it desired the surplus wheat liquidated as quickly as possible. Yet in the autumn of 1935, under the stewardship of McFarland, the Wheat Board took delivery of two-thirds of the prairie wheat crop and the surplus continued to be maintained.[50] Upset that McFarland continued to purchase wheat on behalf of the Board, the government dismissed him and appointed J. R. Murray as the new chair. Under Murray, the Wheat Board's initial delivery price was set below world price levels at 87.5 cents per bushel, and the Wheat Board was ordered to take delivery of grain only when the open market price fell below 90 cents per bushel. In the 1935-36, 1936-37 and 1937-38 crop years, open market prices did not fall below the floor price, and Murray sold the surplus stocks held by the government.[51] Yet the utility of the Wheat Board for other marketing reforms became apparent. From his position as chair, Murray also made efforts to develop alternative markets for Canadian wheat in Europe and reduce Canada's reliance on the British market.[52] In the meantime, the Liberal government selected W. F. A. Turgeon to chair another Royal Commission to investigate the marketing system for wheat.

Turgeon Commission, 1937-38

In his report Justice Turgeon noted that "[a compulsory government Board] was asked for by nearly all of the farmers' organizations, and by a great many of the individual farmers, who appeared before me."[53] The report of the Turgeon Commission (1938), like the Royal Grain Inquiry (1925) and the Commission to Enquire into Trading in Grain Futures (1931), nevertheless argued that the open market system with competitive pricing and futures trading remained the best method of marketing grain. Still, Turgeon had to address the concerns of those farmers who demanded orderly marketing, and especially the position of the prairie wheat pools. In May 1937, Turgeon quizzed pool leaders Paul Bredt and Jack Wesson at great length; in his report he concluded "that the cooperative marketing of wheat is something essentially sound and that it contains possibilities for the future." He also concluded later that "the producer who is antagonistic to speculation may still do his share towards reducing it ... by joining one of the Pools I have suggested."[54] In his conclusions, Turgeon reflected the attitudes of the minister of Agriculture, J. G. Gardiner. Gardiner informed Prime Minister King that he concurred with Turgeon's report and advised that the government introduce legislation to implement Turgeon's recommendations.[55]

Turgeon's solution, however, did not coincide with the position of the pools themselves. These organizations continued to stagger under the debts from the 1929-30 price collapse. They had no interest in returning to the market place and were instead consolidating their position as grain handling enterprises. Following his meetings with the executives of the three prairie pools, Gardiner informed King "they would prefer a Wheat Board handling 100% of the wheat under International Agreements with regard to quotas made between importing and exporting countries." Only when it became clear the government would not consider a compulsory pool did Gardiner obtain limited support from a few "prominent" Board members who acknowledged that a compulsory pool had almost "insurmountable" problems.[56] The wheat pools, however, never re-entered the field of orderly marketing as hoped by the Liberal government, even after the government guaranteed the initial pool payment in the 1939 legislation.[57] Garry Fairbairn writes that "for the Saskatchewan Wheat Pool, however, the era of bold new ventures had been replaced by an era in which mere survival as an elevator company would be tough enough."[58]

Although Turgeon's report recommended the eventual restoration of the competitive market, it recognized the important role the Wheat Board played as an income support mechanism. As a result, his report concluded"

> that under what may be called normal conditions ... the Government should remain out of the grain trade, and our wheat should be marketed by means of the futures market (under proper supervision), and encouragement given to the creation of cooperative wheat marketing associations, or Pools.... But upon the facts before me today, I must say that such a return is not immediately in sight.... For all these reasons (and not withstanding the adverse considerations to which I have referred in relation to government Boards) I do not feel that I can suggest the immediate dissolution of the Canadian Wheat Board. There is a strong possibility that conditions may develop which will require a measure of assistance in the marketing of the coming crop.... I can think of nothing better to suggest than that the Board be maintained to meet any situation which may arise.[59]

The problems Turgeon noted in the world wheat market were the prediction of a bumper crop in Canada and the United States, and a decline in world wheat imports. The result would be a drastic decline in world prices. Turgeon, it was clear, believed the Wheat Board could help subsidize prairie farmers temporarily. The Liberal government conceded as much in 1938. The government set the floor price at 80 cents per bushel when all the evidence suggested this would be too high; it nevertheless defended the decision as improving the overall price of wheat in the export market through orderly marketing. C. D. Howe noted: "our fixed price was imposed to prevent demoralization of the market under heavy country deliveries and is not intended as a bonus to the farmer or an export subsidy." Similarly, the deputy minister of Trade and Commerce, H. Bartons, remarked the floor price not only limited government relief payments but also ensured "more orderly marketing of the crop than would otherwise be likely ... the policy of the Wheat Board is that wheat will be sold at the market price as the market can absorb it."[60] The government's decision led all farmers to deliver their grain to the Board in 1938, and the government lost over $60 million.

The loss shocked the Liberal government. By November 1938, Gardiner con-ceded that such a system could not be maintained. When grain prices were low, he noted, the Wheat Board would receive all of the grain which necessitated a large bureaucracy to market the crop. When the prices were high, however, farmers would not deliver to the Wheat Board but it still maintained its marketing infrastructure.[61] The federal government attempted to eliminate the Wheat Board; but faced with a campaign from western politicians to maintain the voluntary pool, it decided to reform the system instead.[62] The Canada Wheat Board Act was consequently amended: its role as a temporary subsidy mechanism would continue, but the government's liability would be limited. The Board continued to establish a floor price, this time 70 cents per bushel, but the Act limited the Board's purchases to 5,000 bushels per producer. Under these changes, if world prices fell below the floor price, small farmers would continue to receive full benefit of the Wheat Board, while the subsidy to large producers would be reduced. The Board's power to purchase grain was also limited to wheat.[63] Still, the Wheat Board's role as a temporary subsidy system led to a new surplus carry-over of 86.5 million bushels of wheat. It was clear that a more orderly system of marketing would be required if the Board was to continue to offer an alternative market system to farmers.

Single Desk Marketing
World War II

Historians Ian MacPherson and John Herd Thompson have argued that Canadian farmers and politicians "were best prepared to fight the previous war, and they were marching resolutely forward facing backward.[64] Many predicted that wheat prices would inflate rapidly and that the problem of over-supply would disappear. Despite an early upward movement of prices in the 1939-40 crop year, these predictions proved utterly false and a large carry-over developed.[65] From 1939 to 1943 the United Kingdom was the only market for Canadian grain. The British market imported approximately 230 million bushels/year, but in 1941 Canadian wheat stocks available for export amounted to 700 million bushels.[66] Over-supply, consequently, continued to disrupt the Canadian grain trade. The government now had a new rationale for maintaining the Wheat Board; it needed to establish orderly marketing.

Deliveries to the Board early in the war were inconsistent and irregular. In 1939-40, the initial price offered by the Wheat Board was often below the world price, but the Board received 342.4 million bushels (over 70 percent of the marketed crop). The dramatic fluctuation in prices on the competitive market eventually led to the temporary closure of the futures market in May 1940.[67] The Board made most of its sales to the Cereal Imports Committee of the United Kingdom, but still carried a growing surplus of 182 million bushels into the 1940 crop year. The next year prices hovered around, or just below, the Board's initial price. Under these conditions, the Board took delivery of 395.3 million bushels (90 percent of the crop marketed), of which it managed to sell only 141.6 million bushels. Sales of the 1939-40 crop continued, but the surplus had now reached well over 480 million bushels by July 31, 1941. With no market relief in sight, the government was forced to interfere in the market in a manner which, although often demanded by the farmers' organizations, had always been resisted by the federal Liberals.

In a meeting with the cabinet, on January 27, 1941, the Canadian Federation of Agriculture (CFA) had informed the government that:

> For the coming crop year, we believe that the Dominion Government, through the Wheat Board, should undertake to announce the total amount of wheat of which it is prepared to accept delivery. We believe that the amount of wheat which may be delivered by the grower in the crop season 1941-42 should be based on the estimated amount which the Wheat Board may be reasonably expected to market during the 1941-42 crop year, including both export and domestic sales. Having established the maximum amount which may be delivered, we recommend that the individual grower's delivery quota should be based on a policy which will preserve to each farm unit an equitable share of such maximum amount in proportion to the past production of the crop district to which it is located.[68]

The Canadian Federation of Agriculture thus recommended that supply be matched to demand and that producers share equally in the risks of management; in other words they recommended orderly marketing of Canadian wheat. The government followed these parameters closely in defining its 1942 wheat marketing policy: it limited the amount of wheat which could be marketed, although producers could still choose to deliver to either the Wheat Board or the competitive market, and encouraged a reduction in wheat acreage by offering financial incentives for summerfallowing land or shifting to coarse grain production.[69]

Although the CFA recommended that the government increase the initial price, the government maintained its 70 cents per bushel floor price. As a result, world prices advanced beyond the Board price, and only 44 percent of wheat was delivered to the Wheat Board, allowing the Board to reduce its surplus. The initial price rose to 90 cents per bushel in 1942-43, and the Board received 62 percent of the eligible wheat marketed. The good harvest in 1942-43, however, demonstrated the weakness in Canada's grain delivery and marketing system. At the beginning of the 1942 crop year, the terminal and initial elevator system had available only 35 million bushels of storage space. While the government had limited marketing of wheat to 280 million bushels, the total wheat crop was 556 million bushels. After seed requirements and on-farm feed, nearly 200 million bushels of wheat remained on Canadian farms. It was clear that changes were necessary.[70]

Compulsory Board, 1943

By 1943, the Wheat Board had put into place most of the rules for orderly marketing with the exception of ensuring to producers an average price on the year's sales through compulsory pooling. For example, near-record harvests in the 1938-39, 1939-40, and 1940-41 crop years had generated a huge surplus of unsold grain in Canada and placed tremendous strain on the delivery system. The Wheat Board reacted to this problem by implementing delivery quotas and paying farmers to store grain on the farm at the rate charged by the local elevators.[71] Through this system, the Wheat Board provided for an equitable distribution of storage space in the delivery system, and prevented the blockade of storage and rail deliveries which often accompanied the autumn harvest. The success of the system, as well as some of its problems, was demonstrated by the large deliveries of the 1940-41 crop in July 1941. This delivery plugged the elevator network just prior to the 1941-42 crop harvest, and as a result quotas were continued in that year despite the significant reduction in the crop size. Still, the Board managed to get the entire allotted quotas into the system before the end of November, thereby ensuring farmers income for their fall crop.

The Board had not only established delivery quotas: it had also assumed authority for the allocation of grain cars in the west on October 15, 1942.[72] Thus the Board could direct cars to areas where quotas remained unfilled, and to areas where storage facilities were especially taxed by farmers' deliveries. The Wheat Board could also move grains to the terminal market for delivery as needed. If feed grains were in high demand, as they were in 1942-43, the Board moved these grains preferentially over alternative products. Despite the measures of control and regulation, shipments from the initial delivery system to the terminals could not keep pace, and the country elevator system was completely plugged by the shipment of the 1942-43 crop. The Canadian government also faced significant new problems in 1943. World demand for wheat, led by increased imports of feed wheat by the United States, increased in 1942-43 and thus prices began to rise. The United Kingdom, Canada's traditional market, meanwhile faced increased financial pressure and appeared unable to finance additional purchases of war-related commodities.[73] Fearing inflation, and needing to coordinate price, delivery, and shipment of grain, the government finally answered the farmers' organizations' long-standing demand for a compulsory pool.

The Combined Food Board called for the delivery of the designated quantities of Canadian wheat to the United States and the United Kingdom as quickly as transportation allowed. Since the Wheat Board knew how much and what quality of Canadian wheat needed to be delivered during each period, the Board could completely coordinate delivery, shipment, and sale. In order to fulfill its international arrangements and to prevent domestic inflation which might be caused by delivering large quantities internationally, Canada closed the futures market on September 27, 1943 and authorized the Wheat Board to take control of all wheat stocks at the closing price of that day.[74] The decision reflected the government's desire to "order" the Canadian grain trade and deliver grain to the market as the demand dictated. On October 12, 1943, the Canada Wheat Board became a compulsory pool. Its control over the wheat export market has never since been relinquished.

CONCLUSION

Not all farmers supported the idea of orderly marketing, but many farmers, including most farmers' organizations, had long demanded a compulsory pool in Canada. They believed that a system of orderly marketing would bring more reasonable and stable returns to prairie farmers than the competitive open market system. That the Wheat Board became a compulsory pool because of the war and partially reflected the government's desire to control inflation in Canada mattered little to the pool supporters: the Wheat Board as it emerged in 1943 recognized their long campaign to order the grain trade. As H. H. Hannan, president of the Canadian Federation of Agriculture, noted, farmers would not go "back to the planlessness [*sic*] of pre-war days"; instead, he continued, they wanted to improve "the newer, proven policy of systematic coordinated production and orderly, organized, nationally directed and supervised marketing.[75] The farmers who had demanded a Wheat Board had concerns about the economic structure of the grain trade. Years of experimentation and regulation had clearly demonstrated that their concerns could only be addressed by a compulsory pool. Under the compulsory pool, farmers were guaranteed a floor price for their grain and could make economic plans well in advance when the grain was stored on the farm. The grain blockade and the price variations it created had been broken; the difference between small street sellers and large track sellers had been eliminated; specula-tors, and the seasonal variation in prices and the different value for crops delivered at different times, had been removed from the market. The government could deliver grain to customers as demand dictated. All of these programs had been requested by the farmers' organizations; the Wheat Board addressed these problems and thus protected the small family farm. Many farmers, of course, opposed the continuation of the Wheat Board: they believed they held a comparative advantage over their fellow farmers in the marketing of their grain, either by virtue of loca-tion, size, or skill at hedging in the market, and the Wheat Board effectively thwarted those advantages. Once the government had created a compulsory pool, however, it would be difficult to disband it.

NOTES

This article first appeared in *Prairie Forum* 26, no. 1 (2001): 85-105.

1   The most important of these commissions for the purposes of this article are Canada, *Sessional Papers*, 1908, no. 59, "Report of the Royal Commission on the Grain Trade in Canada (1906)," (hereafter cited "Report of the Commission on the Grain Trade, 1906."); *Report of the Saskatchewan Royal Commission on Elevators*, 1910 (hereafter cited *Saskatchewan Elevator Commission*); *Report of the Royal Grain Inquiry Commission*, 1925; *Report of the Commission to Enquire into Trading in Grain Futures*, 1931 (hereafter cited *Report of the Commission on Futures*); and *Report of the Royal Grain Inquiry Commission*, 1938.

2   The two most comprehensive surveys of the Canadian grain trade including the creation of the Canadian Wheat Board are C. F. Wilson, *A Century of Canadian Grain* (Saskatoon: Prairie Books, 1978) and Allan Levine, *The Exchange: 100 Years of Trading Grain in Winnipeg* (Winnipeg: Peguis, 1987). Three old studies remain of tremendous importance. See Harald S. Patton, *Grain Growers' Co-operation in Western Canada* (Cambridge, MA: Harvard University Press, 1928); D. A. MacGibbon, *The Canadian Grain Trade* (Toronto: Macmillan, 1932); Vernon Fowke, *The National Policy and the Wheat Economy* (Toronto: University of Toronto Press, 1957). Economists have focused their attention on the economic benefits of single-desk marketing rather than examine the purpose and intent of the Canadian Wheat Board. Two studies illustrate this perspective. C. A. Carter and R. M. A. Lyons, *The Economics of Single Desk Selling of Western Canadian Grains* (Edmonton: Alberta Agriculture, 1996) argue that the Wheat Board structure is inefficient. Andrew Schmitz and Hartley Furtan, *The Canadian Wheat Board: Marketing in the New Millenium* (Regina: Canadian

Plains Research Center, 2000) argue that the Canadian Wheat Board is an efficient grain-marketing venture. Political scientists and economists have also examined the Wheat Board in the context of Canadian-American relations. See the special issue of the *American Review of Canadian Studies* 28, no. 3 (1998) for articles on this topic. The issue of the wheat board has recently been dealt with in the Federal Court of Canada. See Archibald v. Canada [1997] 3 F.C., 335. Two recent decisions are Alberta v. Canada at http://www.fja.gc.ca/en/cf/1998/vol2/html/1998fca21524.p.en.html and the appeal of Archibald v. Canada at http://www.fja.gc.ca/en/cf/2000/ orig/ html/2000fca26721.o.en.html. The Courts have upheld the constitutionality of the Canadian Wheat Board and its compulsory pool powers in these decisions.

3   David Bercuson and Barry Cooper, *The Monopoly Buying Powers of the Canadian Wheat Board: A Brief History* (Edmonton: Alberta Agriculture, 1997) after a brief synopsis of early grain marketing activity focuses extensively on World War II arguing wartime contingencies and a desire to stop inflation were entirely responsible for the origin of a compulsory Wheat Board. In his best selling attack on the Wheat Board, *Canada's Great Grain Robbery* (Regina: the author, 1998), publicist Don Barron blames the political influence of the Social Gospel and its call to "abolish capitalism" for the Wheat Board. Two other studies, both better informed about the economic complexities of the Canadian grain trade than either of those mentioned above, have linked the origin of the Wheat Board to Canadian political contingencies. See J. E. Rea, "The Wheat Board and The Western Farmer," *Beaver* 77, no. 1 (1997), 1423 and Robert Ankli and Gregory Owen, "The Decline of the Winnipeg Futures Market," *Agricultural History* 56, no. 1 (1982), 272-86. Rea's study, despite its title, is really about the politics of the Wheat Pool movement in Canada prior to 1935.

4   *Report of the Commission into Futures* (1931), 20-21.

5   The mechanisms for selling grain have been discussed in variety of publications. See for example the discussion in Patton, *Grain Growers' Co-operation*, 14-17; MacGibbon, *Canadian Grain Trade*, 102-5; Fowke, *National Policy*, 109-14. Also look at the *Report of the Royal Grain Inquiry*, 1925, 10-19.

6   Interprovincial Council of Grain Growers' and Farmers' Associations, *A Provincial Elevator System* (1909), 2, Saskatchewan Archives Board (hereafter SAB), G 361.1. For a discussion of farmers' and corruption see Patton, *Grain Growers' Co-operation*, 12-19; Fowke, *National Policy*, 121-25 and 138-40; and Robert Irwin, "'The Better Sense of the Farm Population': The Partridge Plan and Grain Marketing in Saskatchewan," *Prairie Forum* 18 (Spring 1993): 36-40.

7   Canada, *Sessional Papers*, 1900, No. 81-81b, "Report of the Royal Commission on the Shipment and Transportation of Grain."

8   SAB, Scott Papers, 40111, F. M. Gates (SGGA Vice-Pres.) to Premier Walter Scott, 20 April 1908.

9   *Report of the Royal Grain Inquiry Commission*, 1925, 15-19.

10   This figure was cited by the *Report of the Commission into Futures* (1931), 38.

11   Walter Davisson, *Pooling Wheat in Canada* (Ottawa: Graphic, 1927), 153-57. See also *Report of the Royal Grain Inquiry Commission*, 1925, 35-36.

12   *Report of the Royal Grain Inquiry Commission*, 1925, 12; D. A. MacGibbon, *Canadian Grain Trade*, 103.

13   As a result, the Board of Grain Commissioners set the maximum fee allowed for

this service at two cents per bushel on wheat, although most companies continued to charge only 1.75 cents per bushel. D. A. MacGibbon, *Canadian Grain Trade*, 99.

14 Fowke, *National Policy*, 112-13.

15 Ibid., 187-88.

16 *Report of the Royal Grain Inquiry Commission*, 1925, 128.

17 *Report of the Commission into Futures*, 1931, 57.

18 Less than one percent of farmers speculated on the futures market according to the Stamp Commission. See *Report of the Commission into Futures*, 1931, 39.

19 Fowke, *National Policy*, 153-65; David Hall, "The Manitoba Grain Act: An Agrarian Magna Carta?" *Prairie Forum* 4 (Spring 1979): 105-20.

20 *Saskatchewan Elevator Commission*, 75-76.

21 See R. D. Colquette, *The First Fifty Years* (Winnipeg: Public Press, 1957); D. S. Spafford, "The Elevator Issue, the Organized Farmers, and the Government, 1908-11," *Saskatchewan History* 15 (1962): 88-96; Robert Irwin, "Farmers and Managerial Capitalism: The Saskatchewan Co-operative Elevator Company," *Agricultural History* 70 (Fall 1996): 626-52; John Everitt, "A Tragic Muddle and a Co-operative Success: The Story of Two Elevator Experiments in Manitoba," *Manitoba History* 18 (1989): 12-24.

22 *Report of the Royal Grain Inquiry Commission*, 1925, 18.

23 Irwin, "'Better Sense of the Farm Population,'" 40-41; SAB, S-B2 1 3, Saskatchewan Grain Growers' Association, Annual Convention Report, 1908, 16-18.

24 Wilson, *A Century of Canadian Grain*, 223-26.

25 Evidence and Proceedings, Royal Grain Inquiry Commission, 29 June 1923, 377-86.

26 SGGA Annual Convention Report, 1908, cited in S. W. Yates, *The Saskatchewan Wheat Pool: Its Origin, Organization, and Progress, 1924-35* (Saskatoon: United Farmers of Canada, n.d.), 16.

27 *Saskatchewan Elevator Commission*, 24-26. In debates surrounding the Saskatchewan government's elevator plans, the failure of the government to pursue this policy was one of the leading objections to the proposal. See SAB, SGGA Annual Convention Report, 1911, 20.

28 P.C. 1604, 11 June 1917, cited in Wilson, *A Century of Canadian Grain*, 95.

29 Wilson, *A Century of Canadian Grain*, 130-36.

30 Ibid., 142. Chapter 7 of Wilson's book provides the best coverage of the first wheat board.

31 Allan Levine suggests this rationale in *The Exchange*, 111.

32 SAB, Dunning Papers, M6 [Q]Y-103-0, 47644-47660, Edward Edgley (farmer) to Dunning, 16 Aug 1922 and Lawrence Banks (farmer) to Dunning, 31 Aug. 1922.

33 SAB, Motherwell Papers, M12 III 133, Motherwell to T. McConica, MP, 24 January 1922.

34 SAB, Dunning Papers, M6 [Q]Y-103-0, 47561, Dunning to W. D. McBean, 25 July 1922.

35 James Stewart and E. W. Riddell, *Report to the Government of Saskatchewan on Wheat Marketing* (Regina: n.p., 1921), 3.

36 Ibid., 15-16.

37 The most recent discussion of the Wheat Pools is Garry Fairbairn, *From Prairie Roots: The Remarkable Story of Saskatchewan Wheat Pool* (Saskatoon: Prairie Books, 1984).

38 Levine, *The Exchange*, 139-41. It was for these reasons that the wheat pools entered the elevator business.

39 Harold Adams Innis (ed.), *The Diary of Alexander James McPhail* (Toronto: University of Toronto Press, 1940), 118.

40 Levine, *The Exchange*, 128. The 100 percent pool called for government legislation to force farmers to market through the Pools. It never received unanimous support (Fowke, *National Policy*, 41). Historian J. E. Rea associated the call for a 100 percent pool with ideologues in the Farmers Union of Canada, Saskatchewan Section who chastised the Central Selling Agency for its use of the futures market in 1925-26. Although never stated explicitly, Rea appears to indicate that had the Wheat Pools become active participants in the futures market and hedged their grain, these crises could have been avoided (Rea, "The Wheat Board and The Western Farmer," 22).

41 Saskatchewan Wheat Pool continued to grapple with the problem of recouping the 1929 initial payment in 1936. It experimented with deducting it from patronage dividends paid to members, and finally absorbed it as a corporate loss. Fairbairn, *From Prairie Roots*, 118. Citing the *Manitoba Free Press*, May 1, 1937, Wilson notes Manitoba Pool Elevators president Paul Bredt demanded the government guarantee the initial pool payment before the Turgeon commission (Wilson, *A Century of Canadian Grain*, 546). A more sympathetic account of the presentation is E. W. Hamilton, *Service at a Cost: A History of Manitoba Pool Elevators* (Saskatoon: Modern Press, 1975), 176-78.

42 For detailed coverage of McFarland's grain price stabilization activities see Wilson, *A Century of Canadian Grain*, 334-448. A brief account is Fowke, *National Policy*, 256-62.

43 Wilson notes the surplus actually peaked in 1934 at nearly 235 million bushels, an amount nearly equivalent to an entire Canadian harvest (Wilson, *A Century of Canadian Grain*, 444).

44 C. E. Wilson, *Canadian Grain Marketing* (Winnipeg: Canadian Grain Institute), 97.

45 McFarland to Bennett, 2 June 1934, in Wilson, *A Century of Canadian Grain*, 460.

46 LAC, RG 17, vol. 3080, file 45-2-6.

47 Fowke, *National Policy*, 264-65. The Liberal party was quite proud of its accomplishments. A Manitoba member informed King: "The Conservative draft provided only for a dictatorial, compulsory method of state-controlled marketing which deprived the farmer of every vestige of freedom in the marketing of his crop." NA, King Papers, vol. 205, 175695-699, J. C. Davis (Liberal-Progressive Election Committee of Manitoba) to King, 13 September 1935.

48 A floor price had been suggested to the federal government by the members of the Winnipeg Grain Exchange and J. R. Murray (soon to head the Wheat Board). Levine, *The Exchange*, 182-83.

49 *The Canadian Wheat Board*, 1935-56 (n.p, 1957), 6-8.

50 Fowke, *National Policy*, 266.

51 Wilson, *A Century of Canadian Grain*, 499-502. Murray was greatly assisted in this task by the failure of the United States in crop in 1936. A memorandum prepared by the Department of Trade and Commerce noted Canada sold 253 million bushels in 1935-36 accounting for a 50 percent share of entire world wheat export trade. LAC, RG 20, vol. 1412, file 146, "Wheat Marketing and Sales Policies and Operations of Canadian Wheat Board," (February 1940).

52   LAC, RG 20, vol. 1932, file 20-40 pt. 1, Murray to J. G. Parmelee, 20 June 1936.

53   *Report of the Royal Grain Inquiry Commission*, 1938, 185.

54   Ibid., 91, 184.

55   LAC, King Papers, vol. 250, 213469-213473,1G. Gardiner, "Memorandum Re: Wheat Marketing," 28 June 1938.

56   Jack Wesson to Gardiner, 7 September 1938, copy in LAC, King Papers, vol. 250, 213496-213498; ibid., 213492-94 and 213512-13, Gardiner to King, 10 September 1938 and 15 November 1938.

57   The Co-operative Wheat Marketing Act (1939) cited in George Britnell, "Dominion Legislation Affecting Western Agriculture, 1939," *Canadian Journal of Economics and Political Science* 6, no. 2 (1940): 276.

58   Fairbairn, *From Prairie Roots*, 115.

59   *Report of the Royal Grain Inquiry Commission*, 1938, 189, 194.

60   LAC, RG 17, vol. 3080, file 45-2-7 (1), C.D. Howe to Skelton, 22 August 1938, and H. Bartons to O. D. Skelton, 25 August 1938.

61   LAC, King Papers, vol. 250, 21351419, Gardiner, "Address to the Saskatchewan Wheat Pool," 3 November 1938.

62   NA, RG 17, vol. 3080, file 45-2-7 (1), Manitoba Legislature Endorsement of Wheat Board, 24 February 1939; John Kendle, *John Bracken: A Political Biography* (Toronto: University of Toronto Press, 1979), 164-70.

63   *The Canadian Wheat Board, 1935-1956*, 16.

64   Ian MacPherson and John Herd Thompson, "An Orderly Reconstruction: Prairie Agriculture in World War Two," *Canadian Papers in Rural History* VI (Gananoque: Langdale, 1984), 11.

65   NA, RG 20, vol. 1412, file 146, "Wheat Marketing and Sales Policies," (February 1940).

66   G. E. Britnell and Vernon Fowke, *Canadian Agriculture in War and Peace, 1935-50* (Stanford, CA: Stanford University Press, 1962), 92-93.

67   The Wheat Board's initial price was 70 cents per bushel. Britnell and Fowke note the traders realized "there could be no appreciable market for Canadian wheat beyond the United Kingdom," in May 1940 and prices fell 30 cents per bushel. Trading was suspended 18 May 1940 with the price at 70 cents per bushel (*The Canadian Wheat Board, 1935-1956*, 18; Britnell and Fowke, *Canadian Agriculture*, 90).

68   Cited in Britnell and Fowke, *Canadian Agriculture*, 95.

69   Wilson, *A Century of Canadian Grain*, 684.

70   *The Canadian Wheat Board, 1935-1956*, 20-29.

71   Ibid., 19-21.

72   Ibid., 31.

73   Britnell and Fowke, *Canadian Agriculture*, 103. Canada would begin selling grain to Britain in return for British assumption of the cost of Canadian troops in Britain. As well, Canada loaned the British substantial sums to purchase Canadian products.

74   Wilson makes it clear that farmers' demands for a pool had little to do with the decisions (Wilson, *A Century of Canadian Grain*, 782); also see MacPherson and Thompson, "An Orderly Reconstruction," 27.

75   H. H. Hannan, "The Future of Agriculture in Canada," *Industrial Canada* (July 1945), cited in MacPherson and Thompson, "An Orderly Reconstruction," 27.

18. "Stuck in Playing the Old Tunes": The *Winnipeg Free Press* and the Canada-United Kingdom Wheat Agreement

Patrick H. Brennan

The death of J. W. Dafoe in January 1944 was a serious blow to the editorial prestige of the *Winnipeg Free Press*, one from which it was never able to recover. A poisonous struggle for editorial control pitting publisher and owner Victor Sifton and Dafoe's longtime managing editor and presumptive heir, George Ferguson, followed. After two years of endless infighting with a man he held in complete contempt, Ferguson concluded he could not preserve the "editorial integrity" of the *Free Press* and resigned, in the process raising more than a few eyebrows among the paper's many admirers in the Ottawa bureaucracy, Liberal party and intellectual circles.[1]

His replacement was Grant Dexter, for most of the previous twenty-two years the *Free Press's* eyes and ears in Ottawa. A superb reporter of the national political scene, he only reluctantly accepted the Dafoe mantle. Unfortunately for Dexter, his employers and the newspaper's editorial reputation, it was soon apparent that he lacked the vision, temperament and administrative skills of a great (or even successful) editor. With the nineteenth-century liberal nostrums so dear to the hearts of both Dexter and Sifton dominating editorial pronouncements, "The Page" became "stuck in playing the old tunes."[2] And as the *Free Press's* editorial centre of gravity drifted farther and farther away from the Canadian mainstream, its longstanding claim to be *the* newspaper voice of the Prairies—a claim which had been eroding during the latter stages of the Dafoe regime—began to sound ever more hollow. No editorial campaign mounted during the Dexter-Sifton regime better illustrates this decline than the lonely crusade against the Canada-United Kingdom Wheat Agreement of 1946.

Dexter's role in this process was especially tragic. Had he been prepared to offer his services elsewhere, he could have stayed put in Ottawa and virtually named his salary. Having worked his way up, literally from copy boy to the pinnacle of his profession, the prospect of becoming editor, the fitting (and deserved) culmination to a long, loyal, and so far distinguished career, must have been appealing and certainly reinforced his Victorian sense of duty. But the overriding consideration for accepting Sifton's offer seems to have been loyalty to Dafoe's memory and the institution of the *Free Press*, the two now being all but indistinguishable in his mind. Dexter had convinced himself that he was the only man who really understood what the *Winnipeg Free Press* stood for and was trying to accomplish, in other words, he was the only man who could carry on the "Dafoe tradition."[3]

Dexter's conceptions of an editor's responsibilities were totally conditioned by his understanding of how Dafoe had operated. Indeed, there was nothing Dexter prided himself on more than having never held an opinion contrary to Dafoe's on any important question.[4] Thus his basic economic ideas were firmly rooted in the nineteenth-century Manchester liberal school. Admittedly, he could depart from these principles for reasons of practical politics—namely at election time to accommodate Liberal back-sliding on issues like the tariff, just as Dafoe himself had often done—but it was always reluctantly and with obvious embarrassment. While it would be a mistake to think that Dexter's small "l" liberal views were pure *laissez faire*, on most questions they were nevertheless a good deal closer to the economic fundamentalism of the Sifton brothers (Clifford, Jr., a Toronto lawyer and businessman, never played more than a peripheral role in *Free Press* operations) than to the mildly progressive ones adopted by Dafoe during his later years when Dexter, of course, had had little personal contact with him. What he had absorbed from Dafoe in undistilled form, however, was an unflinching commitment to internationalism and collective security, convictions which for both men tied in neatly with the free trade gospel.[5] In Dexter's mind, these positions, with their pronounced moral and idealistic overtones, were absolutely unshakeable. When it came to postwar international relations, Dexter's views were forward looking and widely shared. But in the realm of domestic economic and social issues, all too frequently he found himself advocating lines of policy which seemed only marginally relevant to mainstream opinion. CCF leader M. J. Coldwell once characterized the Dexter-edited *Free Press* as "19th centuryish and not very anxious to find out what is happening in this century."[6] Coldwell was hardly an unbiased observer, but the accusation contained more than a grain of truth. Even Edgar Tarr, a prominent Winnipeg businessman and guiding force of the Canadian Institute of International Affairs and as such a longtime confidant of both Dafoe and Dexter, was critical of the paper's ideological "inconsistency" under the latter's editor-ship and in particular its frequent lapses into an "extreme free enterprise tone."[7] While some of this was clearly attributable to Sifton's growing influence over *Free Press* affairs, one is hard pressed to find any position adopted by the newspaper during Dexter's editorial reign which did not reflect ideological positions he had long since worked out.

In attempting to preserve the Dafoe tradition without Dafoe, Dexter had set himself an impossible task. But from the start there were other problems which cast a pall over his editorship. Dexter was a perfectionist and an extraordinarily hard worker. This, and a phenomenal memory—it was said that he could remember the approximate date and subject of every one of the thousands of articles and columns he had written for the *Free Press*—were key elements in his success as a political reporter and had gone a long way to compensate for his self-admitted deficiencies as a writer. Put bluntly, Dexter hated editorial writing and, indeed, editorial work in general, to the point where prolonged bouts of it frayed his nerves and even made him physically sick. Being high strung and a compulsive worrier only made this worse. Furthermore, Dexter possessed practically no talent for administration and displayed little aptitude for dealing with his supporting cast. It was not that he was unfriendly or insensitive; in fact, he could be a very outgoing and considerate man who, when he alienated colleagues, as editors must inevitably do, found this very difficult to handle. Rather, he simply never felt comfortable working closely with others. In Ottawa he had been able to operate on his own, but in Winnipeg, with day-to-day administrative responsibilities and the all-important task of moulding a group of journalists with their distinctive talents and egos into a smoothly functioning editorial team, such deficiencies constituted real liabilities.[8] Simply put, he lacked flexibility; the Dexter way was the Dafoe way, and the Dafoe way, or rather his interpretation of it, was the only way. Unable to accept or even appear to be swayed by counterarguments on relatively minor points, on matters of "principle," which covered a lot of ground with Grant Dexter, there could be no deviation at all. As a consequence, he too often saw constructive dissent, the life blood of an editorial office, as virtually subversive.[9] While it is true that the power of Dafoe's personality coupled with his prestige had invariably resulted in a meeting of minds on the "big" editorial issues along solidly 'Dafoean' lines, Dexter had misunderstood the process by which this agreement had come about.

In the end, all of Dexter's deficiencies were magnified because of his tendency to be unduly influenced by strong, articulate personalities. Without the commonsense wisdom of a Dafoe, Ferguson or Tarr (who died in 1950) to protect him from himself, he fell increasingly under the sway of Victor Sifton. Obsessed by the desire to become as great a publisher as his father, Sifton was a good businessman and quite probably a good publisher as well, but he was at best a mediocre editor, the role into which he was increasingly deter-mined to impose himself. Despite the thin veneer of mildly progressive views which Dafoe had imparted to him, Sifton's values were deeply conservative, even reactionary. Moreover, guided by a political and economic philosophy encompassing little more than Adam Smith and the unseen hand, his intel-lectual horizons on most public issues were, to be generous, limited. At the personal level, Dexter was no match for such a domineering personality who felt no compunction in badgering and bullying him. When early on Sifton demanded that nothing go onto "The Page" without his express approval, a blatant interference in Dexter's editorial prerogatives which his predecessors would have rejected out of hand, Dexter quietly acquiesced. The fact that, for ideological reasons, he and Sifton seldom differed on editorial points anyway neither masked the significance of what had occurred, nor the impression it created among Dexter's staff.[10]

Despite the many constraints imposed from above and from within, by and large Dexter was still able to produce the kind of editorial page he wanted. It was a Dafoe page," and its most important purpose remained unchanged: to educate public opinion and, in so doing, set a policy agenda that the politi-cians would have to address. "Our target," as editorial colleague and confidant Bruce Hutchison later proudly recounted, "was the political establishment in Ottawa."[11] No English-language newspaper in Canada had traditionally had a greater sense of its own self-importance than Dafoe's *Free Press*, and neither Dexter nor Sifton seriously doubted that their written utterances continued to wield real influence in Ottawa's corridors of power, not to mention in homes and offices across the Prairies. The *Free Press's* five-year crusade against the Canada-United Kingdom Wheat Agreement of 1946, and the government and farm organizations championing it, revealed the extent to which ideol-ogy and nostalgia had come to dominate the paper's editorial thinking in the immediate postwar years.

During the "Dirty Thirties" prairie wheat farmers had been impoverished by crop failures, low prices and contracting export markets. While the war brought good crops, improved prices, high demand, and, in 1943, the long sought after introduction of compulsory marketing, western wheat producers feared that their newfound prosperity would be swept away in the postwar recession virtually every Canadian expected. They had spelled this out in unmistakable terms during that June's federal election when Saskatchewan, long a Liberal party bastion, returned only two government members. The farmers' message was not lost on James Gardiner, Agriculture minister since 1935 and self-appointed representative of western agrarian interests in the government of Mackenzie King, who had won personal re-election by a mere twenty-eight votes.[12] Wheat farmers, Gardiner and his colleagues knew, were clamouring for two specific measures: a permanent end to the speculative, private enterprise grain-marketing system—in other words one constructed around the farmer-owned Wheat Pools with the Wheat Board as a compulsory marketing agency—and an enforceable international pricing agreement. Or as Saskatchewan Wheat Pool president J. H. Wesson put it, farmers wanted "a fair price stabilized over a period of time," and to achieve it, were more than willing to forego short-term economic gains.[13] Certainly, if the opinions of their leaders were reflective of the rank and file, and given the democratic structure of the principal farm organizations, there is no reason to doubt they were, the average prairie grain grower not only accepted, but expected, government intervention in the marketplace to achieve these ends.

The federal cabinet considered wheat policy a political minefield and in a decision they lived to regret, granted Gardiner virtually a free hand to find them a safe path out.[14] With little faith in international negotiations or the multilateralist gospel in general, Gardiner favoured a continuation of the bilateral commodity contract system he had set up with the United Kingdom during the war.[15] With fellow westerner James MacKinnon, the minister of Trade and Commerce, merely tagging along for appearances' sake, Gardiner travelled to London in 1945 to sound out the British government. When the latter expressed interest, Gardiner wasted no time in commencing formal negotiations. Simultaneously, he set out to court influential western farm leaders whose support would obviously be crucial when it came to selling any agreement to prairie grain producers. The discussions between British and Canadian government officials were all but completed before 1 January 1946, and by spring, press and radio reports of an imminent Anglo-Canadian wheat deal were almost daily fare in the western media. Thanks to the network of high-level bureaucratic contacts he had built up during his press gallery days, Dexter was extremely well informed about the progress of the negotiations and as a result knew that the British, having sensed the Canadians' eagerness for a deal, had managed to have their terms met almost to the letter.[16]

The agreement was formally announced on 24 July 1946. Canada was to provide the United Kingdom with a minimum of 600 million bushels of wheat during the next four crop years, 160 million during each of 1946-47 and 1947-48 at a fixed price of $1.55 per bushel, and 140 million during 1948-49 and 1949-50 at prices to be negotiated later but in no case to fall below $1.25 and $1.00 per bushel, respectively. Clause 2(b) of the agreement, which would later become a major source of controversy, included a provision that in negotiating prices for the final two years of the contract, the British would "have regard to" any difference between the fixed price of $1.55 in force during the first two years and the so-called "world price" prevailing during the same period. Inserted at the insistence of the Canadian negotiators, it was supposed to ensure that wheat farmers would be compensated for any losses they incurred in the eventuality, then considered highly unlikely, that the world price rose above $1.55 per bushel during the contract's first two years. At least this was the inference drawn by most Canadians, Gardiner included. Even though the world price stood at $2.00 per bushel in July 1946, the sheer magnitude of the sale, representing nearly one and a half years' normal exports, and the simultaneous introduction of several other popular measures successfully calmed any reservations most farmers may have had. And with their support assured, any chance Parliament would examine the deal critically vanished.[17]

In western Canada, the region most directly affected by the announcement, the initial reaction of many of the dailies was positive. Even the Sifton-owned *Saskatoon Star-Phoenix*, after a decent pause, endorsed its terms.[18] Farm leaders, several of whom had just returned from a tour of England and the continent which had left them more anxious than ever over the short-term export prospects for Canadian wheat, saw much to approve. As the editor of the *Western Producer*, the Saskatchewan Wheat Pool's own weekly, concluded:

> The bilateral agreement with Great Britain is better than no agreement at all and if it proves to be merely the first step to a world agreement such as the organized farmers of Canada have advocated for the last twenty years ... so much the better.[19]

As the terms of the deal were debated at farm gatherings that fall, it seemed even more clear that Gardiner's initiative, and especially its income-stabilization aspect, had the overwhelming support of the "organized farmers" on the Prairies. From their perspective, not the least of the pact's advantages was that the sheer size of the contracted amounts effectively guaranteed the government's continuation of the Wheat Board marketing monopoly at least until 1950.[20]

All of this, of course, was antithetical to the position of the *Winnipeg Free Press*. For as long as anyone could remember, the newspaper's editorial page as well as that of its even more widely circulated weekly version, the *Free Press Prairie Farmer*, had fervently supported free trade with no qualifications, and there was no danger that this would change.[21] Of course, being devoted economic liberals, Dexter and Sifton were also strong advocates of private grain marketing, just as Dafoe, too, had been. That Sifton was a social intimate of the most powerful of these interests, starting with the Richardson family, merely served to reinforce this ideological commitment. No institution was more universally despised by prairie farmers, yet found a more receptive hearing at the *Free Press*, than the Winnipeg Grain Exchange. By preaching the old free-market nostrums, Dafoe had lost touch with the thinking of prairie farmers during the 1930s, and with Dexter and Sifton in charge, the chances of the chasm being bridged were nil.

From the Sifton-Dexter perspective, the Canada-United Kingdom Wheat Agreement violated every tenet of sound economics. The first editorial criticism actually appeared while the negotiations were in their final stages, a sort of preemptive strike directed equally at the Liberals and the leaders of the prairie farm organizations. Dexter had minced no words: the rumoured bilateral trade pact was a clear betrayal of economic multilateralism, the American-inspired postwar movement toward unrestricted free trade which had many adherents in Ottawa and a true believer in the editor's chair at the *Free Press*. To "prejudice the most hopeful movement of modern times," an alarmed Dexter warned, "would be little less than an act of treason."[22] In fact, there was considerable opposition to the proposals within the government. While deviation from the multilateralist gospel was not by itself a sin to the economic mandarins and their colleagues at External Affairs, the possibility of American retaliation if Washington were to conclude that Canada was backsliding on its commitment to trade liberalization caused all of them nightmares.[23]

To the surprise of no one, the *Free Press*, along with the other Sifton-owned daily in Saskatchewan, the *Regina Leader-Post*, greeted the formal announcement that a deal had been struck with disdain. Time would show, the former warned, just how bad a bargain it was.[24] But the great majority of prairie grain growers and their spokesmen had little faith in the gospel being preached by the multilateralist evangelists[25] and not surprisingly, given the paper's bias against the Wheat Pools and in favour of the "open market" (a euphemism for the Grain Exchange), they were highly suspicious of the *Free Press's* motives.[26] To Dexter, the leadership of the three Wheat Pools and the Canadian Federation of Agriculture with their dangerous collectivist tendencies were at best dupes in this affair and certainly fair game for criticism. Meanwhile, farmers, in droves, began cancelling subscriptions to the *Free Press Prairie Farmer*, an enormous moneymaker for the Siftons. Yet, as during Dafoe's unpopular Munich stand in 1938, this was one of those occasions where Victor Sifton put principle (and ideology) ahead of the balance sheet, and Dexter was told to press ahead.[27] For his part, Dexter was stubbornly convinced he was right and nothing would have deterred him. If he seemed to be making no headway, it was reassuring to recall that Dafoe had never wavered if a principle was involved and yet had always been borne out in the end.[28] There would be no wavering by the troops either. When Dexter realized that James Gray, his replacement in Ottawa, actually favoured the deal and was not prepared to toe the *Free Press* line in the articles he submitted Dexter promptly fired him.[29] The wheat agreement and the threat it posed to the free market had become Dexter's *idée fixe.*

By the spring of 1947, both editor and publisher had convinced themselves that their attacks were having a marked effect on farm opinion. Yet more objective observers could find no evidence that it was changing, a conclusion which the public utterances of farm leaders and informal polls of farm groups seemed to support.[30] But the *Free Press* had still been "proven" right on one score: the world price of wheat had not fallen.[31] In fact, it had risen steadily and in the early months of 1947 stood at $3.10 per bushel, thereafter declining only gradually. Dexter missed no opportunity to roast the government, and Gardiner in particular, over their "blunder," a delighted in printing running totals of the wheat farmers' mounting "losses" under the contract.

Although Dexter's opposition to the wheat agreement was primarily ideological, his personal dislike for Gardiner played an important secondary role. While Gardiner was generally viewed by western farm organizations as a friend and usually reliable ally in the defence of their interests,[32] the *Free Press* had rarely had anything positive to say about him. Apart from other considerations (and there were several particularly damning ones), his pretensions as a rival western spokesman undoubtedly placed him beyond the pale. Dexter, however, carried the paper's long-established animosity to new lengths. Criticisms of the government were flowing quite freely on this particular issue, and they always implicated Gardiner as the chief culprit. Ever the moralist, Dexter found Gardiner's unconcealed ambition repugnant, and his flagrant patronage dealings sullied his image further in the *Free Press* editor's eyes. Moreover, as Mackenzie King's "minister of elections" for the Prairies, Gardiner's recent record in delivering seats was pretty dismal, no small consideration for a big "L" Liberal like Grant Dexter. Finally, there was the fact that many of the best and brightest among the senior bureaucrats, men whom Dexter placed on a pedestal even if he did not accept the wisdom of all their policies, heartily detested Gardiner as the archetype of the old-style patronage minister: parochial, isolationist and generally unreceptive to new ideas. From Dexter's perspective, this was a crippling indictment.[33]

By the end of the contract's first year, 31 July 1947, the *Free Press* trumpeted that wheat farmers' "losses" had reached a staggering $100 million. This was nothing short of a tragedy, the paper moaned.[34] Even government spokesmen admitted under intense questioning in Parliament that producers had suffered a substantial "paper loss" during 1946-47 and were likely to face another in the coming crop year.

James Garfield Gardiner (1883-1962). When this photograph was taken in 1957, James G. Gardiner had served as federal minister of Agriculture for 22 years; he retired the following year. (Courtesy of the Saskatchewan Archives Board/R-B7146-4.)

Gardiner, meanwhile, hurriedly embarked on a speaking tour of the West in an effort to salvage his reputation and that of the government, dismissing "the gloomy critics of a year ago" and reminding his farm audiences that complete protection was built into the agreement through the "have regard to" clause. Doubts were growing, however, about exactly what Clause 2(b) meant[35] and Gardiner's position had become uncomfortable enough without the continuous heckling from the *Free Press*. First he tried to convince Dexter that his flood of speeches supporting the deal were actually a ploy to trap his own cabinet colleagues into committing themselves to compensation if the British proved "difficult," but the *Free Press* editor would not buy it. Gardiner then adopted another tack, pointing out to the paper's new Ottawa bureau chief, Max Freedman, that no matter what one thought of the agreement and the farm leaders, the former had turned the latter into government supporters "and this was not an asset to be lightly cast aside." But Gardiner, a partisan but pragmatically minded politician, underestimated Dexter's almost religious commitment to his economic principles. As Freedman joked, such a nakedly partisan appeal "won't butter very many parsnips with us," and in this case it was true.[36] The *Free Press's* editorial position was not going to shift an inch. Gardiner, of course, was not the only Liberal left muttering over the damage the paper was doing to the government's reputation in the West. In fact feelings were running hot, as Dexter found out during a one-sided exchange at a dinner party in Ottawa later that summer when an exasperated Jack Pickersgill let him know to his face exactly how the party felt.[37]

Gardiner had no leverage at all with the *Free Press* because, frankly, not only did its editor not like him, he did not need him. The ability to tantalize with "inside" information, the tie that normally bound cabinet ministers to editors, particularly to politically sympathetic ones, was not a factor in the Dexter-Gardiner relationship. While Dexter could now visit Ottawa only infrequently, his extensive network of civil service contacts remained intact. When it came to economic questions, by far the most important of these contacts was John Deutsch. The Saskatchewan-born Deutsch had been a friend of Dexter's for years. Toward the end of the war, worn out by the bureaucratic grind, Deutsch had been persuaded by Sifton to come to Winnipeg as an editorial and feature writer and general "economic advisor." For someone used to the centre of action, journalism was bound to prove to be a pretty mild stimulant, and with Clifford Clark, the deputy minister of Finance, constantly pestering him to return to Ottawa, Deutsch's career as a newspaperman lasted only thirteen months.[38] Thereafter, from his position as director of the International Economic Relations Branch of the Department of Finance, Deutsch was exceptionally well placed to pass on sensitive intelligence on trade matters when this suited his or Clark's purposes.[39] Among the postwar economic mandarins, Deutsch was as orthodox as they came, and this, combined with his commitment to economic multilateralism and suspicion of British intentions toward Canada (the latter was rapidly becoming almost part of the job description for Ottawa's economic advisors), jibed neatly with Dexter's and Hutchison's own thinking. Beyond Deutsch, Dexter could also count on the "Winnipeg Mafia" at the Bank of Canada Research Department: M. C. McQueen, J. Robert Beattie and James Coyne, as well as another Winnipegger at Finance, Mitchell Sharp.[40] And there were many others, too, including the ever helpful Mike Pearson, the newly appointed undersecretary of State for External Affairs who had been a close friend and confidant of Dexter's since their prewar days in London. As these civil servants became more and more disillusioned with the wheat agreement, Dexter's editorial stand earned him a grudging respect, and his

Ottawa men some confidential "briefings," even from those who considered his economic views more than a touch out of date.[41]

By the autumn of 1947, London and Ottawa had agreed on a $2.00 per bushel price for the contract's third year. But unfortunately for Gardiner and the government, the world price would average $2.23 during the 1948-49 crop year.[42] More seriously, until the last minute, the Agriculture minister had been confidently assuring any farmers who would listen that Britain could be counted upon to compensate producers for the "losses" they had incurred as a result of being paid lower than world prices during the first two years of the contact. If that was not enough, he further raised expectations by categorically stating that Finance Minister Abbott's talks in London, talks undertaken, among other reasons, to sort this matter out, had been "very satisfactory for us.[43] Thanks in part to such wishful thinking, questions of the British government's "fair dealing" were now front and centre. This was sufficiently worrisome politically that in announcing the 45 cent per bushel increase, Prime Minister King immediately attempted to defuse the issue:

> Having in mind the magnitude of the agreement and the long-term security which it provides, a precise arithmetical calculation of the difference in price was not suggested [by Canadian negotiators].[44]

Among prairie grain growers, only incorrigible Anglophiles could have swallowed this explanation at face value. Nevertheless, while bitterly disappointed with Britain's "ingratitude," not to mention their less than cooperative attitude in ongoing negotiations to hammer out a world wheat pricing agreement, all the leading farm spokesmen in the West including the three Wheat Pool presidents and H. H. Hannan, the president of the Canadian Federation of Agriculture, continued to defend the market stabilization aspects of the wheat agreement while warning their followers to ignore critics who were plainly fronting for the hated grain exchange. It was typical Wheat Pool rhetoric and all of it had been aired many times before: the interests of 250,000 honest producers pitted against 160 parasitic grain brokers and the "line" elevator companies. In a speech before the Saskatoon Board of Trade at year end, J. H. Wesson bluntly stated the organized farmers' case. Despite the propaganda of the "gentlemen of the Winnipeg Grain Exchange and their supporter, the *Winnipeg Free Press*," the head of the powerful Saskatchewan Wheat Pool categorically asserted, farmers did not want to scrap the British wheat agreement in favour of the open market and a so-called "world price."[45] For their part, the "grain interests" were not exactly raising the level of the debate by running full page ads in the *Western Producer* hysterically accusing the Wheat Pools of supporting the agreement to advance their own monopoly marketing aspirations in the "first step toward communism."[46]

For Dexter the defences hurled up by Hannan and the others were cockeyed and too ridiculous to pass unchallenged. He especially delighted in pointing out that Australian farmers had a compulsory government-run marketing board and still managed to sell their entire crop at the higher world price. What they had wisely avoided, Dexter sarcastically reminded his critics, was the trap of fixed price, bilateral export contracts.[47]

During the ensuing winter, the question became not how much Canadian wheat farmers would receive but whether they would receive anything at all. The United Kingdom was virtually bankrupt and, indeed, for some time had actually been paying for most of its Canadian food imports with dollars from Canadian government loans. However, given Canada's own precarious financial position, continuing this strategy was unthinkable. A solution finally came when London and Ottawa persuaded Washington to allow the diversion of Marshall Plan dollars to "off shore" purchases of foodstuffs, that is, purchased from Canada. But in the meantime, egged on by the economic bureaucrats and the minister of Finance, Douglas Abbott, the *Free Press* had a field day ridiculing the Agriculture minister's apparent ignorance of the cold, hard facts of postwar economic reality. Britain, Dexter's sources all agreed, was a spent force and the American market was the key to Canada's future prosperity. Public opinion toward gallant Britain prevented him from putting the message in print quite so bluntly, but his editorials were clear enough.[48]

The criticism that Gardiner was out of touch when it came to trade matters, uttered privately within the government and publicly by the *Free Press*, was valid enough. But for Gardiner, the Canada-United Kingdom Wheat Agreement was more of an emotional than an intellectual issue anyway. Everyone agreed that prairie grain growers had never been so prosperous; Gardiner pointed to the wheat contract and rested his case.[49] As for the farmers, while there was unquestionably growing dissatisfaction over the price issue, support for the agreement seemed to be holding and assertions by the *Free Press* editor to the contrary drew a sharp rebuttal from the farm press. "Anyone possessed of the slightest capacity for the most elementary analysis can see through him [Dexter] at a glance," scoffed the *Western Producer*. "His sole concern is to eliminate the Wheat Board [and] re-establish the Grain Exchange in all its ancient glory,"[50] which was really not all that far off the mark.

Needless to say such criticism bounced off Dexter, and his attacks continued unabated through the summer and into the autumn of 1948. By now, the arguments were standard enough, rarely changing even in their order of pre-sentation. The lead editorial in the 25 September issue of the paper was typical:

> In this period of high prices, as a result of policies enacted by the Dominion Parliament, at the request of the Wheat Pools, the wheat producers lost vast sums of money. The farmers have not got a shred of an asset to show for these losses. Not a dollar of them lies to their credit in any fund of the Wheat Pools, the Wheat Board or the Dominion Government. Nobody owes the farmer a penny of them. This money has been irretriev-ably lost by reason of the bad judgement of the Wheat Pools' and unwisdom of the Dominon Government, particularly of the Hon. J. G. Gardiner, in accepting their advice. The policy of using the peaks in prices to fill in the hollows has failed in Canada, in a period of unprecedentedly high prices, because the peaks have been deliberately thrown away.[51]

Despite brave talk by Gardiner and other farm spokesmen about "recouped losses" and how Great Britain could be "trusted to keep its word" and "live up to the spirit of the 'have regard to' clause," the real situation was anything but promising. No amount of wishful thinking, or denunciations of the mes-senger—the grain trade's "newspaper stooges"—could mask that for long.[52] Meanwhile, negotiations to fix the price for the fourth year dragged on interminably. Far from recovering any of their first two years' losses during the contract's third year, farmers had "lost" another $30 million, making the cumulative total according to the *Free Press's* calculations a whopping $357 mil-lion. Barring an extension of the agreement, the coming year, 1949-50, would be Canada's last opportunity to invoke Clause 2(b). With the full backing of his cabinet colleagues, Gardiner again took a direct hand in the negotiations, but the end-of-the-year deadline came and went without agreement. The Attlee government saw no need to compromise and the tone of the negotia-tions deteriorated accordingly. At last, on 20 January 1949, Ottawa revealed that the contract price would again be set at $2.00 per bushel. Meanwhile, the question of compensation under Clause 2(b) had been shelved until the agreement's termination some eighteen months hence. In fact, this aspect of the negotiations had gotten nowhere, but at least a potential embarrassment had been delayed until after the next federal election.[53] Thanks to Pearson and Deutsch in particular, Dexter had a virtually complete account of the negotiations and how badly the Canadians had fared. In this light he could see that as far as the British government was concerned, they had met their responsibilities under the "have regard to" clause completely.[54]

The situation was now very confusing. While the support of the Wheat Pools held firm and Gardiner put a positive gloss on it all, frustrated civil servants confided to Hutchison that the only certainty now was an eventual federal payout to placate angry prairie wheat farmers.[55] The unravelling of the Canada-United Kingdom Wheat Agreement had become a "national problem" and therefore a national story. Most major dailies adopted a moderate view, the general feeling being that the government had made an honest error in misjudging postwar price trends, and that the resultant "paper losses" should not be allowed to detract from the tangible benefits Canadian wheat farmers had enjoyed.[56] Only the *Winnipeg Free Press* editorial page, and those of its two sister publications in Saskatchewan, adopted an outright condemnatory position. The *Saskatoon Star-Phoenix*, once a fulsome supporter of the contract, had been brought to heel earlier in the year. Since his appointment in 1946, its editor, Burton Richardson, had managed to maintain a fair degree of editorial independence from the "Winnipeg line," something both Victor Sifton and Dexter professed to expect and respect. However, when Richardson's editorial page emerged as a strong proponent of the wheat agreement and all that it entailed, including the Wheat Board, his days were clearly numbered. By the autumn of 1948, Richardson's failure to reprint the endless attacks on the deal churned out by the *Free Press* while simultaneously publishing his own editorials mocking the "policies of rugged individualism that are still favoured in the sheltered crannies of Winnipeg" had exhausted Dexter's patience and he persuaded Sifton to fire him.[57] Perplexed *Star-Phoenix* readers thus received their copies of the paper on 2 November only to find Richardson's name hastily removed from the masthead. There was no explanation then or later. No chances were taken in choosing a replacement, however; when the appointment came, it was none other than Clifford Sifton, Jr.

The announcement of a federal election for 11 June 1949 brought a toning down of *Free Press* attacks on Gardiner and the Liberal government; it was "the higher wisdom induced by the election in so far as our comment on the wheat contract is concerned," as a bemused Freedman put it.[58] The Liberals, under their new leader, Louis St. Laurent, scored a resounding victory, with Gardiner's home province, Saskatchewan, returning fourteen government members, a dozen more than in 1945 and a result which Gardiner understandably interpreted as a vote of confidence in the wheat contract. Of course with the election safely out of the way, the *Free Press* felt free to resume its attacks.

All that remained now was to negotiate such compensation for the first two years' losses as the British government would agree to, and given a British official's recent boast to Freedman that they planned "to cut us off at the pockets," Canada's prospects looked none too promising.[59] This was certainly the feeling in the East Block. According to Deutsch, the farmers had been duped by Gardiner whose own naivete had made him easy pickings for the unscrupulous British. When farmers finally realized the wheat agreement had blown up in their faces, he fumed, there would be hell to pay politically and obviously the government would bear the brunt. The *Free Press*, Deutsch added approvingly, was now "magnificently vindicated."[60]

Understandably, Dexter now weighed in, repeatedly attacking the folly of the bilateral deal and the duplicity of the Wheat Pools and Gardiner in denying grain growers their deserved profits. Neither the source nor content of these attacks much impressed farm leaders, however. They steadfastly continued to endorse the income-stabilizing aspect of the agreement. The organized farmers' movement, it has to be borne in mind, had always seen the deal as nothing more than a passing stage in the long struggle toward a more stable grain marketing environment built around government subsidies, an international pricing agreement, the Wheat Pools and a compulsory, government-controlled system of "orderly" marketing.[61] Fundamentally, the issue at stake remained the entrenchment of the Wheat Board versus the reestablishment of private grain trading, and it was in that light that officials at the Wheat Pools and Federation of Agriculture saw the *Free Press's* relentless attacks on the wheat contract. Thus, in responding to Dexter's contention that the 1946 agreement had been a plot between the government and the Wheat Pools to defraud prairie farmers, the *Western Producer* gave vent to years of farmer suspicion. To what end did the *Free Press* keep bolstering its "hopeless contention" with "sheer invention and outrageous misrepresentation?"

> There can be only one conclusion. The *Free Press* has not given up hope of destroying the fanners' organizations for the purpose of restoring the grain trade and Grain Exchange of the City of Winnipeg to all the predatory glories of their heyday.[62]

The thankless task of burying Clause 2(b) fell to C. D. Howe, the new minister of Trade and Commerce. Discussions began in London in early May 1950. Rumours in Canada that the extension of the 1946 agreement might be in the works were a pipe dream. At month's end, Howe announced that the United Kingdom would not budge from its view that they had honoured all their obligations, legally and in spirit. For understandable reasons, this was not the most palatable news for the wheat agreement's Canadian champions, and by late autumn, Deutsch's scenario of farmers descending on Ottawa, outraged and bent on compensation, would come to pass.[63]

At last, the *Free Press* really did feel "totally vindicated."[64] During the next two months, Dexter wrote no fewer than seventeen editorials on the subject. Most of them created more heat than light, however, since key parts of the *Free Press's* "factual analysis" were pretty self-serving, as its critics acidly pointed out in their responses.[65] The paper, for example, made much of the "world price" without admitting that it would have been considerably lower than the figures it quoted had 600 million bushels of Canadian wheat not been locked up in a bilateral contract. Similarly, there was obviously no guarantee that the British would have purchased any wheat at all from Canada at other than the discounted (or at the time, at least "protected") price they had received. As it was, contractually tied to Canadian suppliers, the government of the United Kingdom had successfully pressured the Americans to permit them to divert their Marshall Plan dollars to Canada, a major help to Canada's own American-dollar position in 1948.[66]

Predictably, Howe's announcement sparked a free-for-all in Parliament as those involved scrambled to deflect blame elsewhere. The tone of the *Free Press* and the two other Sifton papers was predictably self-righteous. The federal government had bet on sliding postwar prices and lost; the farmers had their much desired "market security" and now were whining about having to pay the price for it.[67] But to other prairie dailies, the Canada-United Kingdom Wheat Agreement had clearly brought economic benefits to the region's chief industry. There was agreement that Britain's failure to live up to the spirit of the contract had rightly left a sour taste in many westerners' mouths, and a few criticisms that Ottawa could have been a tougher negotiator. Over all, however, the prevailing editorial sentiment was that the contract, while flawed, had been good for the country. This was also the official verdict of the Saskatchewan Wheat Pool.[68] Editorially speaking, the farther removed from the Prairies, the more moderate and favourable the assessment. Distance, both geographical and emotional, seemed to bring a clearer perspective.[69]

In February 1951, the prime minister despatched Gardiner to the United Kingdom with, among others, J. H. Wesson in tow, in a final effort to negotiate some kind of settlement of the "losses." This futile initiative was undoubtedly a ploy to help prepare the political groundwork for a modest government compensation package. Of course, the failure of Gardiner's odyssey occasioned another editorial barrage from the *Free Press*. Dexter dismissed the now widespread accusations of British perfidy as a cheap smokescreen designed to obscure the real issue, namely the "straight CCF socialist policy" of government contracting. "From the outset," he pontificated, "the result was inevitable … but the loss will not have been in vain if the producers learn that the only guarantee of selling their wheat for what it is worth is the market."[70] The government promptly allocated the unspent portion of the Canadian loan to Britain, some $65 million, as compensation and closed the books on the wheat agreement with a collective sigh of relief. Press comment was slim and generally critical of the straight politics of the bailout and seemingly endless demands of prairie farmers for more help.[71] The *Leader-Post*, after taking the inevitable slap at the role of the Wheat Pools in the whole "fiasco," offered Gardiner a rare compliment for his role in prying the additional money from the federal treasury. The farm movement, while by no means satisfied with the compensation provided, nevertheless accepted that Ottawa had acted in good faith and, regardless, feared, like their ally Gardiner, that any evidence of division among farmers would simply play into the hands of the Grain Exchange and the *Free Press*.[72] Howe defended the agreement in the House of Commons in light of the circumstances of 1946, calling it "an experiment in stability" and making the valid point that "the time when you must decide whether an agreeement is good or bad is when you sign on the dotted line and we agreed on that dotted line in 1946."[73] Neither observation reflected the private views of most of the senior bureaucrats or even many of his cabinet colleagues, and it seems unlikely they were Howe's, either.[74] Feelings were particularly strong on the question of whether the British government had acted honourably. Hutchison had received an earful from Finance Minister Abbott, the gist of which was that "the have-regard-to clause … meant something in morals and the British [had] ratted on it," and Bank of Canada Governor Graham Towers, Hutchison reported, was equally bitter.[75]

Dexter could not have agreed more with Howe's public statement. From the first rumours that a bilateral contract was in the works, he had used the *Free Press* editorial page to oppose any such measure. No single issue during Dexter's tenure as *Free Press* editor so obsessed him, and the reason was simple: the wheat agreement violated his fundamental economic beliefs. It was the same reason that he fought the Wheat Pools and a compulsory Wheat Board. That the deal seemed to "fail" merely confirmed his opposition; it did not motivate it. Despite the stresses and strains, Dexter's five-year *Free Press* crusade against the contract left him feeling very satisfied. The issue was straightforward, and there were no discomforting doubts about the justice of the cause. Certainly it was emotionally and intellectually reassuring to be able to write the same sort of polemical attacks on the prairie apostates of the liberal economic gospel that he had written for Dafoe on so many earlier occasions.

Still, exactly what had he and the *Free Press* achieved? In terms of influencing government policy, obviously not very much. True, there would be no more fixed-price bilateral wheat contracts with the United Kingdom or anyone else, but the newspaper was pretty far down the list of the factors responsible for that. Within Liberal party circles, the ceaseless criticism, and particularly the *Free Press's* unwillingness to let the whole embarrassing business die after the agreement's expiry on 31 July 1950, left a residue of ill feeling.[76] The tendency of publications with Conservative sympathies and opposition MPs to use material from the *Free Press, a* newspaper which the public identified with the government cause, to give their attacks added credibility had been especially galling to Liberal party loyalists. And certainly one could not point to any evidence that the free-market sermons had wooed many farmers away from their support for the Wheat Pools and the Wheat Board. As the *Western Producer* put it:

> Isolationism is a waning force in the world today. But there is one spot where it still flourishes in full vigor. The editorial office of the *Winnipeg Free Press* which blandly presumes to speak for Western agriculture remains cut off, voluntarily no doubt but obstinately, from any contact with the prevailing currents of thought among Western farmers.[77]

What then had been achieved? Well, a great principle had been upheld, and that would have to—and for Dexter and Sifton, probably did—suffice.

Editorial campaigns of this magnitude took a heavy emotional and physical toll on Dexter, who found the job made for "an intolerable life."[78] Victor Sifton's relentless demands and constant interference in the running of the paper had turned Winnipeg into a pressure cooker for editors. After only two years on the job, Dexter's blood pressure had become a serious concern. By 1950, Sifton was sufficiently alarmed about Dexter's health, and, one might add, disenchanted with his performance as an editor, to consider posting him back to Ottawa, but finding a suitable replacement delayed this move for another four years by which time Dexter's health was broken.

In a revealing letter written shortly after Victor Sifton's death, Dexter outlined his conception of the *Free Press* editor's role in the post-Dafoe era:

> The [editorial] problem is not so much the thinking out of new policies because the *Free Press* has not really had any new policies of a fundamental kind since J. W. made the paper editorially after World War I. The great problem, and it is a very real one, is to apply our fundamental policies to the current, ever-changing, always baffling events as they occur.[79]

That was Dexter's answer to Ferguson's parting warning "to stop looking over your shoulder at the past."[80] He failed to recognize that Dafoe's ideas, if seldom especially new or flexible, had at least been a response to their own times and circumstances. A reluctant editor who took on the responsibility more out of loyalty than ambition, Dexter could see no other way to shape the editorial direction of the paper. He was simply not a man of original thought, nor was he intellectually flexible. Rather, he sifted through others' ideas, retaining the most persuasive of these as his own, and the most persuasive of all had been Dafoe's. When, at age fifty, he assumed the editorship of the *Free Press*, his mind was certainly not empty, but it was closed. And that assessment, as Ferguson later observed, could be applied to Victor Sifton, only more so.[81] Sometimes, as in the case of the country's postwar internationalist impulse, this did not pose a problem. But more often than not it did. In too many instances, change was found wanting because it violated some time-honoured principle, and the *Free Press* was left opposing or, at best, approving change halfheartedly. Official Ottawa, where Dexter had so many friends and was well respected as a man, was not long in noticing this.[82] The penalty, of course, was to be paid by being out of touch with the mood of the paper's readers and of the region for which one aspired to speak. In the case of its response to the Canada-United Kingdom Wheat Agreement, Dexter and Sifton had ensured that while the *Free Press* might fight its editorial battles with the old determination and flare, it would be "stuck playing the old tunes" and be increasingly irrelevant as a consequence.

NOTES

This article first appeared in *Prairie Forum* 18, no. 1 (1993): 77-96.

1   William Christian, ed., *The Idea File of Harold Adams Innis* (Toronto: University of Toronto Press, 1980), 32; Library and Archives of Canada (LAC), MG32B5, Brooke Claxton Papers, vol. 31, George and Mary Ferguson file, Claxton to the Fergusons, 23 May 1946; interviews with Mrs. Alice (Grant) Dexter, James Gray and Burton Richardson; and, Queen's University Archives (QUA), 2142, Grant Dexter Papers, vol. 4, file 29, Ferguson to Dexter, 29 October 1946.

2   Tom Kent, *A Public Purpose: An Experience of Liberal Opposition and Canadian Government* (Montreal and Kingston: McGill-Queen's University Press, 1988), 6.

3   Interviews with Susan Dexter (Dexter's daughter), James Gray and a confidential interview.

4   University of Calgary Archives (UCA), MsC22, Bruce Hutchison Papers, 1.2.7, Dexter to Malone, 10 September 1961.

5   Interviews with Susan Dexter, Wilfrid Eggleston and Wilfred Kesterton; QUA, Dexter Papers, vol. 3, file 26, Dexter to Victor Sifton, 8 April 1944; and, James Gray, *Troublemaker! A Personal History* (Toronto: Macmillan, 1978), 73-75.

6   QUA, Dexter Papers, vol. 3, file 26, Freedman to Dexter, 12 January 1948.

7   LAC, Claxton Papers, vol. 67, Tarr file, Tarr to Claxton, 19 July 1948.

8   Interviews with Eggleston, Gray, Victor Mackie and William Metcalfe; William Metcalfe, *The View from Thirty* (Winnipeg: n.p., 1984), 126-28; and Ruth Wilson Papers (private collection, Toronto), Dexter to Wilson, 18 February 1952. Gray, Mackie and Metcalfe were all senior journalists with the *Free Press* during part of Dexter's tenure as editor.

9   Gray, *Troublemaker*, 191-93 and Metcalfe, *View from Thirty*, 131.

10   Interviews with Gray and Metcalfe; QUA, Dexter Papers, vol. 4, file 28, Dunning to Dexter, 20 May 1946.

11   Bruce Hutchison, *The Far Side of the Street* (Toronto: Macmillan, 1976), 195. See also ibid., 178-89; Peter Demon, *Assignment Ottawa: Seventeen Years in the Press Gallery* (Toronto: General Publishing Co., 1968), 17-18; Gray, *Troublemaker*, 196-97; and LAC, MG30D289, Burton Richardson Papers, vol. 4, file 36, Ferguson to Richardson, 16 February 1951.

12   One of the Saskatchewan Liberal MPs to go down to defeat was, of course, the prime minister.

13   *Western Producer*, 4 April 1946. See also ibid., 3 and 24 May 1945 and 4 July 1946; Vernon C. Fowke, *The National Policy and the Wheat Economy* (Toronto: University of Toronto Press, 1957), 273-76 and D. A. MacGibbon, *The Canadian Grain Trade*, 1931-1951 (Toronto: University of Toronto Press, 1952), 119-20 and 138-39.

14   LAC, Mackenzie King Papers, MG26J, Diaries, 16 July 1946.

15   *Western Producer*, 25 January 1945.

16   QUA, Dexter Papers, vol. 20, file 12, Canada-United Kingdom Wheat Agreement memo, 8 June 1950 and MacGibbon, *Canadian Grain Trade*, 120-21.

17   Fowke, *National Policy*, 277 and MacGibbon, *Canadian Grain* Trade, 121-23.

18   *Saskatoon Star-Phoenix*, 22 August 1946. See also *Edmonton Journal-Bulletin*, 27 July 1946 and *Winnipeg Tribune*, 26 July 1946.

19   *Western Producer*, 12 September 1946. See also ibid., 27 June and 8 and 15 August 1946.

20   Ibid., 8 August 1946. See also ibid., 21 November and 19 December 1946.

21   Interview with Mitchell Sharp; LAC, Richardson Papers, vol. 3, file 29, Ferguson to Richardson, 10 March 1944; and LAC, Clifford Sifton, Jr. Papers, vol. 4, Victor Sifton 1944 file (2), Clifford to Victor Sifton, 23 November 1944 and June-December 1945 file, 6 September 1945.

22   *Winnipeg Free Press*, 19 June 1946. See also ibid., 22, 25, 26 and 27 June and 20 July 1946.

23   Interview with Mitchell Sharp and J. W. Pickersgill; and D. F. Forster, *The Mackenzie King Record*, vol. 3: 1945-1946 (Toronto: University of Toronto Press, 1970), 262-64.

24   *Winnipeg Free Press*, 27 July and 3, 9 and 12 August 1946; *Regina Leader-Post*, 29 July 1946.

25   *Western Producer*, 25 January and 29 November 1945.

26   Gray, *Troublemaker*, 184 and MacGibbon, *Canadian Grain Trade*, 123.

27 Interview with Gray; QUA, Dexter Papers, vol. 7, file 50, Dexter to Malone, 1 May 1961.

28 UCA, Hutchison Papers, 1.2.7, Dexter to Malone, 10 September 1961.

29 QUA, 1022A, John Deutsch Papers, Personal Correspondence, vol. 1, file 5, Sharp to Deutsch, 20 March 1947; Gray, *Troublemaker*, 183, 188 and 191-92; and LAC, Richardson Papers, vol. 3, file 31, MacKinnon to Richardson, 28 January 1947.

30 For instance, a radio poll of farm forum groups conducted in the spring showed that, if the income stabilization aspects of the wheat deal were emphasized, the percentage of farmers approving (and approving with "some reservations") in Alberta, Saskatchewan and Manitoba were, respectively, an astonishing 76% (24%), 68% (24%) and 94% (4%). *Western Producer*, 2 May 1947. See also QUA, Deutsch Papers, Personal Correspondence, vol. 1, file 5, Sharp to Deutsch, 20 March 1947 and *Western Producer*, 6 February and 13 March 1947.

31 In the absence (prior to 1949) of an international wheat pricing agreement between producers and consumers, there was no "world price" as such. The term, or its Canadian synonym, the "Class II price," basically referred to the sum American farmers were receiving for their exports, with "Class I" referring exclusively to the wheat sold to Britain. During this period, all other Canadian exports were being sold by the Wheat Board at or near the Class II price. MacGibbon, Canadian Grain *Trade*, 125.

32 Norman Ward and David Smith, *Jimmy Gardiner: Relentless* Liberal (Toronto: University of Toronto Press, 1990), 258 and 266; and *Western Producer*, 22 March 1945 and 4 July 1946.

33 Interviews with James Coyne and Gray; QUA, 2117, Thomas Crerar Papers, II, vol. 88, Correspondence 1948 file, Tarr to Johnston, 24 March 1948; QUA, Dexter Papers, vol. 5, file 32, Freedman to Dexter, 16 January 1948 and file 37, memo, 30 June 1950; and LAC, Richardson Papers, vol. 4, file 34, Ferguson to Richardson, 8 July 1948.

34 *Winnipeg Free* Press, 21 July 1947.

35 MacGibbon, Canadian *Grain Trade*, 126-27. *Western Producer*, 16 October 1947.

36 QUA, Dexter Papers, vol. 4, file 31, Freedman to Dexter, 20 September 1947. See also ibid., file 30, Dexter to Victor Sifton, 30 May 1947.

37 LAC, Richardson Papers, vol. 3, file 32, Maudie Ferguson to Richardson, 10 September 1947.

38 QUA, Deutsch Papers, Personal Correspondence, vol. 1, file 4, Deutsch to Bennett, 29 March 1945 and Deutsch to Ferguson, 2 April 1946.

39 QUA, Crerar Papers, II, vol. 88, Correspondence 1952 file, Freedman to Hutchison, 21 December 1952 and Dexter Papers, vol. 4, file 31, Freedman to Dexter, 20 September 1947 and vol. 5, file 36, Hutchison to Dexter, 17 October 1950 and file 37, 4 February 1950.

40 Interview with R. B. Bryce (a senior official with the Finance Department during this period).

41 Interview with Sharp; and QUA, Dexter Papers, vol. 4, file 31, Freedman to Dexter, 20 September 1947.

42 MacGibbon, Canadian *Grain Trade*, 127-28.

43 *Western Producer*, 25 September 1947.

44 Ibid., 9 October 1947.

45 Ibid., 11 December 1947. See also ibid., 22 June and 11 and 25 September 1947.

46   Ibid., 25 September 1947.

47   *Winnipeg Free Press*, 16 December 1947. QUA, Dexter Papers, vol. 10, file 12, Canada-United Kingdom Wheat Agreement memo, 8 June 1950, 30-37 and 41-45.

48   *Winnipeg Free Press*, 9 and 12 December 1947. QUA, Dexter Papers, vol. 4, file 31, Freedman to Dexter, 23 November and 5, 8 and 21 December 1947 and vol. 5, file 32, 16 January 1948.

49   Ibid., vol. 4, file 31, Freedman to Dexter, 21 December 1947. Gardener's ambitions to succeed King as party leader combined with the fact that in his efforts to promote the agreement among western farmers he had an unfortunate tendency to portray himself as the farmers' shield against his own government did not exactly endear him to his cabinet colleagues. See LAC, Richardson Papers, vol. 4, file 34, Richardson to Dexter, 8 July 1948. On civil service reaction, interviews with Coyne and Sharp.

50   *Western Producer*, 12 February 1948. See also ibid., 15 January 1948.

51   *Winnipeg Free Press*, 25 September 1948.

52   *Western Producer, 2* December 1948. QUA, Dexter Papers, vol. 10, file 12, Canada-United Kingdom Wheat Agreement memo, 8 June 1950, 53-55 and *Western Producer*, 16 December 1948.

53   QUA, Dexter Papers, vol. 5, file 33, Wilson to Dexter, 28 December 1948. MacGibbon, *Canadian Grain Trade*, 128-31 and LAC, Richardson Papers, vol. 4, file 34, Ferguson to Richardson, 6 July 1948.

54   QUA, Dexter Papers, vol. 5, file 33, memo, 4 December 1948 and file 35, Dexter to Hutchison, 26 January 1949.

55   Ibid., vol. 5, file 35, Hutchison to Dexter, 8 February 1949 and vol. 10, file 12, Canada-United Kingdom Wheat Agreement memo, 8 June 1950, 6468; *Western Producer*, 20 January and 17 February 1949.

56   *Montreal Star*, 21 January and 30 April 1949.

57   *Saskatoon Star-Phoenix*, 29 July 1948. See also ibid., 3, 12 and 20 January 1948 and 9 September 1948; interview with Ben Malkin (*a Free Press* reporter); QUA, Dexter Papers, vol. 5, file 32, Freedman to Dexter, 23 March 1948; and LAC, Richardson Papers, vol. 3, file 21, MacKinnon to Richardson, 28 January 1947 and vol. 4, file 34, Kreutzweiser to Richardson, 20 October 1948.

58   QUA, Dexter Papers, vol. 5, file 35, Freedman to Dexter, 30 April 1949.

59   Ibid., vol. 5, file 37, Hutchison to Dexter, 4 February 1950. See also QUA, Crerar Papers, II, vol. 105, Dexter 1940-50 file, Crerar to Dexter, 7 March 1949 and Dexter Papers, vol. 5, file 37, unsigned memo, 26 November 1949.

60   Ibid., Hutchison to Dexter, 31 January 1950. See also ibid., 4 February 1950.

61   See, for instance, the editorial in the *Western Producer*, 16 February 1950.

62   Ibid., 9 March 1950. See also ibid., 23 March and 6 April 1950 and *Winnipeg Free Press*, 8 December 1949 and 22 February 1950.

63   MacGibbon, *Canadian* Grain Trade, 133 and *Western Producer*, 14 December 1950.

64   *Winnipeg Free Press*, 31 May 1950.

65   *Western Producer*, 17 August 1950.

66   MacGibbon, *Canadian Grain* Trade, 136-37.

67   Regina *Leader-Post* and *Saskatoon Star-Phoenix*, 2 June 1950.

68   *Western Producer*, 8 and 22 June 1950. See also *Edmonton Journal*, 1 June 1950 and *Winnipeg Tribune*, 6 June 1950.

69 Wilfrid Eggleston, "Post Mortem on Wheat," *Saturday Night*, 13 June 1950, 4; *Montreal* Gazette, 1 June 1950; *Montreal Star*, 9 June 1950; and *Toronto Globe and Mail*, 1 June 1950.

70 *Winnipeg Free Press*, 3 March 1951. See also ibid., 16, 19, 26, 27 and 29 March and 2, 4, 5 and 6 April 1951.

71 *Edmonton Journal*, 5 March 1951; *Financial Post*, 3 and 10 March 1951; *Montreal Gazette*, 6 March 1951; *Regina Leader-Post*, 5 March 1951; *Saskatoon Star-Phoenix*, 5 March 1951; and *Toronto Globe and Mail*, 5 March 1951.

72 Editorial in the *Western Producer*, 15 March 1951 and excerpts of Wesson's speech to the Saskatchewan Farmers Union, carried 5 April 1951.

73 MacGibbon, *Canadian Grain Trade*, 144. QUA, Dexter Papers, vol. 6, file 40, Hutchison to Dexter, 2 May 1951.

74 Interview with Sharp.

75 QUA, Dexter Papers, vol. 6, file 40, Hutchison to Dexter, 1 May 1951.

76 QUA, Crerar Papers, III, vol.105, Dexter 1951-59 file, Crerar to Dexter, 30 May 1951.

77 *Western Producer*, 31 May 1951. See also LAC, Richardson Papers, vol. 4, file 36, Ferguson to Richardson, 29 March 1951.

78 UCA, Hutchison Papers, 1.2.3, Hutchison to Dexter, undated (December 1955). Interview with Mackie and Richard Malone (Victor Sifton's assistant and confidant).

79 QUA, Dexter Papers, vol. 7, file 50, Dexter to Malone, 1 May 1961.

80 Ibid., vol. 4, file 29, Ferguson to Dexter, 21 October 1946.

81 LAC, Richardson Papers, vol. 5, file 46, Ferguson to Richardson, 1 May 1961.

82 Ibid., vol. 4, file 36, Ferguson to Richardson, 16 February 1951. See also, QUA, Dexter Papers, vol. 6, file 40, Mackie to Dexter, undated (May 1951).

Index

A

Also in the

HISTORY OF THE PRAIRIE WEST SERIES:

The Early Northwest

ISBN: 978-0-88977-207-6

Immigration & Settlement, 1870–1939

ISBN: 978-0-88977-230-4

Available on-line at

WWW.CPRCPRESS.CA

or through your local bookseller

THE PRAIRIE

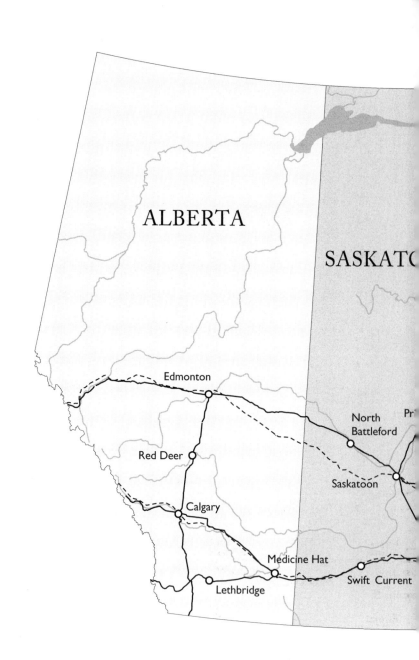

ALBERTA

SASKATO

Edmonton

North
Battleford

Pr

Red Deer

Saskatoon

Calgary

Medicine Hat

Swift Current

Lethbridge